Library of
Davidson College

Beyond The Dyad

Genesis of Behavior

Series Editors: MICHAEL LEWIS
                 Rutgers Medical School
                 University of Medicine and Dentistry of New Jersey
                 New Brunswick, New Jersey

    and LEONARD A. ROSENBLUM
                 Downstate Medical Center
                 Brooklyn, New York

Volume 1   *The Development of Affect*

Volume 2   *The Child and Its Family*

Volume 3   *The Uncommon Child*

Volume 4   *Beyond the Dyad*

Beyond The Dyad

EDITED BY
MICHAEL LEWIS
Rutgers Medical School
University of Medicine and Dentistry of New Jersey
New Brunswick, New Jersey

PLENUM PRESS · NEW YORK AND LONDON

Library of Congress Cataloging in Publication Data

Main entry under title:

Beyond the dyad.

   (Genesis of behavior; v. 4)
   Bibliography: p.
   Includes index.
   1. Social interaction in children—Congresses. 2. Social perception in children—Congresses. 3. Family—Congresses. 4. Child development—Congresses. 5. Psychology, Comparative—Congresses. I. Lewis, Michael, 1937 Jan. 10- II. Series.
BF723.S6B49   1984   115.4′18   83-24639
ISBN 0-306-41446-5

© 1984 Plenum Press, New York
A Division of Plenum Publishing Corporation
233 Spring Street, New York, N.Y. 10013

All rights reserved

No part of this book may be reproduced, stored in a retrieval system, or transmitted, in any form or by any means, electronic, mechanical, photocopying, microfilming, recording, or otherwise, without written permission from the Publisher

Printed in the United States of America

Contributors

PAULA J. BECKMAN, *Department of Special Education, University of Maryland, College Park, Maryland*

JAY BELSKY, *College of Human Development, Pennsylvania State University, University Park, Pennsylvania*

SAUL FEINMAN, *Department of Sociology, University of Wyoming, Laramie, Wyoming*

CANDICE FEIRING, *Department of Pediatrics, Rutgers Medical School—University of Medicine and Dentistry of New Jersey, New Brunswick, New Jersey*

WENDY GAMBLE, *College of Human Development, Pennsylvania State University, University Park, Pennsylvania*

DEVRA G. KLEIMAN, *Department of Zoological Research, National Zoological Park, Smithsonian Institution, Washington, D. C.*

MIRIAM KOTSONIS, *Bell Laboratories, Holmdel, New Jersey*

MICHAEL E. LAMB, *Department of Psychology, University of Utah, Salt Lake City, Utah*

SHARON LANDESMAN-DWYER, *Department of Psychiatry and Behavioral Sciences, University of Washington, Seattle, Washington*

MICHAEL LEWIS, *Department of Pediatrics, Rutgers Medical School—University of Medicine and Dentistry of New Jersey, New Brunswick, New Jersey*

RONALD D. NADLER, *Yerkes Regional Primate Research Center, Emory University, Atlanta, Georgia*

Ross D. Parke, *Department of Psychology, University of Illinois at Urbana-Champaign, Champaign, Illinois*

Elliot Robins, *College of Human Development, Pennsylvania State University, University Park, Pennsylvania*

Zick Rubin, *Department of Psychology, Brandeis University, Waltham, Massachusetts*

Jone Sloman, *Department of Psychology, Brandeis University, Waltham, Massachusetts*

Ross A. Thompson, *Department of Psychology, University of Nebraska, Lincoln, Nebraska*

Barbara R. Tinsley, *Department of Psychology, University of Illinois at Urbana-Champaign, Champaign, Illinois*

Thomas S. Weisner, *Departments of Psychiatry and Anthropology, University of California, Los Angeles, California*

Preface

How are we to understand the complex forces that shape human behavior? A variety of diverse perspectives, drawing on studies of human behavioral ontogeny, as well as humanity's evolutionary heritage, seem to provide the best likelihood of success. It is in an attempt to synthesize such potentially disparate approaches to human development into an integrated whole that we undertake this series on the genesis of behavior.

In many respects, the incredible burgeoning of research in child development over the last decade or two seems like a thousand lines of inquiry spreading outward in an incoherent starburst of effort. The need exists to provide, on an ongoing basis, an arena of discourse within which the threads of continuity between those diverse lines of research on human development can be woven into a fabric of meaning and understanding. Scientists, scholars, and those who attempt to translate their efforts into the practical realities of the care and guidance of infants and children are the audience that we seek to reach. Each requires the opportunity to see—to the degree that our knowledge in given areas permits—various aspects of development in a coherent, integrated fashion. It is hoped that this series—which will bring together research on infant biology, developing infant capacities, animal models, the impact of social, cultural, and familial forces on development, and the distorted products of such forces under certain circumstances—will serve these important social and scientific needs.

Each volume in this series will deal with a single topic that has broad significance for our understanding of human development. Into its focus on a specific area, each volume will bring both empirical and theoretical perspectives and analysis at the many levels of investigation necessary to a balanced appreciation of the complexity of the problem at hand. Thus, each volume will consider the confluence of the genetic, psychological, and neurophysiological factors that influence the individual infant and the dyadic, familial, and societal contexts within which development occurs. Moreover, each volume will bring together the vantage points provided by studies of human infants and pertinent aspects of animal behavior, with particular emphasis on nonhuman primates.

Just as this series will draw on the special expertise and viewpoints of workers in many disciplines, it is our hope that the product of these labors will speak to the needs and interests of a diverse audience, including physiologists, ethologists, sociologists, psychologists, pediatricians, obstetricians, and clinicians and scientists in many related fields. As in years past, we hold to our original objectives in this series of volumes to provide both stimulation and guidance to all among us who are concerned with humans, their past, their present, and their future.

The chapters in this volume are derived from papers presented and discussed at a conference, "Beyond the Dyad," held under the auspices and with the support of the Educational Testing Service in Princeton, New Jersey. The participants in the conference were Paula J. Beckman, Jay Belsky, Saul Feinman, Candice Feiring, Paul Holland, Devra G. Kleiman, Miriam Kotsonis, Michael E. Lamb, Sharon Landesman-Dwyer, Michael Lewis, Ronald D. Nadler, Jane Reinisch, Zick Rubin, and Leonard A. Rosenblum.

MICHAEL LEWIS

Contents

1. Social Influences on Development: An Overview
   MICHAEL LEWIS — 1

2. Is There Social Life beyond the Dyad? A Social-Psychological View of Social Connections in Infancy
   SAUL FEINMAN AND MICHAEL LEWIS — 13

3. The Social Ecology of Childhood: A Cross-Cultural View
   THOMAS S. WEISNER — 43

4. Changing Characteristics of the U. S. Family: Implications for Family Networks, Relationships, and Child Development
   CANDICE FEIRING AND MICHAEL LEWIS — 59

5. Implications of Monogamy for Infant Social Development in Mammals
   DEVRA G. KLEIMAN — 91

6. Biological Contributions to the Maternal Behavior of the Great Apes
   RONALD D. NADLER — 109

7. The Social Network of the Young Child: A Developmental Perspective
   MICHAEL LEWIS, CANDICE FEIRING, AND MIRIAM KOTSONIS — 129

8	*Grandparents as Support and Socialization Agents* BARBARA R. TINSLEY AND ROSS D. PARKE	161
9	*Infants, Mothers, Families, and Strangers* ROSS A. THOMPSON AND MICHAEL E. LAMB	195
10	*How Parents Influence Their Children's Friendships* ZICK RUBIN AND JONE SLOMAN	223
11	*The Determinants of Parental Competence: Toward a Contextual Theory* JAY BELSKY, ELLIOT ROBINS, AND WENDY GAMBLE	251
12	*A Transactional View of Stress in Families of Handicapped Children* PAULA J. BECKMAN	281
13	*Residential Environments and the Social Behavior of Handicapped Individuals* SHARON LANDESMAN-DWYER	299
	Author Index	323
	Subject Index	333

1

Social Influences on Development
An Overview

MICHAEL LEWIS

Historically, consideration of dyadic influences on children's development grows out of two distinct and fundamentally different models. The first is biological in nature and the second is educational. Both models have in common the strong belief that children's development, social as well as cognitive, is influenced primarily by dyadic interaction.

In the biological model, the emphasis on the parent–child interaction, in particular the mother–child dyad, is predicated on the belief that this dyad constitutes a biological unit endowed with and possessing unique characteristics that are essential both for survival and for development. The dyad is conceived of as symbiotic (Rosenblum & Moltz, 1983), as a biological imperative having evolutionary significance. It is often perceived as the critical factor in the young organism's development. Such a view is supported by both animal and human models of development. Animal models, using nonhuman primates or other mammals, demonstrate the importance of the mother–child dyad and of the mother as the single most critical unit in the child's development. Adult males are demonstrated either as absent or as uninterested in the young's development. Data from such studies are impressive in demonstrating the importance of a single adult female's role in her offspring's devel-

---

MICHAEL LEWIS • Department of Pediatrics, Rutgers Medical School—University of Medicine and Dentistry of New Jersey, New Brunswick, New Jersey 08903.

opment and the singularly unimportant role of the biological father. Nevertheless, such models and data often lose sight of the fact that the offspring's development, even though influenced by a mother only, takes place in an environment in which there are large numbers of diverse conspecifics, including younger and older siblings, aunts, adult female and male strangers, and adult males and peers. In the last decade, the importance of other social figures in the child's life have been addressed (Lewis & Feiring, 1978), and although some might still argue for the primary importance of the mother, none would argue that others do not play some role in the child's development.

The teacher–pupil relationship also represents an important dyadic model. This model, focusing on information exchange, holds that the fundamental process through which information is disseminated is a dyadic one, defined as two people: a teacher and a learner. The model of the parent–child dyad, in particular the mother–child, owes much to the teacher–pupil model. The teacher–pupil model of learning focuses on interactions between two members. In the teacher–pupil dyad, it is the adult member, for the most part, who is believed to educate the pupil. Thus, the effects of learning—or in broader terms, the effects of socialization—occur (1) within the dyadic interaction and (2) as a direct consequence of what the adult member does to the child. The influences of the interaction are the influences of direct information and didactic techniques. These, in turn, cause the younger member to change or, in the specific teacher–pupil sense, to learn. Such models of learning, when translated into general dyadic models, clearly emphasize the didactic notion of change. In this view, one member directly influences the other member, and in the most prominent model, the older member influences the younger member. Indeed, the children's learning experience vis-à-vis its social environment is often likened to that of teacher and pupil, and in studies of information exchange between parent and child, the teacher model is used. Thus, the dimensions thought important in the education of children vis-à-vis school are extended to the home and the family in observing mother–child learning situations (Bee, Van Egeren, Streissguth, Nyman, & Leckie, 1969; Hess and Shipman, 1965). Moreover, both reflect, at least in part, the unidirectional feature, in which the teacher influences the pupil. Although recognizing that the dyad may be more interactive and multidirectional (Lewis & Rosenblum, 1974), such models, nevertheless, suggest that the primary mode of socialization is a dyadic one.

The teacher–pupil dyad model suffers from some of the same problems as are found in the mother–child dyad model. Let us explore some of the common difficulties. To begin with, the dyadic learning model neglects the fact that even in the teaching situation the teacher–pupil interaction occurs in the company of others: other students and, at times, even other adults. Thus, to think of the teacher–pupil dyad as the only significant source of information and, therefore, of learning is to negate the role of peers (either older or same-age peers) in the learning process. The classroom in which the teacher–pupil dyad is located is also filled with a multitude of dyads, including the teacher–pupil dyad as well as the pupils' interactions with each other.

A second problem occurs when we focus on the processes involved in either change or learning. Both dyadic models (or didactic effects) tend to focus on direct effects as the process inducing change. Although pupils learn directly from what they are given to read or from what their teachers say, there are other important forms of learning that are not didactic in nature. The existence of alternative forms of learning itself argues against dyadic interactions as the sole or even the predominant model of growth. Observational learning and imitation are examples of such forms of learning, which are not dyadic nor didactic. The process of imitation allows us the opportunity to explore the nondyadic processes that allow the child to imitate anyone it chooses, be it someone in a position of power or someone who acts and/or achieves in a fashion that the imitator wishes to emulate. The process of observational learning likewise allows children to learn from anyone, not just the person directly interacting with them. These alternative forms of learning are particularly important in that observing them allows us to consider units larger than two people and therefore allows for the consideration of interactions beyond the dyad.

Language acquisition is a good example of how learning can take place by more than one method. The example of language acquisition can serve as a starting point for this exploration. Most learning theories relating to the development of language suggest that the mother–child dyad is chiefly responsible for the child's language acquisition (Lewis & Rosenblum, 1977). Although these theories differ considerably in detail, the major model remains that of the mother's speaking to the child as it affects the child's language ability. However, if we embed the mother–child dyad within a family consisting of a father and an older sibling, the influence of the mother talking to the child becomes only one alternative. In the family unit, there are at least two possible influences on the child's language acquisition: already discussed is the significance of the mother's speech (or any other person's speech) on the child. In every case, no matter who comprises the dyad, it is the speech of the older member of the dyad to the younger member that is held responsible for development. The second influence, one that is rarely considered, is the impact of other dyads, which do not include the child directly, on the child's language acquisition. In a mother, father, and child interaction, besides the two dyads that directly include the child (M → C; F → C), there are other dyads that do not include the child: the mother–father dyad, and the parent–older-sibling dyad. Research on indirect effects suggests that the language interaction between the mother and the father may influence the child's language acquisition (Lewis & Feiring, 1982). However, measurement is made increasingly complex by the addition of more than two members and by the addition of indirect effects.

First, it should be noted that particular measures of the dyadic interaction have themselves undergone considerable change. Finally, the parental language interaction may affect each parent's direct interaction with the child. For example, if the parents have an argument, they may be less disposed to talk with the child. Thus, parental language interaction has two types of indirect effects on the infant. Exactly how the language interaction of the mother–father dyad or the parent–older-sibling dyad affects the child is little understood. Nevertheless, these interactions may have significant effects because the young child often witnesses its family members engaging in interactions with each other. Without going into an analysis beyond the dyad it is impossible to study these effects.

MEASUREMENT ISSUES

The focus on dyadic interactions enables the investigator to easily explore the direct consequences of the action of one member on the behavior of another. The ease of measurement of a dyadic interaction, in part, is one reason for its success. The measurement model, as applied to development, is surprisingly simple. One can simply measure how one member of the dyad behaves and also measure, both concurrently and subsequently, how the other member behaves. Thus, children's language development can be studied by measuring some feature of maternal speech at one point in time and also as it relates to infant and child speech both concurrently and subsequently (Freedle & Lewis, 1977). In this fashion, direct influences in a dyadic exchange can be viewed developmentally. Before the early 1960s, mothers' behavior toward their children was measured primarily through clinical ratings or through observation and the use of scales. These scales consisted of categories of behavior believed important, and they were often selected on the basis of psychoanalytic as well as learning theory. Perhaps one of the earliest and most successful schemes of this sort grew out of the Fel's longitudinal study and the work of Sontag and Baker (1958). Here, aspects of maternal behavior ascertained through observation and rating scales were related to changes in intellectual behavior across the first part of the life cycle. Under the influence of the ethologists, such rating-scale procedures became outdated, and new techniques based on direct observation of particular behaviors were initiated. In part, the new systems of micromeasurement (see Cairns, 1979; Lamb, Suomi, & Stephenson, 1979; Lewis & Rosenblum, 1974) owe their origins to a discontent with the use of scales in assessing parent–child interaction. There were many difficulties with these scales, including their relatively weak interobserver reliabilities and their relatively low predictive validity. The scales used to describe the parent–child dyad were often constructs, which, although based on behavioral observation, were removed at least one inference level from the particular behaviors themselves. This discontent with construct variables suggested that a more profitable means of understanding the parent–child dyad would be through the ethological study of particular behaviors as they occurred within the dyad. The introduction of complex coding systems and the use of new technologies such as slow-motion videotape recording much facilitated this new measurement procedure. The basis of the new measurement procedures is the observation of behaviors that occur. No inference about behavior meaning is made; rather, those behaviors that are observable are scored. The advantages of such procedures quickly became apparent; high interobserver agreement on the occurrence of specific behaviors is possible.

The use of a more behavior-oriented methodology resulted in the observation of a large number of maternal and child behaviors. Not only were methodologies that would allow for the accurate observation of such a set of behaviors developed, but statistical methodologies had to be devised so that the large number of data obtained from these observations could be treated. Although many of the difficulties of observation, data management, data analysis, and interpretation remain, many new techniques have emerged. Nevertheless, there are still difficulties in these micro-observational systems. It is quite apparent that the attempt to measure all that occurs during a dyadic interaction is an impossible task because the number of possible behaviors is endless. Moreover, although more complex sequential-time-series analyses of data, in addition to simple counts of behaviors, are available, the use of these techniques, along with deciding what behaviors to measure, requires carefully articulated theories. Without these theories, what behaviors are to be observed and how the data are to be analyzed remain a problem. Although better interobserver reliability has been gained, the validity of these measures—that is, the relationship of the dyadic interaction measures to current as well as subsequent development—is not demonstrably better than with previous techniques. Because of these limitations, some investigators have suggested that the study of dyadic interaction—at least, as it is related to other variables—might best be done through the use of observational scales.

This brief review of the history of the measurement of dyadic interactions should immediately alert us to the major problems we face when we measure the more complete effect of the social environment on the child. However problematic, there has been much success in devising measurement systems for the dyad. This measurement success may act as a motive for the continuing use of the dyadic model as the major unit in development when, in fact, the need for other models has become increasingly clear. How a measurement system including more than two members would work is not yet clear, although several suggestions have been made. Normally, when units larger than a dyad are considered, the measurement system used is restricted to a set of dyadic interactions. Thus, for example, in mother–child–father interaction, the separate mother–child, father–child, and mother–father dyads are each treated separately. Alternatively, the mother–child dyad has been compared when the mother–child dyad are alone versus when they are in the company of another (Lamb, 1979; Pedersen, 1980). Although the new observation systems carry us beyond the concept of the dyad, they nevertheless restrict their analyses to the dyad within the context of other dyads or individuals. Not until methodological issues are addressed will the study of units larger than the dyad become the rule.

How to Study Interactions Larger Than the Dyad

Consideration of the social influences on the development of children from the perspective of units larger than the dyad is necessary. Few would argue (except perhaps those who advocate strong biological models) that the only unit of importance is the mother–child dyad. In studies beyond the dyad, two trends can be observed that are reflected in the chapters in this book. The first is the recognition that one must go beyond the mother–infant or even the father–infant dyad in order to understand development. Such approaches suggest that a large number of other conspecifics need to be considered, including grandparents, uncles and aunts, cousins, siblings, and friends. Studies such as these at least point out that one dyad is insufficient in characterizing the child's development. The second trend in studying interaction is going beyond the dyad toward examining group interactions. Much of our social life, even in early childhood, takes place in groups. Group processes and their measurement allow for study of units larger than the dyad. Approaches that look at group processes do this in a variety of ways. The simplest model is to observe how a dyad is affected by other conspecifics, for example, how mother–child interaction is altered by the presence of the father. Although studies of this type are important, they have only limited application to the study of group interactions. The third trend involves studies of group processes that deal with indirect as well as direct effects. Most recently, the issue of indirect effects has been taken up by Lewis (Lewis & Feiring, 1981; Lewis & Weinraub, 1976) and Parke (Parke, Power, & Gottman, 1979); it is particularly germane to the study of interactions containing more than two members. This topic reappears in several of the following chapters and includes such concepts as social referencing and imitation.

In this book, there are 13 chapters that view the issue of development in children, as well as animals, from a perspective that attempts to go beyond the dyad. In the second chapter, Feinman and Lewis present a social-psychological view of social connections in infancy. They connect issues found in social psychology and discuss their relevance to development. Such concepts as small-group processes, indirect effects, and social referencing are considered. The implication of social behavior beyond the dyad are considered in terms of day care and child abuse.

In the third chapter, Weisner performs an important service by disconfirming our prejudice that the social development of all children requires only a mother–child dyad. Weisner's case studies of other cultures points out in some measure the weakness of the biological notion that the mother–child dyad is by *necessity* the only and most important social relationship. The cross-cultural approach informs us, through examining other cultures, that other relationships have equal importance.

In the chapter on the change in characteristics of the U.S. family, Feiring and Lewis characterize three of the important changes taking place in the United States, including the smaller number of siblings, increased divorces, and working mothers. The implications of these changes for child development cannot be easily appreciated or understood without understanding the effect on family networks. Thus, for example, the effects of increased maternal working on children's development cannot be understood without understanding how this affects the role of the father vis-à-vis child rearing.

The chapter by Kleiman on the implications of monogamy for infant social development in mammals is a careful review of the effects of nonmonogamous relationships on parental care and the dyad. The review of the influence of male parental care on the young forces us away from considering the mother and child as the sole important dyad. Even among nonhuman animals, the influence of the father needs to be recognized. The involvement of the father must weaken the biological argument of the primacy of the mother–child unit. The discussion of indirect effects is also taken up as Kleiman demonstrates that maternal behavior is influenced by other conspecifics surrounding the mother.

In the sixth chapter, "Biological Contributions to the Maternal Behavior of the Great Apes," Nadler helps to balance the issue of the biological importance of the mother–child dyad by the use of naturalistic as well as laboratory studies. Nadler makes clear that, in the wild, for both the orangutan and the great ape, the mother–child dyad is the most important and influential unit in the child's life. These animals are not gregarious by nature and appear to rely on the dyad as a central feature in social behavior. Nevertheless, the maternal gorilla's standing in the group and her relationship to the male gorilla influence her childrearing practices. Thus, the child's relationship to its mother is influenced by group dynamics that involve the mother herself. The indirect effects of others on the mother–child relationship are no better demonstrated than in these studies of the gorilla.

Lewis, Feiring, and Kotsonis, in the next chapter, examine the social network of 3-year-olds. For the first time, data are presented that indicate the complex social network that already exists in the 3-year-old's world. Such a survey serves to point out that from the beginning of life children are embedded in a large and complex social world made up of a variety of people varying in functions they serve in relation to the child. Looking at a large group of 3-year-old children, Lewis, Feiring, and Kotsonis found important differences in children's social networks as a function of their social class, sex, and birth order, and they use these different social structures to account for differences in development. For example, that children of higher socioeconomic status have a larger adult–child ratio than the poor may affect the child's intellectual ability prior to the school experience as well as facilitating the school experience itself.

Chapter 8, by Tinsley and Parke, presents a discussion of grandparents as a part of the parental support system and as socialization agents. The discussion of grandparents draws our attention to an aspect of the social network not yet considered. Because of the overdrawn focus on the mother as the primary caregiver, the significance of others in the child's network has been lost. Nowhere is this more obvious than in the absence of studies on the nature and the role of grandparents. Grandparents now take the place previously assigned to fathers as forgotten parents. To address this imbalance, Tinsley and Parke review the literature on grandparents and relate grandparenting and the role that grandparents play to such factors as social class differences. The discussion of grandparenting also provides a forum for discussing the nature of the social network and the need to consider people other than the mother and the father. Moreover, the effects of grandparents, both maternal and paternal, are delineated within a model that attempts to show the dynamic influence of grandparents and the interactions among three generations.

In "Infants, Mothers, Families, and Strangers," Thompson and Lamb attempt to expand the nature of the attachment relationship between mother and child by considering how the attachment relationship itself undergoes changes as a consequence of the familial experience. Although the direction of the change in the attachment relationship as a function of familial factors remains unclear, the authors do alert us to the need to understand that even attachment is open to and affected by influences outside the dyad. The attachment relationship has focused on the importance of the mother–child dyad in the child's subsequent development (indeed, the attachment relationship between the father and the child is rarely even considered). In this chapter, the consideration of the child's attachment to people other than the mother—in particular, the father—is also considered.

Rubin and Sloman, in their chapter on how parents influence their children's friendships, present a unique example of case study research. In their interviewing of children and parents on the children's friendship, the authors have demonstrated the well-known fact that children's friendships affect the parental relationships (a phenomenon known to parents of adolescents), and also that children's friendships are influenced by their parents. These interactions among children, parents, and peers show that the child's social relationships are not a set of discrete dyadic interactions but are interconnected in a very complex fashion (Harlow, 1969; Hartup, 1979; Lewis, Young, Brooks, & Michalson, 1975).

In Chapter 11, Belsky, Robins, and Gamble focus on the issue of competence. The concept of parental competence is expanded to include both the personal attributes of the parents and the nature of the support system. These authors' view of the child's development as embedded in the capacities of the parent, as well as the social network, points out that any theory that focuses solely on the features of the parent or the child in order to understand the dyad is bound to be inadequate. It is a unique perspective to talk about parental competence by considering factors outside the parents themselves.

In the last two chapters of this book, special attention has been paid to the development of children and adults who are handicapped. We give the handicapped person special attention so that we can examine the dynamics of dyadic and group interaction in groups in which development has been difficult. In a transactional view of stress in families of handicapped children, Beckman allows us to examine family dynamics when the child's development is dysfunctional. Here, the role of the family as a support system and the way in which the relationship between the parents themselves affects the child's development are apparent. Beckman also affords us an opportunity to examine not only how the family affects the child's development but also how children affect their families. Perhaps nowhere is this more evident than in families of handicapped children. The stress on families of handicapped children is intense; as might be expected, family disruptions, such as divorce, are more frequent in such groups.

In Chapter 13, Landesman-Dwyer presents a highly complex naturalistic study of residential environments and the social behavior of handicapped adults. The opportunity to observe the effects of changing environments on the handicapped individual's behavior is not often available, given the methodological problems associated with such studies. Landesman-Dwyer was able to avoid such pitfalls. Of particular importance to our discussion of the dyad is her demonstration that individual differences among handicapped individuals significantly affect and are affected by differences in the social structure. One of the critical aspects of this chapter is the demonstration that social development and social behavior transpire within a context, and that this context has physical as well as social features. Moreover, these physical and social features interact in such a way that the physical environment affects social grouping. Perhaps we should not lose sight of the fact that particular types of dwellings facilitate certain types of social interaction and inhibit others. For example, the close-space living of the poor may facilitate group interactions, whereas the private dwellings of the middle class and the wealthy facilitate more dyadic interactions. The manipulation of environments allows us to examine the social ecology of handicapped adults and the way in which the social ecology (varying from dyadic to group interactions) affects behavior.

All told the 13 chapters that are presented in this volume can be said to focus on at least three of the major factors that are necessary in any attempt to study interaction beyond the dyad. First, they address the issue of considering dyads other than the mother and child, and they look at dyads that include fathers, siblings, peers, teachers, caregivers, and grandparents. This theme runs through most of the chapters and points out the inadequacy of any theory restricted to a single dyad as the sole cause of development. The second feature that many of the chapters have in common is the study of group interactions—if not observing group interactions directly, at least looking at how groups or single individuals affect particular dyads. The final factor, and perhaps the least well studied, although described in some detail, is the issue of indirect effects. Children come to learn from and profit from experiences that are not didactic in nature. Here, perhaps, the chapters as a whole fail to address the processes that underlie group interactions as opposed to dyadic ones. Nevertheless, in several of the chapters, these processes are at least discussed if not studied. With the publication of these chapters, the search for the critical dimensions of social experience is enhanced as we are led from the dyad to larger units of study.

REFERENCES

Bee, H. L., Van Egeren, L. F., Streissguth, A. P., Nyman, B. A., & Leckie, M. S. Social class differences in maternal teaching strategies and speech patterns. *Developmental Psychology*, 1969, 1(6), 726–734.

Cairns, R. B. (Ed.). *The analysis of social interactions: methods, issues and illustrations*. Hillsdale, N. J.: Lawrence Erlbaum, 1979.

Freedle, R., & Lewis, M. Prelinguistic conversations. In M. Lewis & L. Rosenblum (Eds.), *Interaction, conversation and the development of language: The origins of behavior* (Vol. 5). New York: Wiley, 1977.

Harlow, H. F. Age-mate or peer affectional system. In D. S. Lehrman, R. A. Hende, & E. Shaw (Eds.), *Advances in the study of behavior* (Vol 2). New York: Academic Press, 1969.

Hartup, W. W. Two social worlds: Family relations and peer relations. In M. Rutter (Ed.), *Scientific foundations of developmental psychiatry*. London: Heinemann, 1979.

Hess, R. D., & Shipman, V. C. Early experience and the socialization of cognitive modes in children. *Child Development*, 1965, 36, 869–886.

Lamb, M. E. The effects of the social context on dyadic social interaction. In M. E. Lamb, S. J. Suomi, & G. R. Stephenson (Eds.), *Social interaction analysis: Methodological issues*. Madison: University of Wisconsin Press, 1979.

Lamb, M. E., Suomi, S. J., & Stephenson, G. R. *Social interaction analysis: Methodological issues*. Madison: University of Wisconsin Press, 1979.

Lewis, M., & Feiring, C. The child's social world. In R. M. Lerner & G. D. Spanier (Eds.), *Child influences on marital and family interaction: A lifespan perspective*. New York: Academic Press, 1978.

Lewis, M., & Feiring, C. Direct and indirect interactions in social relationships. In L. Lipsitt (Ed.), *Advances in infancy research* (Vol. 1). Norwood, N.J.: Ablex, 1981.

Lewis, M., & Feiring, C. Some American families at dinner. In L. Laosa & I. Sigel (Eds.), *Families as learning environments for children*. New York: Plenum Press, 1982.

Lewis, M., & Rosenblum, L. (Eds.). *The effect of the infant on its caregiver: The origin of behavior* (Vol. 1). New York: Wiley, 1974.

Lewis, M., & Rosenblum, L. (Eds.). *Interaction, conversation, and the development of language: The origins of behavior* (Vol. 5). New York: Wiley, 1977.

Lewis, M., & Weinraub, M. The father's role in the child's social network. In M. E. Lamb (Ed.), *Role of the father in child development*. New York: Wiley, 1976.

Lewis, M., Young, G., Brooks, J.,& Michalson, L. The beginning of friendship. In M. Lewis & L. Rosenblum (Eds.), *Friendship and peer relation*. New York: Wiley, 1975.

Parke, R. D., Power, T. G., & Gottman, J. M. Conceptualizing and quantifying influence patterns in the family triad. In M. E. Lamb, S. J. Suomi, & G. R. Stephenson (Eds.), *Social interaction analysis*. Madison: University of Wisconsin Press, 1979.

Pedersen, F. A. (Ed.). *The father–infant relationship: Observational studies in the family setting.* New York: Praeger, 1980.

Rosenblum, L. A., & Moltz, H. *Symbiosis in parent offspring interaction.* New York: Plenum Press, 1983.

Sontag, L. W., Baker, C. T., & Nelson, U. L. Mental growth and personality development: A longitudinal study. *Monographs of the Society for Research in Child Development,* 1958, 23(68), 1–143.

2

Is There Social Life beyond the Dyad?

A Social-Psychological View of Social Connections in Infancy

SAUL FEINMAN AND MICHAEL LEWIS

The study of infants' social relations has focused primarily on dyadic interaction. Early research concentrated on the mother–infant and the stranger–infant dyads. During the past decade, the view of infants' social network has expanded to include relationships with fathers, infant peers, substitute caregivers, siblings, and grandparents. There has been increasing concern in the last few years about extradyadic social forces, as evidenced by research on second-order and indirect effects, and by consideration of the infant as a member of a family system (Belsky, 1981; Lewis & Rosenblum, 1979; Lewis & Weinraub, 1976; Pedersen, Anderson, & Cain, 1980). Nevertheless, in comparison with research on social relations in adults and older children, the study of infant sociability

---

SAUL FEINMAN • Department of Sociology, University of Wyoming, Laramie, Wyoming 82071. MICHAEL LEWIS • Department of Pediatrics, Rutgers Medical School—University of Medicine and Dentistry of New Jersey, New Brunswick, New Jersey 08903. Preparation of this paper and the collection and analysis of some of the social referencing data reported in it were supported by NIMH National Research Service Award IF32MH07625-01, NIMH Grant 1RO3MH35384-O1A1, and a Basic Research Grant from the College of Arts and Sciences of the University of Wyoming to the first author. The second author was supported by BEH Grant 300-77-0307.

remains dominated by a dyadic focus. This chapter considers the various ways in which social connections beyond the dyad affect the infant.

There seem to be several major reasons why infant social research has concentrated on dyadic interchange. First, dyads are the building blocks of sociability. If we could select only one focus for research on interpersonal relations, dyadic interaction would be the appropriate choice. Second, dyadic data are logistically easier to collect, analyze, and interpret. Third, adult–infant interaction has been conceptualized as a teacher–pupil relationship, in which the caregiver teaches and the infant learns. Fourth, the immaturity of infants' cognitive skills has suggested that more complex social relationships are beyond their grasp. Fifth, the frequent recruitment of middle-class participants may encourage the belief that infants in general interact mainly within the nuclear family and especially with the mother. Lower-class children live in larger families (Chilman, 1975), interact more often with siblings (Brossard & Boll, 1955; Brossard & Sanger, 1952), and have more contact with adults other than the mother (Tulkin & Kagan, 1972) who often act as substitute caregivers (Gans, 1962; Stack, 1974; Young & Wilmott, 1957). Finally, the concentration of research on the mother–infant dyad may have been a realistic response to the social isolation of mother and infant 20–30 years ago. When mothers and infants stay at home, their social experiences are probably limited. The increased labor-force participation of women with young children and the more frequent utilization of substitute caregivers in infancy have thrust the mother and infant into more complex social relationships, opening new avenues for extradyadic influence.

Although the dyadic approach to infant sociability is understandable and appropriate, its explanatory power is limited by the importance of social forces beyond the dyad. Dyadic interaction itself is influenced by the dyad partners' relationships with other people, by their social positions within the society, and by the way in which others influence their perceptions of each other. In addition, social interaction often occurs in triads and larger groups (e.g., extended families and day-care centers). Dyadic analysis is a reasonable first step toward an understanding of such higher level interaction. But just as dyadic interaction is more than the sum of the two participants, higher level interaction is more than the sum of the component dyadic relationships.

One guideline for investigating the extradyadic social forces that affect infants is the elaborately developed framework utilized by social scientists to study social connections within small groups, social institutions, and societies. Social-psychological study of interaction in families, juries, work groups, and experimental laboratory groups; sociological investigation of the effects of family systems, educational institutions, and occupational structures on interpersonal behavior; and cultural anthropological considerations of extended kinship networks and political alliances—all represent research traditions that focus on social contexts beyond the dyad. Although much understanding of infant behavior is founded on the premise that infants' mental activity and information processing are markedly different from those of adults and verbal children, there are some dimensions of infant response that may be explained by theories previously developed to explain the behavior of older humans.

A Classification of Extradyadic Social Influences

This chapter considers *four* types of extradyadic social factors that are likely to influence infant behavior. This classification is derived from research and theory on interpesonal behavior in older humans, but it does not contravene earlier catalogs of indirect influence on infant behavior (Bronfenbrenner, 1974; Lewis & Weinraub, 1976; Parke, 1979). For each of these social forces, we discuss general considerations about human interaction and specific applications to infants:

1. The physical presence of other people to form triads and larger groups (e.g., families and play groups).
2. The extent to which a dyad is isolated from or integrated within social networks.
3. The social experiences and relationships that each dyad partner has outside the dyad itself.
4. The ways in which others modify dyad partners' perceptions of each other, thus influencing dyadic interaction.

The Presence of Other People

The presence of additional individuals significantly affects social interaction through two major avenues. First, the new people can modify interaction within each dyad. Second, the presence of three or more individuals creates opportunities for higher level interaction, as when several individuals converse together as a group.

The interest of behavioral scientists in the effects of numbers dates back at least as far as the turn-of-the-century theorist Georg Simmel (see Wolff, 1950) who suggested that the entrance of a third person, transposing the dyad into a triad, greatly alters the interactive structure. In triads, coalition formation is possible, as two individuals can ally in order to gain an advantage over the third. Indeed, interaction in three-person groups often results in coalition formation in the pursuit of profit (Caplow, 1968; Gamson, 1961; Komorita & Moore, 1976; Miller, 1981). Alliances are especially common between two lower power actors who join forces to counter the more powerful actor, who would otherwise dominate the group and acquire most of the available rewards. Simmel also suggested that when there is discord between two persons, the third individual may either act as a mediator to resolve the differences or take advantage of the rift in order to increase his or her own profit. Indeed, the third person may actively drive a wedge between the other two.

Although the transition from dyad to triad is especially significant, the general issue of how group size influences interaction has also received much social-psychological attention. As group size increases, the number of potential dyadic relationships increases. In an n-member group, the addition of one person increases the number of dyads by n. Furthermore, there is greater potential for simultaneous interaction among three or more people, as when group members work together on a common task. But larger group size is also correlated with decreased individual participation. Because time is, and sociability may be, a limited good (Granovetter, 1971; Nelson, 1966), an increase in group size decreases the average participation of each member in both adults' and children's groups (Bales & Borgatta, 1967; Dawe, 1934; Hare, 1952; Indik, 1965; Williams & Mattson, 1942).

Small groups are inclined to become internally stratified so that activity is controlled more by some individuals than by others (Burke, 1974; Goetsch & McFarland, 1980; Savin-Williams, 1979). As group size increases, interaction comes to be controlled by a smaller percentage of individuals; that is, Mischels iron law of oligarchy operates (Bales, 1970; Bales, Strodbeck, Mills, & Roseborough, 1951; Mayhew & Levinger, 1974, cited in Webster, 1975). Whereas 50% of the individuals contribute 61% of the interaction in 4-person groups, only 20% of the group members contribute 61% of the interaction in 10-person groups (Bales, 1970).

Families and kinship networks provide important small-group interaction in all societies; even societies with the simplest social structure organize social life around kin-based groupings (Van den Berghe, 1978). Research on families has often focused on the distribution of power in families and conjugal relationships (Centers, Raven, & Rodrigues, 1971; Cromwell & Cromwell, 1978; Cromwell & Olson, 1975; Falbo & Peplau, 1980; Salamon & Keim, 1979). Blood and Wolfe's (1960) study of how the husband–wife power balance is altered by the entry of children into the family system reflects an interest in how the presence of additional group members affects dyadic interaction (see also Lewis & Feiring, 1981).

Psychiatrically oriented research on family process has considered the question of whether some interaction patterns can facilitate the development of schizophrenia, psychosomatic disturbances, and phobias in children. Rather than focusing exclusively on each dyad within the family, investigators have also considered the interplay among dyadic relationships and higher level interaction *per se*. Indeed, it is in such interplay that the origin of children's psychological disturbances appears to reside. Competing parents may each attempt to form a coalition with the child against the other parent (Broderick & Pulliam-Krager, 1979; Minuchin, 1974). Parents may confront the child with conflicting interpretations of situations (Sojit, 1971). In a somewhat different solution to family distress, parents may resolve their disagreement by forming an alliance with each other and scapegoating the child, who is then perceived as the source of marital discord (Vogel & Bell, 1960).

The understanding of family interaction requires consideration not only of dyadic relationships but also of higher level interaction and the interplay among dyads. A parallel situation can be seen in cognitive balance theory (Festinger, 1964; Heider, 1946). Person P's psychological comfort with persons O and R is a function not only of whether P likes O and R, but also of whether R and O like each other. When P likes O and R, but O and R despise each other, P is likely to be uncomfortable in the group. A similar discomfort affects the child whose parents do not care for each other.

Generally, behavioral research on small-group interaction focuses on (1) the effect of group size on individual participation; (2) the distribution and use of power within the group; and (3) complex higher level interaction involving three or more people at one time. How does this approach correspond with previous infancy research, and what does it suggest for future investigations?

Information about extradyadic forces in infancy can be derived from earlier studies that were not explicitly directed toward such issues. Investigations of the effect of the mothers' proximity on infant–stranger interaction (Ainsworth, Blehar, Waters, & Wall, 1978; Campos, Emde, Gaensbauer, & Henderson, 1975; Feinman, 1980; Morgan & Ricciuti, 1969) were, after all, concerned with the effect of a third person on dyadic interaction. The proximity of the mother, who can be thought of as the infant's higher power coalition partner, increases the infant's comfort with the stranger, who is obviously more powerful than the infant. Lower power actors, be they developing infants or developing nations, are more comfortable in confronting a powerful and unknown entity when a friendly and powerful ally is present.

Investigations of the effects of birth order on caregiver–infant interaction (Jacobs & Moss, 1976; Lewis & Kreitzberg, 1979; Thoman, Barnett, & Leiderman, 1971) have also considered extradyadic influences.

Mothers' less frequent interaction with later-borns than with firstborns is consistent with the small-group finding that the quantity of interaction per dyad decreases with increasing group size. Later-borns receive less attention because they are members of larger family groups than are firstborns. Similarly, the decrease of maternal attentiveness to and playfulness with firstborn offspring following the birth of a second infant reflects the effect of group size on dyadic interaction (Feiring & Lewis, 1982; Kendrick & Dunn, 1980). The finding that later-born infants spaced at least six years from the previous sibling receive about as much attention as do firstborns (Lewis & Kreitzberg, 1979) could be explained by the absence of older siblings from the household for the hours during which they are attending school. For about six hours each weekday, the distantly spaced later-born is a member of a smaller group and can receive more parental attention. The principle that interaction between two people intensifies as group size diminishes is evident here.

Research on indirect effects considers how the presence of one parent affects the infant's interaction with the other parent (Clarke-Stewart, 1978; Lamb, 1976, 1978; Lytton, 1979; Pedersen et al., 1980). Infants interact more frequently with a parent when alone with that parent than when both parents are present. Similarly, if one dyad in a three-person group does not interact, then interaction in the other two dyads should intensify. This pattern is reflected in the finding that infants interact more with each parent when the parents do not communicate with each other than when they do (Pedersen et al., 1980).

Similarly, 2-year-olds approach their mothers less often when in a triad of mother, infant, and older sibling than when with only the mother (Samuels, 1980). Infants between 10 and 14 months directed 36% more behaviors to peers and 174% more behaviors to peers' mothers when their own mothers were absent than when they were present (Field, 1979). Infants' response to variations in group size does appear to resemble that of older humans.

Although current infant research allows us to consider some of the issues relevant to extradyadic influence, such data have very limited utility in examining (1) infants' triadic or higher level interaction, (2) the distribution of power within groups with infant members, and (3) infants' behavior in coalition formation. Several methodological practices and conceptual barriers have led to this situation. First, many studies of infants in groups use data collection techniques and procedures that inhibit higher level interaction. Mothers and fathers, or parents and strangers, are asked to refrain from conversing with each other (Cohen & Campos, 1974; Lamb, 1978), or mothers are requested not to initiate interaction with their infants in peer studies (Hay, Nash, & Pedersen, 1981; Pastor, 1981).

Second, although research on older humans often focuses on task-oriented groups in which individuals interact to solve a common problem or engage in a common activity, analogous situations are infrequently studied for groups with infant members. For example, the effect of group size on stratification cannot be evaluated when an experimental procedure blocks some channels of dyadic interchange, or when there is no collective activity. Research strategies currently used to investigate second-order effects have been constructed to be effective in determining how dyadic interaction is affected by a third person. These techniques are not effective, though, for studying triadic interaction *per se*. A recent exception is found in the study by Lewis and Feiring (1979) of interaction in families at dinner. All dyadic channels were left open, and a common activity—eating dinner—engaged the members of the family group. Such data allow us to identify and describe the stratification structure in the group. In families with infants and young children, each parent initiated a higher proportion of communication than did each older child, who initiated more communication than each younger child. As family size increased, the ratio of the percentage of interaction initiated by both parents increased, compared with the percentage of family members who were parents. As in adult groups, interactional inequality in families with very young children increases as group size increases.

Investigations of infants in larger groups have tended to collect dyadic data and to ignore higher level interaction (e.g., Field's 1979 study of cooperative day care in which 12 infants and 6 mothers interacted). The same situation could have been studied with an eye toward larger scale interaction, sociometric preferences among mothers and infants, and variations among infants in social participation rates and ability to control valued resources (Is there stratification among 1-year-olds?). The difference between investigations of infants in groups and studies of preschoolers' play groups (Goldman, 1981; McGrew, 1972; Smith, 1974) resides largely in the dyadic focus of the former and the group (as well as dyadic) focus of the latter. Recent exceptions can be found in Frankel and Arbel's investigation (1980a,b) of dominance hierarchies and relationships in groups of 2-year-olds, and in the study by Lakin, Lakin, and Costanzo (1979) of group processes in $1\frac{1}{2}$- to $3\frac{1}{2}$-year-olds. It is probably not coincidental that both of these studies are based on Israeli data, given the long-standing interest in collective and group behavior in that society.

A major conceptual barrier arises if we fail to think of infants in terms of power, conflict, and coalition formation. A power exchange and conflict perspective characterizes much research on adults (Blau, 1964; Cook & Emerson, 1978) but is not as often applied to the interaction of parents and very young children (Trivers, 1974). But mother–infant interaction can be discussed with regard to control (Schaffer & Crook, 1980) and coerciveness (Martin, 1981). Infants possess many features of low-power actors in that they have little control over what others do to them and cannot prevent bigger humans from invading their personal space, touching them, and moving them to other places. The response of infants to unfamiliar people may be conceptualized in terms of power because more positive reactions occur with regard to strangers who are apparently less powerful, for example, children and shorter people (Brooks & Lewis, 1976; Feinman, 1980; Weinraub & Putney, 1978), and when infants can control the strangers' actions (Leavitt, 1980).

Do infants initiate coalition formation? Can they become partners in an alliance if an adult or an older child takes the first step? These questions are quite reasonable ones if we substitute the term *attachment* for the terms *alliance* or *coalition*. In the alliance of artist and patron, the wealthy patron nurtures the artist's career. Analogously, parent and infant ally (attach) to further the infant's well-being and maturation. Furthermore, each parent may try to enhance her or his power by forming a closer relationship with the infant, who can become a pawn in the social dynamics of paradoxical messages and competition. A parent can either mediate or aggravate a conflict between the infant and the other parent. In such disagreements, the parent might form a coalition with one against the other. Indeed, young children are more likely to comply with one parent's request if it is supported by the other parent (Lytton, 1979). Perhaps similar effects also occur within the families of even younger children. It may be that social psychological theories oriented to power, conflict, and alliance formation can effectively explain behavior in groups that have infant members, just as such theories have aided in the understanding of the group behavior of older humans.

The Integration of the Dyad within Social Networks

The dyad's integration within or isolation from social networks of kinship, friendship, and community affects dyadic interaction in several ways. First, social connectedness enhances the probability that the dyad will be drawn into higher level interaction.

Second, social integration provides support and aid in everyday activities and in emergency situations. Individuals in socially isolated dyads do not have other people to call on when situational demands exceed the resources within the dyad. The finding that triads adapt better than dyads to isolation is a case in point (Smith & Haythorn, 1972).

Third, extradyadic networks provide opportunities for social comparison, the process by which individuals evaluate their opinions and abilities through comparison with those of other people (Festinger, 1954). For example, a physician's ability to care for patients is enhanced by social connections within professional networks. In a study of the diffusion of medical innovation, Coleman, Katz, and Menzel (1966) found that physicians who were well integrated within professional networks of other physicians were more likely to prescribe new drugs. Thus, physicians' interaction with patients is affected by their understanding of how to treat their patients' ailments, and such understanding is influenced through social connections with other physicians.

Social integration within a community—whether of scholars, parents, or neighbors—provides information, social comparison, support and assistance, and higher level interaction (Erikson, 1976; Gans, 1962; Stack, 1974). In 1972, the Buffalo Creek community, a collection of neighboring settlements in a West Virginia coal-mining hollow, was greatly damaged by a flood (Erikson, 1976). In addition to leaving a path of death, injury, and property damage, the flood disrupted the social integration of the community. The flood's physical effects displaced residents from their homes, physically isolating them from the family and friends near whom they had lived. Government assistance in providing new shelter exacerbated this disruption by neglecting to reestablish prior patterns of neighboring in the reconstruction of the physical community. Community integration was weakened in that the frequency of contacts that provide support and social comparison was diminished. People no longer lived in proximity to their friends and neighbors. In emergency situations, such as a flood, social support and social comparison become especially important. The psychological recovery of the Buffalo Creek community was severely hindered by the disintegration of its social community as well as by the destruction of its physical community.

It is the community, with its broad networks of social ties, that is able to shoulder the responsibility for convincing people that it is safe to venture forth without excessive anxiety into a precarious world (Erikson, 1976). This task of reality construction exceeds the capabilities of isolated dyads by a considerable margin. Although dyadic relationships, such as those within nuclear families, were maintained in the aftermath of the Buffalo Creek flood, the tearing of the fabric of community ties left individuals socially isolated from sources of social support and reality construction.

Connectedness with broader social networks can affect infant–caregiver dyads as well. Social ties provide information that can influence the caregiver's judgment of the appropriateness of particular interactive patterns with the infant. In societies lacking mass media and professional child-care advisers, integration within a network of kith and kin is the only channel through which such information can flow. Even when alternative information sources exist, interpersonal networks can affect parent–infant interaction. Despite the nuclear family orientation of industrial societies, parents and children tend to be in contact with friends and relatives. For example, two-thirds of the elderly people surveyed in a study of six industrial societies reported that they lived either with a grown child or within 10 minutes of that household (Shanas, 1973). In addition to providing the infant with opportunities to interact within a wider social network (Weinraub, Brooks, & Lewis, 1977), relatives and friends can be important sources of advice about child care.

Social integration also facilitates support systems that supplement the dyad's capacity to cope with everyday demands, for example, the distribution of an infant's care among several people (Stack, 1974; Turnbull, 1961; Young & Wilmott, 1957). Social isolation and weaker ties with extended kin and the community are salient features of child-abusing parents (Garbarino, 1977; Parke & Collmer, 1975; Spinetta & Rigler, 1972), suggesting that social isolation renders the dyad more vulnerable to stress. Garbarino and Sherman (1980) found that children who resided in neighborhoods with higher rates of child abuse had fewer contacts with people beyond the nuclear family than did children in lower risk neighborhoods. Lessened social integration appears to diminish the resources and the social support available to the dyad.

Furthermore, socially integrated parents who abuse their children can learn, through social comparison, that their behavior is neither considered appropriate nor practiced by other people. The isolated child-abuser lacks such opportunities. Within the realm of acceptable child-care practices, parents who interact beyond the nuclear family have multiple bases for assessing their own and their infants' behavior because they can engage in comparative discussion and can also observe other infants.

The effects of extradyadic ties lie in their structure as well as their quantity (Granovetter, 1971, 1973). Strong ties within close-knit networks are the major providers of social support. The individual who has acquaintances but no close friends or relatives is unlikely to receive aid. But weak ties within loose-knit networks expose the individual to a more divergent range of sources of social comparison. Close-knit networks tend to become isolated groups characterized by conformity. Loose-knit networks provide a wider variety of opinions because the people with whom one individual is connected are unlikely to be connected with each other. Investigation of the effects of the extensiveness and structure of caregivers' social connections beyond the dyad are likely to be of considerable value in clarifying variations among parents with regard to their socialization goals and strategies.

The mother–infant dyad appears to be more socially integrated in the 1980s than it was 20 years ago. As more mothers work outside the home, and more young children experience substitute care, they come to interact with new people and to develop opportunities for social support and social comparison. The establishment of cooperative day-care centers in which all parents participate some but not all of the time (as in the group studied by Field, 1979) may (1) enable parents to compare their infants with other infants and (2) activate potential support and comparison systems with other parents. One result of recent alterations in the role of women in some industrial and industrializing societies seems to be the greater social integration of infants and mothers.

Social Experiences and Relationships outside the Dyad

Most individuals occupy more than one position in a society. A woman may be a mother, a wife, a carpenter, and a member of the local art guild. An infant may be a son or daughter, a sibling, a member of a day-care center, and a pediatric patient. What individuals do, how they are treated, and what they learn outside a particular dyad can affect their interaction within the dyad. Such effects can derive from (1) the existence of upper limits on time, sociability, and commitment; (2) the impact of values and attitudes learned in other social contexts; (3) the conflicts caused by mutually exclusive role expectations; and (4) the impact of success or satisfaction in one relationship on other interactions.

Because time clearly is a limited good, the hours devoted to interaction in one social position cannot be utilized in another position. A previous section of this chapter focused on the way in which an increase in dyadic-interaction matrix size within one social position—as when group size increases—diminishes the available interaction time per dyad. Here, we are concerned with the effects of increasing the number of social roles that a person plays. For example, employed mothers have less time than full-time housewives to interact with their children (Moore & Hofferth, 1979), and men who work long hours report having less time for their families (Pleck, Staines, & Lang, 1980). There may also be qualitative differences in how people interact as a function of the quantity of time available. Spouses in dual-career marriages appear to adapt to the smaller amount of nonwork time spent together by making adjustments that enable them to maintain affective bonds (Aldous, 1978).

Sociability and love are also limited goods to the extent that each individual has a maximum "carrying capacity" for social ties and affect (Granovetter, 1971). Such limits may underlie the development of jealousy over love and sex in group marriages (Constantine & Constantine,

1973). Similarly, the finding of less intense affect per dyad in loose-knit networks with many members than in close-knit networks with fewer individuals implies the existence of a finite "lump of sociability" (Granovetter, 1971). The concept of the limited good may also characterize the individual's commitment to various social positions if investment in one takes away from the emotional valuation that can be devoted to another, as suggested by the findings that working mothers value parenthood less than do unemployed mothers (Lamb, 1982), and that families trade off between children and the wife's employment (Cramer, 1980).

The affective relationships that infants form with day-care providers (Cummings, 1980; Ricciuti, 1974) do not appear to diminish affect for their parents (Belsky & Steinberg, 1978; Lamb, 1982). The failure to find (with the exception of Blehar's study of 1974) a reduction in day-care infants' affect toward their parents suggests that the sum of affect devoted to parents and supplementary caregivers does not approach the infant's potential. On the other hand, infant–older-sibling agonism (Abramovitch, Corter, & Lando, 1979) and the negative correlation between firstborns' affect to the mother and to the infant sibling (Dunn & Kendrick, 1981) may indicate that siblings compete over limited parental resources. The decrement in mothers' interaction with firstborns once a second child is born (Kendrick & Dunn, 1980) and the diminished level of interaction between mother and later-borns compared with firstborns (Lewis & Krietzberg, 1979) would suggest that the limits of mothers' time, if not affect, are being approached when she cares for several children.

Variation in the effects of maternal employment on mother–infant interaction (Hock & Clinger, 1980; Lamb, 1982; Schubert, Bradley-Johnson, & Nuttal, 1980) may derive from the multifaceted nature of working outside the home. Not only are employed mothers engaged in another social position, but their motivations for working do vary, and their infants are usually placed in substitute care. Lamb (1982) reported that employed mothers valued parenthood less than did full-time housewives, and that 1-year-olds whose mothers valued parenthood less were more likely to be insecurely attached. Employed mothers who valued parenthood highly had more secure relationships with their infants. Although there certainly are other forces (e.g., quality and continuity of substitute care—Anderson, 1980) that determine the effects of maternal employment on mother–infant interation, the mother's distribution of her time, affect, and commitment among various roles appears to be a significant factor.

The values and attitudes that are associated with particular social positions often infiltrate the person's other roles, guiding performance in social positions other than the one in which they were learned. Thus, work values can diffuse into and influence child rearing. Kohn's (1959) analysis of the greater middle-class desire for self-control, curiosity, and independence in their children's behavior revealed that the character of the father's work predicted child-rearing values. Orientation to children's self-control was more often found when the father's work had little supervision and was more involved with people than with physical objects. The underlying principle is that values, attitudes, and behavioral inclinations learned in one social position diffuse into the individual's interaction with partners in other roles.

It is not particularly surprising to find that infants become friendlier with peers after having previous experience with same-age infants (Becker, 1977; Lewis, Young, Brooks, & Michalson, 1975). But infants also seem to generalize from learning in one social position to behavior in another. Infants' prior interaction with older children affects their later behavior with agemates (Snow, Jacklin, & Maccoby, 1981; Vandell, Wilson, & Whalen, 1981). Later-borns are less friendly than firstborns to peers, a finding suggesting that interaction with siblings may affect peer relationships. The generally agonistic quality of later-borns' interaction with older siblings and, in particular, the tendency of older children to direct and dominate infants' behavior (Abramovitch et al., 1979) may diminish later-borns' enthusiasm for such interactions. A disinclination for interaction with older children may generalize to infant peers as well, resulting in later-borns' lesser sociability with peers. Just as adults' work values may influence their child-rearing behavior, infants' experiences with older siblings may generalize to modify interaction with peers as well as with older children. Similarly, infants who had play-group experience with other infants interacted more actively with their parents (Vandell, 1979). Perhaps the more egalitarian nature of peer relationships carried over into the infants' expectations about interaction with adults as well.

Status assigned to individuals as a result of occupying particular social positions can influence their other relationships if such status generalizes and diffuses from one role to another. For example, interaction among jurors is affected by social class; middle-class jurors are more likely than lower-class jurors to initiate interaction (Strodbeck, James, & Hawkins, 1957). The status associated with being a middle-class person diffuses into and affects behavior in the social position of jury member. One source of the effects of diffuse status characteristics (Berger, Fisek, Norman, & Zelditch, 1977) appears to be the tendency of higher status individuals to think more positively of themselves. Indeed, middle-class people tend to have higher self-esteem and to perceive themselves are having greater power (Neal & Seeman, 1964; Rosenberg, 1979).

The effect of social class on self-concept and perceived powerfulness would seem to be relevant to parent–infant interaction. Lewis and Wilson (1972) and Field and Pawlby (1980) found that middle-class mothers engaged in more distal interactions (e.g., vocalization) with their infants than did working-class mothers. Similarly, middle-class mothers vocalized, attempted to entertain, and responded more often to their 10-month-olds (Tulkin & Kagan, 1972). A broader impact of social class can be noted in the finding of Farran and Ramey (1980) that lower-class mothers were less involved with their 6-month-olds. Thus, mothers' social class position within the society is correlated with their interaction with infants. It has been proposed that lower-class mothers are especially likely to perceive themselves as unable to effect changes in their infants' behavior (Lewis & Wilson, 1972; Tulkin & Kagan, 1972). Perhaps such beliefs stem from the general perceptions of ineffectiveness and poor self-image found in lower-class communities with regard to the world outside (Liebow, 1967). The mother may view the infant as someone who controls her, rather than as a person whose behavior can be effectively modified. The position of "lower-class person" may lead mothers to adopt a less positive attitude toward themselves, which then affects their interactions with their infants.

Because an individual's multiple roles often possess conflicting expectations, using the behavioral norms for Position A while playing the role of Position B can be disturbing to role partners in the latter position. For example, a college professor who lectures not only to students but to spouse and children as well is likely to cause discomfort within the family. Similarly, Whyte (1943) found that, in a lower-class Italian community in Boston, the expectations for the role of "friend" and the role of "politician" were contradictory, in that political advancement interfered with the maintenance of old friendship bonds. Along the same line, whereas American adolescents experience contradictory expectations from parents and peers (Coleman, 1961), the greater compatibility of parents' and peers' values in the USSR diminishes such conflicts for Soviet youth (Bronfenbrenner, 1970).

It is, of course, unreasonable to expect infants to be able to make sense of abstract concepts such as multiple roles and the division of labor. Yet infants do interact with people who possess differing interpersonal styles. The finding that year-old infants engage in more play with their fathers than with their mothers (Lamb, 1977) suggests that they distinguish among people not only according to physical features, but also with regard to behavioral style. Indeed, infants' formation of differential expectations about various individuals or classes of people has been noted by Edwards and Lewis (1979), Lamb (1981), and Lewis and Feiring (1978).

Reward satisfaction in one position can modify other interaction in two different ways. First, the rewards associated with two social positions can sometimes serve as functional alternatives (Merton, 1968) in that one is a viable substitute for the other. Much social commerce is characterized by social exchanges in which individuals depend on each other for material resources, such as shelter and food, and social rewards, such as affiliation and love. Active involvement in multiple roles can diminish the individual's dependency on any particular dyadic partner for the satisfaction of these needs. As Emerson (1962) has noted, A's power over B is a direct function of B's dependency on A for rewards. Therefore, a common strategy used by lower power actors—whether they be individuals or nations—is to seek out alternative sources of the desired reward (Webster, 1975). Similarly, people with multidimensional interests can derive self-esteem from one area of their lives when it cannot be obtained from another realm of activity (Rosenberg, 1979).

Infants who are extensively socially connected appear to be less vulnerable when one primary caregiver is absent. It is a salient feature of investigations of long-term separation (Heinicke & Westheimer, 1966) that the children studied typically reside in nuclear families that do not practice multiple caregiving. The absence of alternative caregivers necessitated the placement of the children in residential nurseries when their mothers had to be hospitalized. If these children had lived within extended families with alternative caregivers, it is unlikely that they would have been separated from their homes and from other kin during the absence of a primary caregiver. Children who have multiple caregivers may, indeed, be less dependent on any one particular person for their physical and emotional needs (Lewis, 1982). When one person cannot care for the child, other caregivers are still available. Somewhat similarly, the accessibility of peers when rhesus monkey infants are isolated from their mothers appears to provide substitute sources of affect—as in the together–together monkeys who become strongly attached to each other rather than to their absent mothers—and may even facilitate appropriate social development (Suomi & Harlow, 1975).

Experiences in one social position can also function as prerequisites or facilitators of effective performance in another role, rather than as functional alternatives. The mother's ability to interact effectively with her children may be strengthened by the father's provision of physical, financial, and emotional support to her (Lewis & Weinraub, 1976; Parke, 1979). Adequate functioning of the mother–father relationship would then be a facilitator of effective interaction between mother and young child. Similar effects appear to have been found for infants as well as for older humans. For example, the quality and continuity of substitute caregiving may be a prerequisite for the prevention of the potential negative effects of day care and maternal separation on mother–infant attachment (Anderson, 1980). Similarly, the nature of mother–infant interaction can affect infants' willingness to explore contacts with new infant peers (Easterbrooks & Lamb, 1979). Clearly, one prerequisite for successful relationships with peers must be the willingness to leave the mother's proximity to interact with other infants. Easterbrooks and Lamb found that infants whose attachment to their mothers inhibited exploration were less friendly to peers. Thus, mother–infant interaction can produce an interpersonal style that either facilitates or inhibits infant peer relationships.

Social Influence on Perceptions of Dyadic Partners: Social Referencing

Dyadic interaction is greatly influenced by the perceptions that the dyadic partners have of each other. Because social influence often modifies individuals' responses to situations (Asch, 1951; Sherif, 1958), it is reasonable to expect that perceptions of dyadic partners can be affected by others' opinions. Similarly, people evaluate their own abilities, opinions, and relationships through social comparison (Festinger, 1954). Other people and groups provide points of reference—hence the term *reference group* (Merton & Rossi, 1968)—by which individuals understand the world around them, and themselves as well. In particular, perceptions of one's dyadic partner can be shaped by the opinions expressed by people outside the dyad.

Social influence, social comparison, and reference group phenomena derive from a common reality-construction process that can be termed *social referencing*—the request, receipt, and use of one's perception of other persons' interpretations of the situation to form one's own understanding of that situation (Feinman, 1982). Human response often proceeds in a multistage process in which (1) the stimulus is sensed, (2) the sensation is interpreted, and (3) action is then based on constructed meaning. The construction of reality is a social process because other people can influence the individual's interpretation of the stimulus.

At times, a person may truly not be able to interpret a situation and will turn to others for cognitive clarification. In this process of *internalization* (Kelman, 1961), the individual's interpretation—and not merely his or her behavior—is modified by others' opinions. A case in point is the mother who, being unsure about her 6-month-old's social behavior, asks the pediatrician to assess the infant's social competency and accepts that interpretation because she truly believes it to be accurate. At other times, reality is clear-cut, but the individual's behavior shifts in *compliance* with another's interpretation in order to gain rewards and avoid punishments. Compliance underlies the behavior of the 4-year-old who, while holding firm in the belief that her 3-month-old sibling is crying at night in order to keep everyone awake, expresses verbal agreement with her father's interpretation of the crying as unintentional. In such cases, the individual's interpretation of the situation has not changed, although her or his behavior has been altered.

A distinction can also be made between *direct* and *indirect* influence. A father who is uncertain about his 12-month-old's fear of large stuffed toys can directly elicit a second opinion by asking his friend—who just happens to be an infant researcher—for an evaluation. Alternatively, he can indirectly observe the way his friend behaves on viewing the infant's reactions to the toys. In the first case—direct influence—the friend provides a verbal report of her evaluation of the behavior. In the case of indirect influence, the friend's interpretation is inferred from her actions and nonverbal expressions in reaction to observing the child's behavior.

It is commonly found that an adult is more likely to help out in an emergency when alone with the victim than when others are also present (Latané & Darley, 1970; Latané & Rodin, 1969). One explanation of this inhibiting effect of groups is that the individual's perception of the victim depends on the reactions of other people. When others in the room fail to act as if there is an emergency, the individual is not likely to believe that there is a need for aid. Whereas 64% of those subjects who were alone helped, only 7% helped when in the company of a confederate who did not respond to the crash of a workman in an adjacent area. In comparison, 40% of those with another naive subject helped the workman (Latané & Rodin, 1969). Some of the naive subjects may, indeed, have emitted cues that suggested that the situation was an emergency.

Because the subjects in such studies are usually engaged in cognitive or questionnaire-answering tasks when the emergency occurs, they are unlikely to receive the interpretive cues provided by the facial expressions of others in the room. To consider the importance of facial cues in defining the situation, Darley, Teger, and Lewis (1973) presented an emergency to subjects who were either (1) alone, (2) back-to-back with another subject whose face could not be seen; or (3) seated face-to-face with another subject. When the subjects heard a crash and a cry of pain from the next room, 90% of those working alone helped the victim, compared with only 20% of those with another person whose facial responses could not be seen. But 80% of those who were face-to-face with another person helped the victim. Those in the back-to-back condition could not see the startled faces of their fellow subjects. But the subjects who were face-to-face could infer the other's definition of the situation as an emergency via her or his facial reactions to the accident. Thus, the subjects' interaction with the victim (i.e., whether or not they helped him) was directed by a social referencing process in which they were influenced by other subjects' perceptions of the victim's condition.

Although investigations of second-order effects (Clarke-Stewart, 1978; Lamb, 1978; Lewis & Feiring, 1981; Parke, 1979) have indicated that, for example, the mother's presence affects father–infant interaction, research has not focused on how the infant's perception of the father is influenced by the mother. Nor has the even more intriguing question of whether the infant's behavior toward another person influences the parent's interpretation of that person been considered. Nonetheless, there is evidence that suggests that the cognitive and social skills needed for social referencing develop during the second semester of the first year of life (Feinman, 1982; Klinnert, Campos, Sorce, Emde, & Svejda, 1983). Social referencing can occur if perception is based, at least in part, on constructionist activities involving interpretation of meaning (i.e., appraisal). Activities that appear to indicate the evaluation of the probable consequences of an event increase in frequency between 6 and 12 months (Piaget, 1952). Similarly, the tendency for action to await evaluation (e.g., when reaching for an object is delayed until after observation) seems to become more pronounced around 9 months (Schaffer, Greenwood, & Parry, 1972).

Receptivity to social sources of information about situations appears to emerge between 6 and 12 months. First of all, second-semester infants often look toward their caregivers when encountering a new person or object (Feinman, 1980; Gunnar, 1980; Haviland & Lewis, 1975; Rheingold & Eckerman, 1973). Such behavior could represent the request and receipt of information about the new event. Second, although very young infants do replicate familiar behaviors (Jacobson, 1979; Killen & Uzgiris, 1981; Lewis, 1979), the imitation of novel behaviors emerges later in the first year (Eckerman & Whatley, 1977; Eckerman, Whatley, & McGhee, 1979; Kaye, 1971). Third, beginning around 6 months, infants appear to distinguish among and react appropriately to emotional expressions (Charlesworth & Kreutzer, 1973). In light of infants' limited word comprehension, sensitivity to nonverbal affective cues is probably the major avenue through which social referencing can proceed during the second half-year.

There are a few direct hints that infants engage in social referencing. Seven to fifteen-month-olds were more likely to move toward or away from a stranger after looking at or touching the mother (Feinman, 1980). Perhaps variation in the mothers' nonverbal affective cues modified the infants' interpretation of the stranger. Similarly, 15-month-olds were friendlier to a female stranger when the stranger had been seen interacting positively with the mother than when she had interacted with another person or with no one (Lewis & Feiring, 1981). The mother's favorable behavior to the stranger when the stranger was more positive to the mother may have influenced the infant's understanding of the stranger. Although such results are, of course, subject to alternative explanations, they are consistent with social referencing.

The ability of mothers to socially influence their 1-year-old infants' responses to depth was indicated in a recent study reported by Klinnert et al. (1983). When 12-month-olds were placed on the shallow side of a modified visual cliff with an apparent drop of 12 inches, their mothers' facial expressions influenced the infants' responses. Whereas 14 of the 19 infants whose mothers smiled when the infant confronted the midpoint traversed the deep side, none of the 17 infants whose mothers exhibited a fear face crossed. Thus, infants' responses to moderate depth was affected by their mothers' facial expression. It is important to note, though, that a social referencing effect on infants' willingness to cross the deep side did not appear when a standard visual cliff with an apparent $3\frac{1}{2}$-foot drop was utilized.

Infants' Social Referencing and Interaction with Strangers

We have recently investigated the question of whether social referencing can influence the infant's response to an unfamiliar person (Feinman & Lewis, 1981, 1982, 1983). In the study, 87 ten-month-olds were approached by a stranger while their mothers provided nonverbally positive or neutral messages each time the stranger paused during the approach. The mother spoke either directly to her infant about the stranger, or indirectly by letting the infant observe her speak to the stranger. There was also a control condition with no maternal communication. After the approach, the stranger sat next to the infant for a one-minute period but did not initiate interaction. The mother gave a toy to her infant when the stranger sat down.

Analyses of variance were performed to investigate the impact of the affect (positive or neutral) and the direction of the mother's message (direct to the infant, indirect to the stranger), as well as infant temperament, on infant smiling, proximity, and toy offers to the stranger during the one-minute period. Because the length and the affect of the mother's message were positively correlated, the number of words expressed by the mother was included as a covariate in all analyses. Temperament was measured by the Revised Infant Temperament Questionnaire (Carey & McDevitt, 1978). The infants were grouped into an "easy" category, consisting of Carey and McDevitt's easy and intermediate-low classifications, and a "difficult" category, consisting of the intermediate-high, difficult, and slow-to-warm-up classifications.

Infants whose mothers spoke directly to them smiled at the stranger more often in the positive than in the neutral condition. Affect did not influence smiling when the mother spoke to the stranger. Although the mother's affect had no impact on difficult infants' proximity to the stranger, easy infants were more often proximate to the stranger in the positive than in the neutral affect condition. Similarly, infants offered the toy to the stranger more often in the positive condition than in the neutral condition, but only when the mothers spoke directly to infants of easy temperament.

When mothers communicated directly, especially with easy-temperament infants, the results were consistent with the social referencing prediction that the mother's positive emotion would generate a more favorable interpretation of the stranger by the infant. Could it be, though, that the infant did not understand the mother's communication as information about the stranger? Perhaps the mother's affect modified the infant's mood generally, while not influencing her or his interpretation of the stranger. In mood modification, the more positive the mother's affect, the more favorable the infant's mood becomes, and, therefore, the friendlier he or she will be to the stranger and to other people and objects. But in social referencing, only the behavior to the stranger—and not to other people and objects—is influenced by the mother's message.

In order to compare the social referencing and the mood modification hypotheses we reran the analyses of variance, using measures of infant behavior to the stranger minus behavior to the mother as the dependent criteria. In mood modification, the impact of the mother's message on infant smiles, proximity, and toy offers to the stranger would also be found for behaviors to the mother. Consequently, when behavior to the mother is subtracted from that to the stranger, the significant effects reported above for behaviors to the stranger would not be detected. But in social referencing, the affect of the mother's message would influence behavior to the stranger but not to the mother. Therefore, the frequency of friendly behavior to the stranger minus that to the mother would be greater in the positive condition than in the neutral condition. The results of these analyses were not substantially different from those reported above for behavior to the stranger, suggesting that the mother's message did not modify the infant's overall mood. Rather, it appears that the infants did engage in social referencing by utilizing their mothers' messages as information about the stranger specifically.

The finding that the mother's affect did not influence her infant's behavior when the mother spoke to the stranger implies that, at 10 months, direct social referencing is more effective than indirect social referencing. A 10-month-old may not attend to the mother's affective tone when she speaks to the stranger, failing to understand that such vocalization can provide information about the mother's opinion of the stranger. Or perhaps the infant interprets maternal greeting of the stranger as a threat to the mother's emotional availability. Understanding that Person A's behavior to Person B can be used to infer the former's attitude toward the latter seems to be a skill that develops at a later time.

The results indicated that easy-temperament infants engaged in social referencing more effectively than did difficult infants. Because the Carey and McDevitt (1978) Infant Temperament Questionnaire is multidimensional, further analysis considered whether one of these dimensions could account for this difference. The dimension that tapped the infants' initial response to new situations provided the best discrimination among infants for direct social referencing. Infants rated as accepting of new situations were more influenced by their mothers' affect. Perhaps the infants who were relaxed when meeting a stranger in an unfamiliar room were more receptive to their mothers' messages. Or this dimension may predict individual differences in social referencing because the mother's communication about the stranger was perceived by the infant as being novel, as it provided new information about the situation.

THE SIGNIFICANCE OF SOCIAL CONNECTIONS BEYOND THE DYAD

Social science research can be conceptually dichotomized into micro- and macrolevels of analysis. In this chapter, we have focused on the microlevel of interpersonal behavior in triads, families, and small groups. It is important, though, to note that macrolevel forces generated by the social institutions and social structure of societies (and even of world systems composed of interdependent societies) shape the extradyadic interpersonal environment within which dyadic interaction proceeds.

For example, the effect of infant day care on parent–infant interaction is a microlevel issue. A concern with macroissues leads to the consideration of larger social forces, such as changes in the roles of women and men, and in the sex composition of the labor force, which have increased the availability and the perceived desirability of substitute caregiving facilities. Similarly, the investigation of the father's influence on the mother–infant dyad can inquire into the social forces that determine the availability of the father to interact with the infant and the mother. Differences in subsistence economy strongly influence whether the father's workplace is close to the home, as in many agrarian societies, or far removed, as in hunting and gathering societies, where males are absent for several days at a time in search of large game (Lenski & Lenski, 1982).

Research on adults and older children has looked at sociability not only with respect to dyadic interaction, but with regard to social influences on the dyad and the relationship of the individual with social groups. Although infants should not be viewed as "little adults" (as was sometimes done in earlier societies—Aries, 1962), neither should infant research be isolated from the mainstream of research about human sociability. We have suggested in this chapter that concepts and hypotheses developed to explain extradyadic influence on the social behavior of older humans can also be profitably applied to the study of infants.

The study of human sociability logically begins with the dyad because it is the basic unit of social interaction. Nonetheless, social life is more than the sum of dyadic interchanges; there *is* social life beyond the dyad. Infant researchers should by no means abandon dyadic investigation. But we should expand our vistas and also investigate (1) social forces that influence the dyad and (2) triadic and higher level interaction. Because infants do exist within an intricate network of interpersonal relationships and social structures, the widening of our sights to include extradyadic forces will yield a fuller and more accurate view of infant sociability.

ACKNOWLEDGMENTS

Discussions with Gary Hampe were extremely helpful in formulating the applications of social-psychological theory to the study of infant social behavior.

REFERENCES

Abramovitch, R., Corter, C., & Lando, B. Sibling interaction in the home. *Child Development*, 1979, *50*, 1189–1196.
Ainsworth, M. D., Blehar, M. C., Waters, E., & Wall, S. *Patterns of attachment*. Hillsdale, N.J.: Lawrence Erlbaum, 1978.
Aldous, J. *Family careers: Developmental change in families*. New York: Wiley, 1978.

Anderson, C. W. Attachment in daily separation: Reconceptualizing day care and maternal employment issues. *Child Development*, 1980, *51*, 242–245.
Aries, P. *Centuries of childhood*. New York: Random House, 1962.
Asch, S. E. Effects of group pressure on the modification and distortion of judgements. In H. Geutzkow (Ed.), *Groups, leadership, and men*. Pittsburgh: Carnegie, 1951.
Bales, R. F. *Personality and interpersonal behavior*. New York: Holt, Rinehart and Winston, 1970.
Bales, R. F., & Borgatta, E. F. Size of group as a factor in the interaction profile. In A. P. Hare, E. F. Borgatta, & R. F. Bales (Eds.), *Small groups*. New York: Knopf, 1967.
Bales, R. F., Strodbeck, F., Mills, T. M., & Roseborough, M. Channels of communication in small groups. *American Sociological Review*, 1951, *16*, 461–468.
Becker, J. M. T. A learning analysis of the development of peer-oriented behavior in nine-month-old infants. *Developmental Psychology*, 1977, *13*, 481–491.
Belsky, J. Early human experience: A family perspective. *Developmental Psychology*, 1981, *17*, 3–23.
Belsky, J., & Steinberg, L. D. The effects of day care: A critical review. *Child Development*, 1978, *49*, 929–949.
Berger, J., Fisek, J., Norman, R. Z., & Zelditch, M., Jr. *Status characteristics and social interaction: An expectation states approach*. New York: Elsevier, 1977.
Blau, P. *Power and exchange in social life*. New York: Wiley, 1964.
Blehar, M. C. Anxious attachment and defensive reactions associated with day care. *Child Development*, 1974, *45*, 683–692.
Blood, R. O. Jr., & Wolfe, D. M. *Husbands and wives: The dynamics of family living*. Glencoe, Ill.: Free Press, 1960.
Broderick, C. B., & Pulliam-Krager, H. Family process and child outcomes. In W. R. Burr, R. Hill, F. I. Nye, & I. Reiss (Eds.), *Contemporary theories about the family* (Vol. 1). New York: Free Press, 1979.
Bronfenbrenner, U. *Two worlds of childhood: U.S. and U.S.S.R.* New York: Russell Sage, 1970.
Bronfenbrenner, U. Developmental research, public policy, and the ecology of childhood. *Child Development*, 1974, *45*, 1–5.
Brooks, J., & Lewis, M. Infants' responses to strangers: Midget, adult, and child. *Child Development*, 1976, *47*, 323–332.
Brossard, J. H. S., & Boll, E. S. Personality types in the large family. *Child Development*, 1955, *26*, 71–78.
Brossard, J. H. S., & Sanger, W. The large family system—A research report. *American Sociological Review*, 1952, *17*, 3–9.
Burke, P. J. Participation and leadership in small groups. *American Sociological Review*, 1974, *39*, 832–843.
Campos, J. J., Emde, R. N., Gaensbauer, T., & Henderson, C. Cardiac and behavioral interrelationships in the reactions of infants to strangers. *Developmental Psychology*, 1975, *11*, 589–601.
Caplow, T. *Two against one: Coalitions in triads*. Englewood Cliffs, N.J.: Prentice-Hall, 1968.
Carey, W. B., & McDevitt, S. C. Revision of the Infant Temperament Questionnaire. *Pediatrics*, 1978, *61*, 735–739.
Centers, R., Raven, B. H., & Rodrigues, A. Conjugal power structure: A reexamination. *American Sociological Review*, 1971, *36*, 264–278.
Charlesworth, W. R., & Kreutzer, M. A. Facial expressions of infants and children. In P. Ekman (Ed.), *Darwin and facial expression*. New York: Academic Press, 1973.
Chilman, C. S. Families in poverty in the early 1970s: Rates, associated factors, some implications. *Journal of Marriage and the Family*, 1975, *37*, 49–60.

Clarke-Stewart, K. A. And daddy makes three: The father's impact on mother and young child. *Child Development*, 1978, *49*, 466–478.
Cohen, L. J., & Campos, J. J. Father, mother, and stranger as elicitors of attachment behavior in infancy. *Developmental Psychology*, 1974, *10*, 146–154.
Coleman, J. S. *The adolescent society*. Glencoe, Ill.: Free Press, 1961.
Coleman, J. S., Katz, E., & Menzel, H. *Medical innovation: A diffusion study*. Indianapolis: Bobbs-Merrill, 1966.
Constantine, L. L., & Constantine, J. M. *Group marriage: A study of contemporary multilateral marriage*. New York: Macmillan, 1973.
Cook, K. S., & Emerson, R. M. Power, equity, and commitment in exchange networks. *American Sociological Review*, 1978, *43*, 721–739.
Cramer, J. C. Fertility and female employment: Problems of causal direction. *American Sociological Review*, 1980, *45*, 397–432.
Cromwell, R. E., & Olson, D. H. (Eds.). *Power in families*. New York: Wiley, 1975.
Cromwell, V. L., & Cromwell, R. E. Perceived dominance in decision-making and conflict resolution among Anglo, Black, and Chicano couples. *Journal of Marriage and the Family*, 1978, *40*, 749–759.
Cummings, E. M. Caregiver stability and day care. *Developmental Psychology*, 1980, *16*, 31–37.
Darley, J. M., Teger, A. I., & Lewis, L. D. Do groups always inhibit individuals' responses to potential emergencies? *Journal of Personality and Social Psychology*, 1973, *26*, 395–399.
Dawe, H. C. The influence of size of kindergarten group upon performance. *Child Development*, 1934, *5*, 295–303.
Dunn, J., & Kendrick, C. Interaction between young siblings: Association with the interaction between mother and firstborn child. *Developmental Psychology*, 1981, *17*, 336–343.
Easterbrooks, M. A., & Lamb, M. E. The relationship between quality of infant–mother attachment and infant competence in initial encounters with peers. *Child Development*, 1979, *50*, 380–387.
Eckerman, C. O., & Whatley, J. L. Toys and social interaction between infant peers. *Child Development*, 1977, *48*, 1645–1656.
Eckerman, C. O., Whatley, J. L., & McGhee, L. J. Approaching and contacting the object another manipulates: A social skill of the 1 year old. *Developmental Psychology*, 1979, *15*, 585–593.
Edwards, C. P., & Lewis, M. Young children's concepts of social relations: Social functions and social objects. In M. Lewis & L. A. Rosenblum (Eds.), *The child and its family*. New York: Plenum Press, 1979.
Emerson, R. M. Power-dependence relations: Two experiments. *American Sociological Review*, 1962, *27*, 31–41.
Erikson, K. T. *Everything in its path: Destruction of community in the Buffalo Creek flood*. New York: Simon and Schuster, 1976.
Falbo, T., & Peplau, L. A. Power strategies in intimate relationships. *Journal of Personality and Social Psychology*, 1980, *38*, 618–628.
Farran, D. C., & Ramey, C. T. Social class differences in dyadic involvement during infancy. *Child Development*, 1980, *51*, 254–257.
Feinman, S. Infant response to race, size, proximity, and movement of strangers. *Infant Behavior and Development*, 1980, *3*, 187–204.
Feinman, S. Social referencing in infancy. *Merrill-Palmer Quarterly*, 1982, *28*, 445–470.
Feinman, S., & Lewis, M. Social referencing and second order effects in ten-month-old infants. In J. Rubenstein (Chair), *Conceptualizing second order effects in infancy*. Symposium presented at the Biennial Meeting of the Society for Research in Child Development, Boston, April 1981.

Feinman, S., & Lewis, M. *Infant temperament and social referencing.* Presented at meeting of the International Conference on Infant Studies, Austin, Texas, March 1982.

Feinman, S., & Lewis, M. Social referencing at 10 months: A second order effect on infants' responses to strangers. *Child Development*, 1983, 54, 878–887.

Feiring, C., & Lewis, M. Early mother–child interaction: Families with only and firstborn children. In G. L. Fox (Ed.), *The childbearing decision.* Beverly Hills, Calif.: Sage, 1982.

Festinger, L. A theory of social comparison processes. *Human Relations*, 1954, 7, 17–40.

Festinger, L. *A theory of cognitive dissonance.* Stanford, Calif.: Stanford University Press, 1964.

Field, T. M. Infant behaviors directed towards peers and adults in the presence and absence of mother. *Infant Behavior and Development*, 1979, 2, 47–54.

Field, T. M., & Pawlby, S. Early face-to-face interaction of British and American working-class and middle-class mother–infant dyads. *Child Development*, 1980, 51, 250–253.

Frankel, D. G., & Arbel, T. Group formation by two-year-olds. *International Journal of Behavioral Development*, 1980, 3, 287–298. (a)

Frankel, D. G., & Arbel, T. Social organization of young children's groups: A comparison of 2 models of dominance relations. *Psychological Reports*, 1980, 46, 911–915. (b)

Gamson, W. A. A theory of coalition formation. *American Sociological Review*, 1961, 26, 565–573.

Gans, H. J. *The urban villagers.* New York: Free Press, 1962.

Garbarino, J. The human ecology of child maltreatment: A conceptual model for research. *Journal of Marriage and the Family*, 1977, 39, 721–727.

Garbarino, J., & Sherman, D. High-risk neighborhoods and high-risk families: The human ecology of child maltreatment. *Child Development*, 1980, 51, 188–198.

Goetsch, G. G., & McFarland, D. D. Models of the distribution of acts in small discussion groups. *Social Psychology Quarterly*, 1980, 43, 173–183.

Goldman, J. A. Social participation of preschool children in same-versus mixed-age groups. *Child Development*, 1981, 52, 644–650.

Granovetter, M. S. *Child-rearing, weak ties, and socialism: A conjecture.* Paper presented at the meeting of the American Sociological Association, Denver, September 1971.

Granovetter, M. S. The strength of weak ties. *American Journal of Sociology*, 1973, 78, 1360–1380.

Gunnar, M. R. Control, warning signals, and distress in infancy. *Developmental Psychology*, 1980, 16, 281–289.

Hare, A. P. Interaction and consensus in different sized groups. *American Sociological Review*, 1952, 17, 261–267.

Haviland, J., & Lewis, M. *Infants' greeting patterns to strangers.* Paper presented at the meeting of the Human Ethology Session of the Animal Behavior Society, Wilmington, N.C., May 1975.

Hay, D. F., Nash, A., & Pedersen, J. Responses of six-month-olds to the distress of their peers. *Child Development*, 1981, 52, 1071–1075.

Heider, F. Attitudes and cognitive organization. *Journal of Psychology*, 1946, 21, 107–112.

Heinicke, C., & Westheimer, I. *Brief separations.* New York: International Universities Press, 1966.

Hock, E., & Clinger, J. B. Behavior toward mother and stranger of infants who have experienced group day care, individual day care, or exclusive maternal care. *Journal of Genetic Psychology*, 1980, 137, 49–61.

Indik, B. P. Organization size and member participation: Some empirical tests of alternatives. *Human Relations*, 1965, 18, 339–350.

Jacobs, B. S., & Moss, H. A. Birth order and sex of sibling as determinants of mother–infant interaction. *Child Development*, 1976, 47, 315–322.

Jacobson, S. W. Matching behavior in the young infant. *Child Development*, 1979, *50*, 425–430.

Kaye, K. *Learning by imitation in infants and young children.* Paper presented at the Biennial Meeting of the Society for Research in Child Development, Minneapolis, March 1971.

Kelman, H. Processes of opinion change. *Public Opinion Quarterly*, 1961, *25*, 57–78.

Kendrick, C., & Dunn, J. Caring for a second baby: Effects on interaction between mother and first born. *Developmental Psychology*, 1980, *16*, 303–311.

Killen, M., & Uzgiris, I. C. Imitation of actions with objects: The role of social meaning. *Journal of Genetic Psychology*, 1981, *138*, 219–229.

Klinnert, M., Campos, J. J., Sorce, J., Emde, R. N., & Svejda, M. J. Social referencing: An important appraisal process in human infancy. In R. Plutchik & H. Kellerman (Eds.), *The emotions* (Vol. 2). New York: Academic Press, 1983.

Kohn, M. L. Social class and parental values. *American Journal of Sociology*, 1959, *64*, 337–351.

Komorita, S. S., & Moore, D. Theories and processes of coalition formation. *Journal of Personality and Social Psychology*, 1976, *33*, 371–381.

Lakin, M., Lakin, M. G., & Costanzo, P. R. Group processes in early childhood: A dimension of human development. *International Journal of Behavioral Development*, 1979, *2*, 171–183.

Lamb, M. E. Twelve-month-olds and their parents: Interactions in a laboratory playroom. *Developmental Psychology*, 1976, *12*, 237–244.

Lamb, M. E. Father–infant and mother–infant interaction in the first year of life. *Child Development*, 1977, *48*, 167–181.

Lamb, M. E. Infant social cognition and "second-order" effects. *Infant Behavior and Development*, 1978, *1*, 1–10.

Lamb, M. E. The development of social expectations in the first year of life. In M. E. Lamb & L. R. Sherrod (Eds.), *Infant social cognition: Theoretical and empirical considerations.* Hillsdale, N.J.: Lawrence Erlbaum, 1981.

Lamb, M. E. Maternal employment and child development: A review. In M. E. Lamb (Ed.), *Nontraditional families: Parenting and childrearing.* Hillsdale, N.J.: Lawrence Erlbaum, 1982.

Latané, B., & Darley, J. M. *The unresponsive bystander: Why doesn't he help?* New York: Appleton Century Crofts, 1970.

Latané, B., & Rodin, J. A lady in distress: Inhibiting effects of friends and strangers on bystander intervention. *Journal of Experimental Social Psychology*, 1969, *5*, 189–202.

Leavitt, M. J. Contingent feedback, familiarization, and infant affect: How a stranger becomes a friend. *Developmental Psychology*, 1980, *16*, 425–432.

Lenski, G., & Lenski, J. *Human societies* (4th ed.). New York: McGraw-Hill, 1982.

Lewis, M. The social network systems model: Toward a theory of social development. In T. Field (Ed.), *Review in human development.* New York: Wiley, 1982.

Lewis, M. *Issues in the study of imitation.* Paper presented at the Biennial Meeting of the Society for Research in Child Development, San Francisco, March 1979.

Lewis, M., & Feiring, C. The child's social world. In R. M. Lerner & G. B. Spanier (Eds.), *Child influences on marital and family interaction: A lifespan perspective.* New York: Academic Press, 1978.

Lewis, M., & Feiring, C. *Some American families at dinner.* Paper presented at a conference on the Family as a Learning Environment, Educational Testing Service, Princeton, N.J., November 1979.

Lewis, M., & Feiring, C. Direct and indirect interactions in social relationships. In L. Lipsitt (Ed.), *Advances in infancy research.* New York: Ablex, 1981.

Lewis, M., & Kreitzberg, V. S. Effects of birth order and spacing on mother–infant interaction. *Developmental Psychology*, 1979, *15*, 617–625.

Lewis, M., & Rosenblum, L. A. (Eds.). *The child and its family.* New York: Plenum Press, 1979.

Lewis, M., & Weinraub, M. A. The father's role in the child's social network. In M. E. Lamb (Ed.), *The role of the father in child development.* New York: Wiley, 1976.

Lewis, M., & Wilson, C. D. Infant development in lower-class American families. *Human Development,* 1972, *15,* 112–127.

Lewis, M., Young, G., Brooks, J., & Michalson, L. The beginning of friendship. In M. Lewis & L. A. Rosenblum (Eds.), *Friendship and peer relations.* New York: Wiley, 1975.

Liebow, E. *Tally's corner.* Boston, Little, Brown, 1967.

Lytton, H. Disciplinary encounters between young boys and their mothers and fathers: Is there a contingency system? *Developmental Psychology,* 1979, *15,* 256–268.

Martin, J. A. A longitudinal study of the consequences of early mother–infant interaction: A microanalytic approach. *Monographs of the Society for Research in Child Development,* 1981, *46,* (3, Serial No. 190).

Mayhew, B., & Levinger, R. L. On the emergence of oligarchy in human interaction. Unpublished manuscript, Temple University, 1974.

McGrew, W. C. *An ethological study of children's behavior.* New York: Academic Press, 1972.

Merton, R. K. *Social theory and social structure.* New York: Free Press, 1968.

Merton, R. K., & Rossi, A. S. Contributions to the theory of reference group behavior. In R. K. Merton, *Social theory and social structure.* New York: Free Press, 1968.

Miller, C. E. Coalition formation in triads: Effects of liking and resources. *Personality and Social Psychology Bulletin,* 1981, *7,* 296–301.

Minuchin, S. *Familes and family therapy.* Cambridge: Harvard University Press, 1974.

Moore, K. A., & Hofferth, S. L. Women and their children. In R. E. Smith (Ed.), *The subtle revolution: Women at work.* Washington, D.C.: The Urban Institute, 1979.

Morgan, G. A., & Ricciuti, H. N. Infants' responses to strangers during the first year. In B. M. Foss (Ed.), *Determinants of infant behavior* (Vol. 4). London: Methuen, 1969.

Neal, A., & Seeman, M. Organization and powerlessness: A test of the mediation hypothesis. *American Sociological Review,* 1964, *29,* 216–225.

Nelson, J. I. Clique contacts and family orientation. *American Sociological Review,* 1966, *31,* 663–672.

Parke, R. D. Conceptualization of the effects of fathers in infancy. In J. D. Osofsky (Ed.), *Handbook of infant development.* New York: Wiley, 1979.

Parke, R. D., & Collmer, C. W. Child abuse: An interdisciplinary analysis. In E. M. Hetherington (Ed.), *Review of child development research* (Vol. 5). Chicago: University of Chicago Press, 1975.

Pastor, D. L. The quality of mother–infant attachment and its relationship with toddlers' initial sociability with peers. *Developmental Psychology,* 1981, *17,* 326–335.

Pedersen, F. A., Anderson, B. J., & Cain, R. L. Jr. Parent–infant and husband–wife interaction at age five months. In F. A. Pedersen (Ed.), *The father–infant relationship.* New York: Praeger, 1980.

Piaget, J. *The origins of intelligence in children.* New York: International Universities Press, 1952.

Pleck, J., Staines, G. L., & Lang, L. Conflicts between work and family life. *Monthly Labor Review,* 1980, *103*(3), 29–31.

Rheingold, H. L., & Eckerman, C. O. Fear of the stranger: A critical review. In H. W. Reese (Ed.), *Advances in child development and behavior.* New York: Academic Press, 1973.

Ricciuti, H. N. Fear and the development of attachments in the first year of life. In M. Lewis & L. A. Rosenblum (Eds.), *The origins of fear.* New York: Wiley, 1974.

Rosenberg, M. *Conceiving the self.* New York: Basic Books, 1979.

Salamon, S., & Keim, A. M. Land ownership and women's power in a Midwestern farming community. *Journal of Marriage and the Family*, 1979, *41*, 109–119.

Samuels, H. R. The effect of an older sibling on infant locomotor exploration of a new environment. *Child Development*, 1980, *51*, 607–609.

Savin-Williams, R. C. Dominance hierarchies in groups of early adolescents. *Child Development*, 1979, *50*, 923–932.

Schaffer, H. R., & Crook, C. K. Child compliance and maternal control techniques. *Developmental Psychology*, 1980, *16*, 54–61.

Schaffer, H. R., Greenwood, A., & Parry, M. H. The onset of wariness. *Child Development*, 1972, *43*, 165–175.

Schubert, J. B., Bradley-Johnson, S., & Nuttal, J. Mother–infant communication and maternal employment. *Child Development*, 1980, *51*, 246–249.

Shanas, E. Family-kin networks and aging in cross-cultural perspective. *Journal of Marriage and the Family*, 1973, *35*, 505–511.

Sherif, M. Group influences upon the formation of norms and attitudes. In E. Maccoby, T. Newcomb, & E. Hartley (Eds.), *Readings in social psychology*. New York: Holt, Rinehart and Winston, 1958.

Smith, P. K. Social and situational determinants of fear in the playgroup. In M. Lewis & L. A. Rosenblum (Eds.), *The origins of fear*. New York: Wiley, 1974.

Smith, S., & Haythorn, W. W. Effects of compatibility, crowding, and leadership seniority on stress, anxiety, hostility, and annoyance in isolated groups. *Journal of Personality and Social Psychology*, 1972, *22*, 67–79.

Snow, M. E., Jacklin, C. N., & Maccoby, E. E. Birth-order differences in peer sociability at thirty-three months. *Child Development*, 1981, *52*, 589–595.

Sojit, C. M. The double-bind hypothesis and the parents of schizophrenics. *Family Process*, 1971, *10*, 53–75.

Spinetta, J. J., & Rigler, D. The child abusing parent: A psychological review. *Psychological Bulletin*, 1972, *77*, 296–304.

Stack, C. B. *All our kin: Strategies for survival in a black community*. New York: Harper and Row, 1974.

Strodbeck, F. L., James, R. M., & Hawkins, C. Social status in jury deliberations. *American Sociological Review*, 1957, *22*, 714–719.

Suomi, S. J., & Harlow, H. F. The role and reason of peer relationships in Rhesus monkeys. In M. Lewis & L. A. Rosenblum (Eds.), *Friendship and peer relations*. New York: Wiley, 1975.

Thoman, E. B., Barnett, C. R., & Leiderman, P. H. Feeding behaviors of newborn infants as a function of parity of the mother. *Child Development*, 1971, *42*, 1471–1483.

Trivers, R. L. Parent–offspring conflict. *American Zoologist*, 1974, *14*, 249–264.

Tulkin, S. R., & Kagan, J. Mother–child interaction in the first year of life. *Child Development*, 1972, *43*, 31–41.

Turnbull, C. *The forest people*. New York: Simon and Schuster, 1961.

Vandell, D. L. Effects of a playgroup experience on mother–son and father–son interaction. *Developmental Psychology*, 1979, *15*, 379–385.

Vandell, D. L., Wilson, K. S., & Whalen, W. T. Birth-order and social experience differences in infant–peer interaction. *Developmental Psychology*, 1981, *17*, 438–445.

Van den Berghe, P. L. *Man in society* (2nd ed.). New York: Elsevier, 1978.

Vogel, E. F., & Bell, N. W. The emotionally disturbed child as a family scapegoat. In N. W. Bell & E. F. Vogel (Eds.), *A modern introduction to the family*. New York: Free Press, 1960.

Webster, M. A. *Actions and actors*. Cambridge, Mass.: Winthrop, 1975.

Weinraub, M. A., & Putney, E. The effects of height on infants' social responses to unfamiliar persons. *Child Development*, 1978, *49*, 598–603.

Weinraub, M. A., Brooks, J., & Lewis, M. The social network: A reconsideration of the concept of attachment. *Human Development*, 1977, *20*, 31–47.
Whyte, W. F. *Street corner society*. Chicago: University of Chicago Press, 1943.
Williams, R. M., & Mattson, M. L. The effect of social groupings upon the language of preschool children. *Child Development*, 1942, *13*, 233–245.
Wolff, K. H. *The sociology of Georg Simmel*. New York: Free Press, 1950.
Young, M., & Wilmott, P. *Family and kinship in East London*. Baltimore: Penguin, 1957.

3

The Social Ecology of Childhood
A Cross-Cultural View

THOMAS S. WEISNER

INTRODUCTION

Imagine that you could arrange for a satellite to circle the earth with extremely accurate high-resolution cameras and sound equipment on board, and that you could direct this satellite to hover over the culture areas of the world. Imagine that you could program your satellite to randomly sample communities, and within each community systematically select households with boys and girls of different ages, and that you could then monitor the members of those households throughout their day. Your cameras would record whom these family members are with; how far they stray from their home; what work they do; and with whom they do it. It would tell you the characteristics of the work and play groups, as well as the domestic group surrounding the child. It would tell you the kinds of physical stimulation the children are exposed to and how the children react and interact with their physical surroundings and social settings. Imagine that the microphones in your satellite could systematically sample what the children talk about; could study the content of their speech and with whom the children talk; and could observe the context of speech activities.

You would be systematically sampling the social ecology (Bronfenbrenner, 1979), or *cultural niche* (Super & Harkness, 1980), of the children.

THOMAS S. WEISNER • Departments of Psychiatry and Anthropology, University of California, Los Angeles, California 90024. Preparation of this chapter was assisted by the Department of Psychiatry and Biobehavioral Sciences, University of California, Los Angeles.

The satellite would allow us to know something that the existing cross-cultural record of the social ecology of childhood and development allows us only partial, fleeting glimpses of at the present time: the systematic assessment of the social milieu within which children grow up around the world. Until the cross-cultural and anthropological material on the social ecology of child development can equal such a detailed record, what is available as a substitute? We have materials in the ethnographic record, a bit of which is coded in the Human Relations Area Files for residence patterns, child-rearing practices, socialization norms, economy and ecology, and scores of other variables (e.g., Barry & Schlegel, 1980). Among the difficulties of interpreting these data is that they are limited to a certain period in human cultural evolution—from the late nineteenth century to the mid-twentieth century. This is the period when colonial contact—and subsequent systematic study by anthropologists) "froze" the evolution of societies into the subsistence types, locations, and sizes that they are described in. Although I will use an abstracted notion of ideal-type cultures, it should be clear from the outset that even this view and this time period are not completely accurate. Obviously, cultural change, urbanization, the spread of Western schooling, Islam and other factors have heavily influenced what the satellite would see. Nonetheless, the imprint of that early- to mid-twentieth-century way of life is still an extremely powerful influence on most peoples of the world, and I will use it as the "ethnographic present."

Even with our satellites, our microphones, and our sampling procedures, we would still need two other things before we could make use of our data in research on the cultural niche and its role in human development. First, we would need members of each social community to discuss our films with us; to give us an interpretation of their meaning; and to point to the things that seem important to the members of the community. We would need to initiate a dialogue with an insider, a participant, to define the relevant contexts; the folk meaning; and the emotional and affective sentiments attached to what we find on our film. I see this as an essential dialogue between a cultural insider and an ethnographer/outsider. The usefulness of the social ecology samples depends on both the folk view and the scientific, Western view.

The second element that would still be needed are some theories, some hypotheses, some lists of the things that are going to make a difference in the social ecology of childhood. We would face an impossible data selection as well as interpretation problem if simply presented with all the evidence from our imaginary satellite. We have also to imagine what it is that will made a difference for families and children, to focus in on those things, and to test for their effects. This chapter outlines what cross-cultural research has shown some of those characteristics of social ecology to be, and then illustrates their effects with a brief example from East Africa.

THE CULTURAL NICHE

An important use of anthropological data is to extend our knowledge of the range of variation in children's behaviors or in the environments of children's experience. Super and Harkness (1980) argued that this is not the most important contribution. The comparative view also reveals behavior as it has adapted to a cultural-ecological niche. Anthropologists characteristically provide interpretations of childhood and parenting (individual differences, cognitive development, social-behavioral development, and so on) in the context of ecological circumstances and household and domestic environments. Anthropology hardly deserves original credit for this idea. It is a venerable one in developmental research in many fields, although a review of most research studies in developmental psychology journals would not lead one to believe that this was so. No doubt anthropologists have been "forced" into this view as much because of the dramatic differences in the cultural ecologies that they find in other parts of the world as because of a clearly defined theoretical approach. In any event, the study of human development in different cultures has certainly not led to a unified theory. Although anthropologists continue the tradition of cultural accounts of families and children in context, comparative work has focused on the kinds of social-ecological features likely to affect social relations within the child's world; on how they have their effects; and on ways in which these can be measured. John and Beatrice Whiting, Caroline Edwards, Lawrence Baldwin, I, and others—drawing on comparative cross-cultural research on the family and childhood (see the reviews in Munroe, Munroe, & Whiting, 1981)—prepared an inventory of the features of the cultural niche that are likely to affect children's social development (Whiting, Edwards et al., in preparation), and these are listed in Figure 1.

These conditions are closely interrelated, and the list of 14 topics is somewhat arbitrary. Some can be included under other topics (chores under division of labor, for instance), and some probably deserve independent status, although incorporated under a broader heading (bilingualism or class differences are here included under heterogeneity, for instance). The first group of measures (work, subsistence, health and community demography, and safety) influence the protective care of infants and young children (e.g., LeVine, 1977). Work and task pressures faced by parents and children structure requirements for compliance, obedience, discipline, and aggression training, as well as types of af-

1. The characteristics of the *work cycle,* including wage work, tending crops and animals; distance of fields from the home; transhumance.
2. The *health status and demographic characteristics* of the community, including mortality, availability of health care, birth control.
3. Overall *community safety* (including dangers other than health and mortality), such as dangers from motor vehicles, wild animals, intra- and intercommunity violence and warfare.
4. The *division of labor* by age and sex at adolescence and in adulthood, including the relative importance of various activities for subsistence and prestige.
5. The *daily routine* and activities of children by age and sex at each period of the day; number of hours spent on various tasks and in different settings; those with whom these activities are done.
6. The *chores that children do,* beginning as toddlers, through adolescence.
7. Child-rearing and *child-care tasks* as a special component of family tasks.
8. The composition of *children's play groups* by age, sex, and kinship category (siblings, cousins, relatives, and nonrelatives).
9. The autonomy, independence, and *role of women in the community.*
10. Institutionalized *women's support groups,* including both formal and nonformal groups, such as work groups; church clubs; mutual aid societies; co-wife relationships.
11. *The role of the father* in child care as a special issue in parental child care (much as sibling care of other children is of special interest).
12. Various sources of child "stimulation" including the child's contact with the media; the outside world; toys; *available sources of cultural influence* on children, both literate and oral.
13. *Parental sources of information* concerning child health; nutrition; new methods of subsistence activities; new methods of child care; and novel or contrastive beliefs about childhood available to the community.
14. Measures of *community heterogeneity,* including the presence of subethnic communities; bilingualism; subcastes; social class differences; social solidarity; the role of minorities; group oppression and lack of community commitment among some subgroups; information on migration; the number of generations families have lived in the community; and overall estimates of the heterogeneity and changeability of the community setting.

FIG. 1. Fourteen cross-culturally relevant conditions in the social ecology of child development.

fective expression in the family (e.g., Minturn & Lambert, 1964; Munroe et al., 1981). The next group (the division of labor, the daily routine, chores, child and sibling caretaking, and children's play groups) includes what are probably the most powerful features influencing the *personnel* likely to be present around children, as well as the task and activity contexts within which children learn (Katz & Konner, 1981; Weisner & Gallimore, 1977; Whiting, 1980; Whiting & Whiting, 1975).

The next set of measures affects the *role of women* in the community as the primary socializers of young children (women's autonomy, women's support groups, and the role of the father). The models that mothers provide for their sons and daughters are in many ways the products of their subsistence roles and their social support groups outside of the home. The final set of factors (sources of information and stimulation and the heterogeneity and variability within the community) provides measures of the range of *cultural alternatives* beyond the local community, which can influence parents' and children's folk views of their world.

It would take us far beyond the bounds of this essay to give extensive illustrations of studies that have found relationships between human development and each of these social-ecological features. Any one of them could be the topic of a review. However, were our satellite to be in operation, these would be the features of the social and physical landscape around families and children that deserve our attention. These conditions produce local adaptation pressures on families and shape their cultural beliefs.

How Do Ecocultural Niche Pressures Affect Families?

How do the conditions of the niche produce effects on dyadic relations in families? John and Beatrice Whiting have proposed that what we think of as "cultural-ecological" influences have their proximate effects through the decisions made by the members of the household as to the places that children are sent and "the company children keep" when they are in those local settings (B. Whiting, 1980, 1983; B. Whiting *et al.*, in preparation). Parents may not be actively and directly involved in the tuition of their children, in behavior change, child discipline, and so forth, but parents usually do structure the child's daily schedule and thereby place the child in the environments where he or she will, in turn, learn with others. From this point of view, it is the *manipulation of and control over settings—especially personnel and scarce resources—by parents* that produces ecocultural niche influences on children. Who gets into those settings; why are they there; and what do children have to do while they are there? These are parental decisions that affect children's social behavior indirectly, through shaping their learning environments. Such parental decisions about learning environments can be systematically compared across cultures.

The folk views that parents and children have about such learning environments—one's perceptions of the company one keeps, for instance—is also of demonstrable importance in determining the effect of social situations on child development. In a recent study of nonparental caretaking, for example, Weisner, Gallimore, and Tharp (1982) asked children in Hawaii about their own views concerning their involvement in sibling caretaking. The children's opinions were compared with those of trained observers; a careful set of criteria was used for judging whether sibling caretaking was occurring. There was a statistically significant concordance between the two sets of reports. On this topic (judging participation in the child caretaking hierarchy), the children's views often agreed with the observers'. Observer and child reports did disagree, but usually only in certain situations. Thus far, anthropology has surprisingly little to tell us about the child's-eye view (cf. Schwartz, 1981) of his or her own social and learning environment. The impact of such a view, and of children's culture more generally, is of profound importance in the study of the social context "beyond the dyad."

Boehm (1982), Chibnick (1981), Cancian (1972), Fjellman (1976), and others have described some of the decision-making patterns that are probably operating in parental responses to the changing pressures of their ecocultural niche. "Precognitive" decisions are those based on biosocial predispositions and on related cultural patterns that are so basic and pervasive that they are rarely consciously articulated. Other kinds of decisions involve calculations of energy cost, degree of risk, reduction of uncertainty, and the degree of novelty that results from making a certain choice. Familial adaptation to the community niche does not necessarily involve "rational" or optimum or "satisficing" calculations. Rather, these mechanisms appear to be a mixture of intended and unintended consequences of decisions that are shaped by prior histories, cultural constraints, and the limits of our "folk logic" and ways of thought (e.g., Shweder & D'Andrade, 1980).

A Descriptive Illustration: Kisa

It is not possible in this brief chapter to cover in detail the influence of each of these domains (work loads, social support, child-care systems; and so forth) or to review extensively the cross-cultural research behind each one (see Munroe et al., 1981). Instead, I would like to illustrate some of these social-ecological effects on human development by describing a series of studies on the effects of urban migration on children's social behavior, cognitive functioning, and parenting among the Abaluyia of Western Kenya, East Africa. The strategy of this research included the systematic assessment of the social ecology, in an attempt to understand changes in child rearing and children's social behavior resulting from rural-urban migration (Sangree, 1966; Wagner, 1949; Weisner, 1976a,b, 1979; Weisner & Abbott, 1977; Weisner & Ross, 1977).

The Abaluyia of Western Kenya are a Bantu group of horticulturalists living in dispersed settlements in homesteads of patrilineally related kin. Horticulture is the primary subsistence activity, along with wage labor inside and outside the community. There is also some supplemental care of sheep, goats, and cattle, as well as some shopkeeping activities. Horticultural work of some kind goes on for much of the year, as the many small (three- to seven-acre) plots of land are intensively cultivated, often with two crops per year.

Although child mortality and health status have improved dramatically during the past two generations, approximately 20% of the infants born in the community do not survive to adolescence. A private health clinic and government and church missions all provide health care, but it is costly and not easily available. Traditional medical techniques are also used. Members of the community feel that intercommunity violence has increased, and there is more fear of theft and robbery than before. At the same time, children and adults roam widely over an area of several square miles in daytime and evening hours with complete freedom and seldom meet people they do not know on the many paths branching off the main dirt road. This main road carries some auto and infrequent bus traffic. Children almost always find relatives, kin, or acquaintances wherever they go.

Children participate actively in the tasks and chores of the homestead, including horticultural work, tending livestock, domestic chores, and child care. About 60% of the children in the community attend primary school, which takes from four to six hours of the day. The work load is fairly heavy, although not grueling. Horticultural tasks, petty trading, child care, getting fuel and water, cooking, baby tending, house repair—all of these are constant and demanding tasks. Men and women both participate in horticultural work, although the women are almost exclusively responsible for domestic activities in infant care and most of the management of children in the homestead. The children characteristically are surrounded by multiage, multisex groups of siblings and cousins throughout the day, and they do many tasks in the context of such a group.

The women marry into their new husbands' exogamous Kisa Abaluyia patrilineal clans from their own natal clans, which are located anywhere from 2 to 20 miles away. The women's natal homes are close enough so that there is a constant interchange of relatives between the natal and the married home. The women can find support from other kinswomen, and they visit home fairly frequently. At the same time, however, they do not own property, nor do they directly inherit substantial independent economic resources. In the public-political domain outside their husbands' homesteads, they are restricted in their activities and their autonomy. Within the domestic and economic management domain, however, they have a fair amount of independent authority and decision-making power. The women are also active in community, church, and social affairs, although seldom in the legal-political realm. There are also informal collective women's work groups and mutual aid societies. Overall, Abaluyia married women have a formal role of dependence on their husbands' clan, but informally, there are a number of areas where they make autonomous decisions and influence community affairs.

Abaluyia men seldom engage in infant or child care and are only very rarely observed performing any chores related to direct care. However, they are rather warm and involved in interactions with their young children, especially at the toddler and early childhood ages, and they are not the distant father figures sometimes portrayed in ethnographic studies. The role of the father in the daily experience of the child, however, is nonetheless very limited, as the fathers are not regularly in and around the homestead interacting with their children. They are involved at leisure and rest times of their choosing. The mothers and the children do not routinely and regularly have the fathers as interactive models present in the home during most of the day. Older siblings (usually girls), grandparents, or aunts play this role more often.

The Kisa area is in extensive contact with the world outside the local community. Many homesteads have radios. National newspapers are delivered by daily bus to the village shops, and there is a steady stream of national and regional cultural influence on the local community. Children who attend school learn the national standardized school curriculum in English and their own language (Luluyia). At any one time, some 40% of the adult working-age men are way from the community seeking wage labor in cities or on large farms; a survey indicated that 94% of all of the adult men in Kisa had worked away from the home community at least once in their lives. Thus, the Abaluyia are highly involved in the world outside their local villages, and they are influenced by their national economy and the world economy beyond it. Because Kenya is a former British colony, English language, culture, and economics heavily affect these new contacts.

Kisa is a culturally homogeneous community, with Luluyia as the common first language. English and Kiswahili are also widely spoken. Differences in income and in educational and occupational status are growing, as some families are resourceful or fortunate enough to obtain sources of outside wage income. This rapid change is in addition to the normal vagaries of family demography, which find some parents gaining more land or inheriting more cattle than others. The community is becoming more socially and economically heterogeneous.

Leaving aside the many inevitable differences in detail, this sketch of a horticultural-pastoral mixed economy infiltrated by wage labor and under increasing outside influence describes the characteristic ecocultural niche for a plurality of the world's families and children today (cf. Leiderman & Leiderman, 1977). It is likely that even given the rapid, and potentially tragic, changes affecting this type of ecology because of population pressure, the pattern is likely to persist for the next two or three generations.

THE URBAN CONTRAST: NAIROBI

Nairobi presents a kind of environment dramatically different from Kisa. Although men from Kisa leave to seek wage employment 230 miles away in Nairobi, their families continue to move back and forth, commuting between city and country in a duolocal residence pattern. Most Nairobi men's wives, families, and some of their children visit them periodically. The men work at wage jobs that take them away from their small urban rooms $5\frac{1}{2}$ days a week. They leave early in the morning and come back near dark. Unskilled work is often inconsistent or seasonal, and there is a constant concern over being laid off.

Relatively few Abaluyia women, whether married or unmarried, work in Nairobi. Very few live there alone. The vast majority of women in Nairobi are staying with their husbands, visiting other relatives, or attending school. If not working, the women stay fairly close to their husbands' rented urban rooms during most of the day. They do not have the characteristic tasks and chores to perform that they have in the rural area. Health services are more readily available.

Nairobi is perceived as a dangerous place. Women and children do not stray far from home unless they are visiting other kin; except in areas where there are relatives or fellow tribeswomen to visit with, people are quite cautious and fearful. Children do not stray very far from the home and certainly cannot roam around easily in most of the housing areas where nonelite Africans live. Although the children help with the domestic routine, they have far fewer chores and tasks to do and spend far more time at home, idling with their mother, than they would in the rural area.

Most older children are sent to school in the rural areas. One of the important differences in the urban cast of characters that results from migration is that older children are very seldom sent to stay in Nairobi for lengthy periods of time. It is costly for them to stay in town, and they are needed for rural chores; parents also do not want their children to be exposed to the temptations of city life. They also prefer that their children go to school in the local community so that they learn in their indigenous language as well as in English in the early years of their education. Thus, the women usually come to stay in Nairobi for a few months at a time with children 8 years and younger. Children in school who are 10 or older are usually sent to schools in the rural area. Older children are not typically available in many families to provide sibling caretaking and to do the other chores and tasks in the home that they would be doing in the rural community.

The fathers are probably around the home more in Nairobi than in the rural area, especially on weekends. The men vary in the preferences they have for how to arrange their family between rural and urban households. Some men prefer their wives to be with them as much as possible and still maintain their rural homestead, and some wives prefer to be in town as well. Some men definitely do not want their wives with them and prefer that they remain on their rural homestead and prefer not to have any children in town. These men invest more in travel to the rural areas on weekends and in holiday periods to visit their families. Other men prefer to invest in having their families visit Nairobi and live with them. In general, these decisions are related to the income available to the father to spend on travel and food and rent—income that is usually insufficient to keep the family in town. The decision also depends to some extent on education and type of occupation. Men with highly skilled occcupations, stable incomes, and more education more often have their spouse with them in town. In general, the men have some conjugal family members and some consanguineal kin (brothers or cousins) with them in Nairobi perhaps 30%–40% of the time, with wide variation in individual cases.

Nairobi provides a dramatic increase in the available sources of outside stimulation and influence on parents and children alike. However, most fathers and mothers tend to avoid intertribal contact and activities that they consider too atypical and nontraditional. Nairobi is, of course, an incredibly heterogeneous community compared to Kisa. Europeans, East Asians, and every African tribal community in the country jam together in a city of well over 1 million population. There are enormous and visible class differences, as well as multilingualism, surrounding every family.

Child Rearing and Family Relationships in Community Families

Family members were constantly commuting back and forth between Kisa and Nairobi. This natural experiment in varying social situations and parenting permitted a controlled comparison of the social-ecological influences of city and country life on both children and parents. I began by locating every man ($N = 24$) living in one housing area of Nairobi who was a member of one of the major subclans in the Kisa area. I then was able to match each urban-resident man with a rural-resident man from the same clan of approximately the same age and education, and with a similar kinship status in the clan. This resulted in 24 matched pairs of households: half living in Nairobi and half living in Kisa. There is frequent mobility between the two locations, and none of the urban families intended to remain permanently in Nairobi.

Systematic observations of children's social behavior and parents' socialization practices were undertaken both in the urban setting and in the rural setting for each family that had one or more children between the ages of 2 and 8 (Weisner, 1979; B. Whiting, 1968). In addition, the parents were interviewed about their child-rearing practices, their residential patterns, their economic status, their modernity, their self-reported psychophysiological stress, and other topics.

There were remarkably few consistent differences between the rural and the urban families on most of these measures. There were no pronounced cultural-ideological differences between the two samples. Nor were there differences in the language used in the house or in the overall acculturation level. In other words, the matching process largely equalized the sample on dimensions other than rural-urban residence, personnel in the household, and local ecology.

The matching also controlled for some ecological features and emphasized certain others. The influence of new knowledge and attitudes, for example, appears not to have been important within the sample, but the personnel present around the child did vary systematically, because older children (10+) were much less frequently present in the city. Similarly, chore and task pressure was far heavier in the rural areas for women and children than it was in the city. The women in Nairobi were far more often around their children without the buffer of available older children. At the same time, the women had fewer social supports available in town and were more economically dependent on their husbands.

The Abaluyia children proved to be fairly similar in the *types* of social behaviors that they were observed to show toward each other. The Spearman rank-order correlation between the mean urban and rural scores for the 15 most frequent social behaviors was .56, significant at the .05 level. The 7 most frequent social behaviors included seeking sociability from others, social play, talking with other children, nonverbally sitting with or following others, seeking approval or praise, seeking material goods or food from others, and offering sociability or offering interaction toward others. These same 7 behaviors were most frequent in both city and country. But the *relative* rank-order and mean proportions of these behaviors differed significantly between the two settings.

There were a number of differences in mean scores between rural and urban settings in the proportion of different kinds of social behaviors. The children in rural settings engaged in more sociable and friendly interactions with one another than did the children in the city. This was true of both boys and girls. Similarly, the children in the rural settings *offered* sociable and friendly behavior to others more often than in the city. The city children *sought* interaction from others more often than did the rural children. The urban children attempted to gain the attention of and interact with their mothers nearly 20% of the time, compared to less than 9% in the rural setting. The urban-resident children sought attention or praise from their mothers more often than the rural children. The urban children sought help, comfort, or permission to do things from their mothers 10.5% of the time, compared to only 3.7% for the rural children.

The urban children (older or younger, boys or girls) were more dominant and aggressive; there were more disruptive interactions in those small urban rooms crowded with mothers and younger children. Such behaviors include annoying, hitting or attempting to hit others, teasing, insulting and threatening. Disruptive and aggressive behaviors were virtually all directed toward siblings or other children in the domestic group rather than toward the mothers. The households in which older children, usually an important part of the sibling caretaking group, were missing, were more prone to disruptive behaviors. The city mothers—who were home with their younger children without task or chore pressures, with greater child-care responsibilities, and without older children to diffuse these caretaking responsibilities—changed their pattern of child care. In the absence of the rural-ecological setting constraints, *dependency seeking* and *aggression/disruption* increased.

The pattern of findings from this series of studies shows the important role of the social ecology or cultural niche in predicting differences in sociability, disruption, and the role of the mother and the peers in caretaking. But some urban-rural ecological features had little or no effect at all; and some others were *culturally* invariant (that is, the Abaluyia were rather homogenous on the trait, whether in the city or in the country). There were clearly panculturally invariant (Lewis & Ban, 1977) behaviors in this sample, unaffected by urban-rural differences, although this finding was not directly demonstrated by this study alone. It is very likely that most effects beyond dyadic interactions are like this: their effects are variable across behavioral domains; are mediated by maturational invariance; and vary by individual difference and cultural characteristics.

The series of Kenya studies shows many of these kinds of effects: the increased disruption observed between child and child and between mother and child in the city families was due to the indirect effects of the separation of the sibling group, as well as to changes in the daily routines, tasks, and activity settings of the children in the two samples. Many of the more obvious urban-rural contrasts did not, in fact, lead to behavioral differences—things like urban media exposure, linguistic diversity and stimulation, and the "bright lights" of Nairobi. Less direct variables that were related to superordinate niche pressures (such as a change in the subsistence base or in the sibling group) led in turn to domestic group changes, which in turn led to behavioral differences in child–child and child–mother dyads.

The study of the econiche effects that influence dyadic interactions will often produce statistical differences within a population that are of theoretical significance, yet that, from a cross-econiche or comparative point of view, prove to be trivial in magnitude. For example, a consistent urban-rural difference for Abaluyia mothers, when their educational level or years in the city were controlled for, showed that they instructed and verbally taught their children more often. The mean difference in dyadic proportion scores was statistically significant. But the *magnitude* of the mean proportion of dyadic interaction was far more revealing of a powerful cultural effect on verbal teaching: the mean for urban mothers to children was under 1%, and the corresponding rural mean was near 0. The point, from a comparative perspective, is that the statistically significant dyadic difference was overshadowed by the rare amount of direct mother–child verbal instruction altogether in Abaluyia society, regardless of the urban or rural location. Compared to American children, for instance, they can barely be said to be "doing" the behavior. There is simply very little overt verbal teaching in this culture between parents and children, in comparison to middle-class American dyads. When reporting naturalistic or experimental dyadic behavioral differences observed in American culture, an analysis of effects on behavior beyond the dyad should include a report of the relative magnitude of the dyadic behavioral difference, or test score, in comparison to similar scores from other niches. Were this reporting practice to become routinely accepted, I believe that many findings similar to the Kenya mother–child teaching example would be found—namely, a theoretically important intracultural difference in the context of far larger cross-cultural differences, produced in other niches (Weisner *et al.*, 1983).

The same point can be made for *activity setting* differences, as well as for direct interactional measures between dyads, such as in the Kenya study reported here. Large-magnitude cross-cultural differences in settings in which children find themselves will place in context the theoretically important yet perhaps substantively quite modest differences between settings found in Western cultures.

THE DIRECTION-OF-EFFECTS PROBLEM

Objections to cultural-ecological theories that move beyond dyadic relations often turn on the question of direction of influence. There is an unnecessary association of social-ecological views with versions of environmental determinism. The family, in this view, is no more than the passive clay for environmental effects, acting as a haven from the extrapolitical-economic world impinging on it at every point. But studying ecological relationships beyond the dyad, and how they affect the family, need not necessarily imply a unidirectional determinism (cf. Bell & Harper, 1977; Weisner, 1982). Families can be vigorous and proactive discoverers of innovations that can work in their local niche. They are active decision makers about the allocations of their family resources and personnel to activity settings. Dyadic relations are reactive to the local environment in creative, innovative ways, not as puppetlike targets of environmental pressures. Taking niche effects as a part of the context for dyadic interactions is a liberating view, for it is ultimately the larger cultural stage, and the goals and ideals that the actors bring to it, that gives meaning to any set of social relationships.

A DEMOCRATIC RELATIVISM

Our satellite views of the cultural niches in which families function around the world provide us with a humbling reminder of how historically and ecologically limited the view is from within any one culture. It also alerts us to the grave dangers of making invidious comparisons regarding developmental outcomes, modes of thought, differences in sex roles, beliefs about disease and illness, beliefs about stages or types of childhood, and ideas of what is valuable for children—without taking into account the larger context. In a democratic society, the nurturance and understanding of diversity, rather than the production of conformity, should remain a fundamental social value. Similarly, it should be a scientific goal to understand the processes of family and human development by studying children in as wide a range of diverse ecocultural niches as we can find in our own country—and throughout the world.

REFERENCES

Barry, H., & Schlegel, A. (Eds.). *Cross-cultural samples and codes.* Pittsburgh: University of Pittsburgh Press, 1980.
Bell, R., & Harper, L. Child effects on adults. Hillsdale, N.J.: Erlbaum, 1977.

Boehm, C. A fresh outlook on cultural selection. *American Anthropologist*, 1982, *84*, 105–125.
Bronfenbrenner, U. *The ecology of human development: Experiments by nature and design.* Cambridge: Harvard University Press, 1979.
Cancian, F. *Change and uncertainty in a peasant economy: The Maya corn farmers of Zincantan.* Stanford, Calif.: Stanford University Press, 1972.
Chibnick, M. The evolution of cultural rules. *Journal of Anthropological Research*, 1981, *37*(3), 256–268.
Fjellman, S. Natural and unnatural decision making: A critique of decision theory. *Ethos*, 1976, *4*(1), 73–94.
Katz, M., & Konner, M. The role of the father: An anthropological perspective. In M. E. Lamb (Ed.), *The role of the father in child development* (2nd ed.). New York: Wiley, 1981.
Leiderman, P. H., & Leiderman, G. F. Economic change and infant care in an East African agricultural community. In P. Leiderman, S. Tulkin, & A. Rosenfeld (Eds.), *Culture and infancy.* New York: Academic Press, 1977.
LeVine, R. Child rearing as cultural adaptation. In P. H. Leiderman, S. Tulkin, & A. Rosenfeld (Eds.), *Culture and infancy.* New York: Academic Press, 1977.
Lewis, M., & Ban, P. Variance and invariance in the mother–infant interaction: A cross-cultural study. In P. Leiderman, S. Tulkin, & A. Rosenfeld (Eds.), *Culture and infancy.* New York: Academic Press, 1977.
Minturn, L., & Lambert, W. *Mothers of six cultures: Antecedents of childrearing.* New York: Wiley, 1964.
Munroe, L., Munroe, R., & Whiting, B. *Handbook of cross-cultural human development.* New York: Garland, 1981.
Sangree, W. H. *Age, prayer and politics in Tiriki, Kenya.* London: Oxford Press, 1966.
Schwartz, T. The acquisition of culture. *Ethos*, 1981, *9*(1), 4–17.
Shweder, R. A., & D'Andrade, R. G. The systematic distortion hypothesis. In R. A. Shweder (Ed.), *New directions for methodology of behavioral science: Fallible judgment in behavioral research.* San Francisco: Jossey-Bass, 1980.
Super, C., & Harkness, S. (Eds.). *New directions for child development: Anthropological perspectives on child development.* San Francisco: Jossey-Bass, 1980.
Wagner, G. *The Bantu of western Kenya.* London: Oxford Press, 1949.
Weisner, T. The structure of sociability: Urban migration and urban–rural ties in Kenya. *Urban Anthropology*, 1976, *5*(2), 199–223. (a)
Weisner, T. Urban–rural differences in African children's performance on cognitive and memory tasks. *Ethos*, 1976, *4*(2), 223–250. (b)
Weisner, T. Urban–rural differences in sociable and disruptive behaviors of Kenya children. *Ethnology*, 1979, *18*(2), 153–172.
Weisner, T. As we choose: Family life styles, social class, and compliance. In J. G. Kennedy & R. B. Edgerton (Eds.), *Culture and ecology. Eclectic perspectives* (Special publication No. 15). Washington, D. C.: American Anthropological Association, 1982.
Weisner, T., & Abbott, S. Women, modernity, and stress: Three contrasting contexts for change in East Africa. *Journal of Anthropological Research*, 1977, *33*(4), 421–451.
Weisner, T., & Gallimore, R. My brother's keeper: Child and sibling caretaking. *Current Anthropology*, 1977, *18*, 169–190.
Weisner, T., & Ross, M. The rural–urban migrant network in Kenya: Some general implications. *American Ethnologist*, 1977, *4*(2), 359–375.
Weisner, T., Bausano, M., & Kornfein, M. Putting family ideals into practice: Pro-naturalism in conventional and nonconventional California families. *Ethos*, *11*(4), 1983.
Weisner, T., Gallimore, R., & Tharp, R. Concordance between ethnographer and folk perspectives: Observed performance and self-ascription of sibling caretaking roles. *Human Organization*, *41*(3), 1982, 237–244.

Whiting, B. B. *Transcultural code for social interaction*. Unpublished manuscript, 1968.
Whiting, B. B. Culture and social behavior: A model for the development of social behavior. *Ethos*, 1980, *8*(2), 95–116.
Whiting, B. B. The genesis of prosocial behavior. In D. Bridgeman (Ed.), *The nature of prosocial development: Interdisciplinary theories of strategy*. New York: Academic Press, 1983.
Whiting, B. B., Edwards, C., *et al*. The company they keep: The genesis of gender role behavior, in preparation.
Whiting, B. B., & Whiting, J. W. M. *Children of six cultures: A psycho-cultural analysis*. Cambridge: Harvard University Press, 1975.

4

Changing Characteristics of the U.S. Family
Implications for Family Networks, Relationships, and Child Development

CANDICE FEIRING AND MICHAEL LEWIS

INTRODUCTION

From the moment of birth the child is embedded in a large social network, the fabric of which is made up of many people, functions, and situations (Lewis & Feiring, 1978, 1979). This network includes, among others, parents, siblings, relatives, teachers, and peers. For the purposes of the present discussion, we shall focus on that part of the social network known as the family and will concentrate on the changing demographic characteristics of the family in the United States and the implications that these changing family characteristics have for the development of the child.

In the discussion that follows, the family is viewed as a social system (Feiring & Lewis, 1978; Lewis, 1982), possessing the characteristics of a social system but distinguishable from other systems by its goals, functions, and social climate. The family is defined by its members, the culture, and the community within which it exists. It provides the opportunity for intimate social interaction for its members and is usually

---

CANDICE FEIRING AND MICHAEL LEWIS • Department of Pediatrics, Rutgers Medical School—University of Medicine and Dentistry of New Jersey, New Brunswick, New Jersey 08903.

the first environment (and system) in which the child experiences intimate social interaction. The family as a social system may be characterized by a particular set of elements (e.g., parents and children) that are interdependent and that function with the purpose of socializing and raising children. Typically, families consist of persons occupying at least three different social roles: wife–mother, husband–father, and child–sibling. Of course, some families may not have one of these categories for part or all of the family's existence, and other roles exist, for example, grandmother–mother.

The family is just one of the many social systems within which the child must learn to function. It is, however, the most basic of all social institutions. The family system is the most basic because it is the first system to which the organism must adapt and because it helps to regulate and control the adaptive systems themselves. For example, Sander (1977) has suggested that the sleep–wake cycle in newborns is regulated by the mother–infant interaction. The interface between the family system and the child's subsequent social behavior has been widely discussed (Harlow & Harlow, 1969; Lewis & Rosenblum, 1975). Because the family is the primary center of the child's social life, it is the social system on which we focus most of our interest here. For most cultures, the family is also the primary institution responsible for maintaining and perpetuating societal goals, and for transforming these goals into directives for the new individual. Thus, the family is the core mechanism for socialization and is the society's adaptational unit that changes through history. By studying changes in the family structure, its goals, and its values, we can come to a broad appreciation of the types of socialization practices generated and thus to a greater understanding of the types of child-rearing activities of a particular culture.

The embeddedness of a child in a family, of a family in society, and of all social systems in history is obvious. However, the interface between these levels of analysis is practically unexplored (Spanier, Lerner, & Aquilino, 1978). The interrelationships among children, their families, the social network, and history have, for the most part, been neglected in theory and research. What we do have is information about children, families, and the change in families over time, although this information comes from disciplines that utilize different units and levels of analysis. Drawing from information provided by demographers, economists, sociologists, and psychologists, we can begin to integrate our knowledge and to speculate about how changes in the family over the past few decades might influence the individual child's development, in both the present and the future.

On the macroanalytic level, three changes within families in the United States are considered here. Each of these has received considerable attention from demographers and sociologists: (1) a drop in the fertility rate; (2) an increase in the number of working mothers of young children; and (3) an increase in the number of divorced and unwed mothers. Lower fertility rates suggest that the average family size will be smaller. An increase in the number of working mothers indicates an evolution in parental functions within the family. An increase in the number of divorced and unwed mothers signifies that single-parent families have become and will continue to be a more prevalent family structure.

On the microanalytic level, we focus on issues pertaining to the quantity and the quality of interaction between family members and the child. Finally, a connection between the macro- and microlevels is made, in regard to how quantity and quality of interaction may be affected by the changing characteristics of the American family. For example, a drop in fertility rate indicates families that are smaller in size. Smaller families mean fewer children, and fewer children mean that the parents may have more time and energy to spend with each child.

In the discussion to follow, we consider the three demographic issues cited above. The difficulty in predicting the possible consequences of these demographic changes for family functioning and child development involves two major points. The first has to do with the issue of quantity versus quality of interaction between family members, and the second has to do with families as systems.

Quantity versus Quality of Interaction in Families

The young child spends a good deal of time within the context of the family. However, the amount of time that the child spends in the family setting is not necessarily indicative of the extent or the quality of interaction between parent and child. Available evidence suggests that even when the parent is in the same room as the child, interaction is infrequent (Clarke-Stewart, 1973). Furthermore, the amount of time that parents and children spend together is a poor predictor of the nature of the child's relationship with its father or mother (e.g., Ban & Lewis, 1974; Feldman, 1974; Pederson & Robson, 1969; Schaffer & Emerson, 1964). It seems that, for the emotional and intellectual growth of the child, far more important than the quantity of time spent with the parents is the quality of interaction (cf. Bossard & Boll, 1966; Clarke-Stewart, 1978; Lewis & Goldberg, 1969). For example, the quality of interaction between parent and child is more predictive of the child's intellectual ability than the amount of contact. Lewis and Coates (1980) have shown that, for a group of mothers, the amount of behavior, such as vocalization or touching, was far less related to the child's intellectual performance than were measures of responsivity and interaction. Moreover, Coates and Lewis (1980) have shown that the nature of the early mother–infant interaction rather than amount of maternal contact is predictive of school-age children's cognitive and achievement behavior. These data, as well as other investigations (Belsky, 1980; Lamb, 1976b; Sroufe & Water, 1977), underscore the importance of the nature rather than the amount of the parent–child interaction.

In terms of emotional development, the same case can be made. The quality of the interaction, such as the parents' responsiveness and sensitivity to the child's needs, is predictive of a secure child–parent relationship (Ainsworth, Blehar, Waters, & Wall, 1978). The security of parent–child relationship may affect the child's social development inasmuch as secure children are more willing than insecure children to interact with other persons (Easterbrooks & Lamb, 1979; Lewis & Schaeffer, 1981). Consequently, the quality of time spent in parent–child interaction, rather than the quantity of time, may reveal more about the nature of the child's development. In our discussion of changes in selected characteristics of the American family, it is necessary to pay particular attention to how these demographic changes may affect the nature of the parent–child interaction, as well as to how these changes may affect the amount of interaction.

The Family as a Social System

The family within which the child functions operates as a social system, and the implications of this conceptualization have been discussed in detail elsewhere (Feiring & Lewis, 1978; Lewis, 1982). However, several important aspects of family systems must be considered, as they illustrate the complexity of the functioning of social systems and the difficulty of making simple predictions about how demographic trends influence family functioning and child development. Briefly, we will examine the following aspects of social systems that indicate families' complexity in functioning: (1) the interdependence of elements and subsystems; (2) nonadditivity; and (3) steady states.

A system is a set of interrelated elements in which each element influences and is influenced by each other element (Monane, 1967; Taylor, 1975; Von Bertalanffy, 1967). Said differently, the defining elements are simultaneously the subject and the object of influence. Structurally, a system is usually composed of interdependent subsystems. For example, within the family, the elements, or the family members (parents and siblings), can be conceptualized into parent–child, child–sibling, and parent–parent subsystems. These child–family-member subsystems of the child's social network can be characterized by their interdependent nature. Pedersen (1975), for example, demonstrated the ways in which the child–parent and the parent–parent subsystems mutually influence each other. He found a high interrelationship for the families of boys between father–infant play and the effect of the father's esteem for the mother as a "mother," as well as the amount of tension and conflict in the marriage. In regard to the demographic trends discussed in this paper, it should be noted that the interdependent nature of the family system is one aspect of the system that makes it difficult to predict the impact on the child of changes in demographic characteristics. For example, how the mother's work status affects the child's cognitive development may depend on the father's view of his wife's role (the parent subsystem), which, in turn, may influence the mother's behavior with her child (the parent–child subsystem).

Social systems also possess the quality of nonadditivity. Knowing everything about the subsystems of a system will not tell us everything about the system as a whole. Any subsystem operates quite differently within the system from the way it does in isolation. For example, the nature of the interaction in the parent–child subsystem in isolation is different from the interaction when it is embedded in a larger system. Lamb (1976a, 1977) reported that infants interact more with both parents in isolated dyads than when the parent–child dyads are embedded in the entire family system of two parents and child. In addition, the parents are far less interactive with the child when the spouse is present. Clarke-Stewart (1978, 1980) found that the quality and the quantity of the mother–child interaction changed when this subsystem was embedded in the mother–child–father triad. The mothers in this study initiated less talk and play with the child (quantity) and were also less engaging, reinforcing, and responsive in their child-directed interactions (quality) when the father was present than when the mothers were alone with the child. Pedersen et al. (Pedersen, Yarrow, Anderson, & Cain, 1979; Pedersen, Anderson, & Cain, 1980) reported that whereas the father-husband frequently divides his behavior between child and spouse in a three-person setting, the mother-wife spends much more time in a dyadic interaction with the child.

Another issue pertaining to the nonadditive nature of systems concerns the operation of indirect effects within systems larger than a dyad. Relatively little is known about the processes that do not include direct interaction but that influence the nature and the form of future interactions and relationships. Children acquire information in a variety of indirect ways that are not necessarily characterized by immediate interaction with people. Indirect effects may be categorized into two general types (Lewis & Feiring, 1981). In the first, indirect effects are those sets of interactions that affect the child but that involve events that occur in the absence of the child. *In the absence of* refers to interactions among members of the child's social network when all the members are not present. For example, the nature of the husband–wife relationship when the husband and wife are alone has been shown to be related to future parent–child interactions (Pedersen, 1980). The second type of indirect effect refers to the influences that occur in the presence of the child but that do not involve the child as an active participant in a social interaction. For example, in the context of the mother–father–child triad, the child may gather information about appropriate sex-role behavior from observation of but not participation in the wife–husband interaction (Lewis & Feiring, 1982).

The nonadditive nature of the family system and the operation of indirect effects must be taken into account when considering demographic trends and their impact on child outcomes. For example, the increased number of single-parent families may indicate that although the ratio of dyadic to triadic father–child interaction increases, the frequency of father–child interaction decreases (as he sees the child only on weekends) and the duration of dyad time increases. Furthermore, the indirect effect of stress in husband–wife interactions may influence parent–child interaction. As can be seen from this example, the task of prediction is difficult in light of the complex, nonadditive nature of the subsystems as well as the operation of indirect effects.

Social systems are also characterized by the process of steady states. A steady state is a balance process whereby a system maintains a viable relationship between its elements and its environment. The balance denoted by the term *steady state* is not the static one implied by the terms *homeostasis* or *equilibrium*, which denote a process whereby adjustment maintains a system at a given level. Rather, *steady state* describes the process whereby a system maintains itself while always changing to some degree. Steady states are characterized by a high degree of complexity and organization and involve an interplay of stability and flexibility in a social system or network.

In the social network system, the child functions as a member of a social network with ever-changing members and environment. For example, when the 2-year-old firstborn child has a mother who returns to work, the structure of its social network changes. The firstborn must modify its mode of functioning in order to cope with its mother's change in availability. Although the child's goals of learning to adapt to its environment remain the same, the child's contact with people and actual activities may alter. For example, if the mother returns to work, she may be less frequently available for dyadic interaction, although the duration of interactions may increase. Meanwhile, siblings, relatives, and other adult caregivers may take on some of the functions previously performed by the mother. Thus, the goal of socialization remains the same, but the persons performing certain activities may change. The child adapts, as does the family system, to changing circumstances while certain functions and goals are maintained. The way in which demographic trends influence child outcomes must be related to the adaptability and the flexibility of the child and the family system.

In discussing the consequences of smaller families, working mothers, and single parents on child development, we will try to address the issues of quality and quantity of interaction and the complexity of the family as a social system. These issues are critical in any attempts at prediction and must be incorporated into an analysis of each demographic trend and its possible impact on child development. The constraints on our ability to make predictions are related to the complexity of the phenomena being examined, as well as to the lack of relevant information on the subject. We are well aware of these constraints and will attempt to address them in our discussion.

Decline in Fertility: Decrease in the Average Size of the American Family[1]

According to the data collected and analyzed by demographers, there has been a decrease in the fertility rate in the last few decades. If today's woman adopted a birth rate comparable to a woman of her own age in 1960, she would produce 3.6 children in her lifetime, but only 1.8 children if she adopted the birth rate of 1978. Thus, a woman of childbearing age had half as many children in 1978 as in 1960. A more conservative estimate of birth rate, and the one preferred by demographers, is the number of lifetime births expected for wives from 18 to 34 years of age. The statistics given in Table 1 indicate a shift from 1967 to 1979 so that wives were more likely to have two children than three or more children. It is also interesting to note that the figures suggest a twofold increase in the number of one-child families from 1967 to 1979.

The lifetime number of births per woman in the United States is approximately two-thirds what it was in the baby boom period. It appears that now the "typical or average" American nuclear family is comprised of two parents and two children rather than two parents and

---

[1] The population statistics for this section are taken from U.S. Population Bulletins and *Statistical Abstract of the United States*, U.S. Department of Commerce, Bureau of the Census, 1980.

TABLE 1. Percentage of Lifetime Births to Wives
18–34 Years Old in 1967 and 1979

Number of births	Wives 18–24 years of age		Wives 25–29 years of age		Wives 30–34 years of age	
	1967	1979	1967	1979	1967	1979
0	1.3	5.2	2.5	5.3	6.1	6.4
1	6.1	11.8	5.1	12.9	5.5	12.6
2	37.1	54.9	29.3	51.9	18.8	46.2
3	29.8	20.9	33.5	21.3	14.9	22.4
4+	25.7	7.1	29.9	8.6	55.3	12.4

three children. The economic costs of raising a child may be one factor influencing the declining birth rate (Reed & McIntosh, 1972). The direct cost of raising a child in the United States to the age of 18 years was roughly $35,261 in 1977 for a low cost estimate and $53,605 for a moderate cost estimate (Espenshade, 1977). These direct cost estimates include out-of-pocket expenses for clothing, food, health care, and education but do not include the cost of a college education or opportunity costs (i.e., the income that spouses forgo by staying home and raising children rather than working). That the economic cost is a prime consideration in the decision to have smaller families is suggested by research that reveals that the most common reason mentioned by wives for wanting no more than their desired number of children was the economic cost (Bulatao & Arnold, 1977; Hoffman, 1977).

The trend evidenced by the data presented in Table 1 may be taken as an overall measure of shrinking family size. It has been suggested that a current decline in household size is mainly due to the long-term decline in the fertility rate; that is, households are smaller, in part, because they are populated by fewer children. The growing number of smaller households also reflects an increasing number of single-parent families as well as an increase in young adults living away from their parental home and elderly persons living alone. Consequently, the average family may have fewer adults as well as fewer children, so that the changing characteristics of social networks may represent more than simply a reduction in fertility.

The current trend in birth rate and size of household indicates that children are more likely to be members of smaller families. Research on family size suggests that the child who is a member of a small family is different in its cognitive, social, and personality characteristics from the child who is a member of a large family. Perhaps the most well-known finding is the negative relationship that exists between family size and intelligence. A review of the literature has shown that the negative correlation between intelligence and family size is reliable (Terhune, 1974). The correlation is small, ranging in size from $-.20$ to $-.30$, reducing to $-.12$ to $-.16$ when social class is controlled.

Zajonc and Markus's (1975, 1977) recent work on the relationship between family configuration and intelligence indicates that the influence of parents' and siblings' intellectual levels on the child's cognitive growth is mediated by diverse processes of social interaction in the family. Their confluence model is based on the assumption that cognitive growth is influenced by the quality of a child's immediate intellectual environment (in most instances, the family). The intellectual environment is assessed by averaging the absolute intellectual levels in a given family. Increasing family size is thus linked to lower intellectual levels because the addition of a newborn to the family lowers the average intellectual level from what it was previously. The extent to which some portion of the intellectual growth of the child is determined by an interaction with the intellectual level of parents and siblings is the extent to which larger families will be associated with lower intellectual levels because the larger the family, the larger the proportion of individuals with low absolute intelligence.

The inverse relationship between family size and intellectual performance finds support in the research literature (Belmont & Marolla, 1973; Marjoribanks & Walberg, 1975; Zajonc, 1976; Zajonc, Markus, & Markus, 1979). Unfortunately, little attention has been given to defining *intellectual environment* in terms of factors such as the amount of time that family members spend with each other and the form, content, and quality of the interaction (Svanum & Bringle, 1980). Little information exists on how family size influences the kinds of family processes and structures that have consequences for the child's intellectual development. Some data indicate that a child who is a member of a small family spends more time in interaction with its parents (Lewis & Feiring, 1982; Sears, Maccoby, & Levin, 1957) and that the quality of the intellectual environment is different for this child and for the child who is a member of a larger family.

One of the most frequently invoked explanations of the influences of family size is the amount of attention that the parents devote to each child. Great emphasis is placed on this factor in interpreting effects on children's intelligence, the explanation being that interaction with parents provides intellectual stimulation for the child. The data indicate that the larger the group, the less time each person has for interaction. Several investigators report that infants interact more with either parent in isolated dyads than when these parent–child dyads are embedded in a three-element system (Clarke-Stewart, 1980; Lamb, 1977; Pedersen et al., 1980). Parents are far less interactive with the child when their spouse is present, and the addition of more children reduces the interaction even further (Lewis & Feiring, 1982). Moreover, the quality of child care decreases with family size. It has been reported that infant care and child management are negatively affected by increasing family size (cf. Douglas & Blomfield, 1958; Douglas & Simpson, 1964).

In our work on the relationship between family size and family interaction (Lewis & Feiring, 1982), we have found that the kinds of activities that family members engage in change as a function of the number of children. In a sample of 15 families, with 5 families in each size of 3, 4, and 5, we examined the behavior of the parents and the target children, who were 3 years of age. In our study, the parents' interactions with the target child were affected by family size: information seeking and nurturance behavior declined with increasing family size. With increased family size, both the father and the mother were observed to spend less time asking questions and expressing positive affect toward the target child, probably because their other children also wanted and needed their attention. The target child's behavior toward its parents also showed the effect of family size. With increasing family size, information seeking declined from the child to either parent. Here it is important to note that the child's cognitive skills were related to its tendency to seek information from its parents. Specifically, the number of questions that the target child asked its mother ($r = .58, p < .05$) and its father ($r = .51, p < .10$) was positively related to the Stanford-Binet intelligence score. Consequently, given the research findings cited above concerning family size, family functioning, and cognitive development, it is reasonable to expect a modest increase in the intellectual level of children as small families come to predominate (Terhune, 1974). For this sample, we also found a significant negative relationship between the 3-year-old child's Stanford-Binet score and family size.

In addition to smaller families' being associated with higher "intellectual" levels in children, the research literature indicates that high aspirations and need for achievement are consistently associated with small as compared to large families (Rehberg & Westby, 1967; Rosen, 1964; Turner, 1962). The tendency for children from small families to have greater ambition or drive may be partly the result of their parents' tendency to aspire more for their children and to stress personal achievement. Douglas and Simpson (1964) found several indications that parental interest in children's schoolwork declined with increasing family size in both the working and the middle class. Eysenck and Cookson (1970) reported a negative correlation between family size and teachers' ratings of parents' interest in their children's schoolwork. Although the magnitude of the associations found in these studies is small, the results seem consistently to indicate that, as family size increases, parents become less interested in their child's schooling. Consequently, one might expect that the type of parent–child interaction in smaller families is more oriented toward achievement-related content. Also, with respect to quantity of time, parents in larger families may not have as much time to become interested in their children's attainments. Time is a limited resource, and per child allocations may decrease as family size increases.

Family size probably also affects the personality characteristics of the child. Children from small families tend to have higher self-esteem and self-confidence than children from larger families. Sears (1970) suggested that self-esteem is engendered by parental expressions of admiration and acceptance and that these parental expressions must be divided over the number of children in the family. Data collected by Sears on public-school children show a small negative association between family size and self-esteem. In addition to feeling good about themselves, children from smaller families may also be more sociable (Landis, 1955) as well as more popular (Borod, Grossman, & Eisenman, 1971) than children from larger families. With increased family size, there is also an increased probability of antisocial behavior, such as misdemeanors, delinquency, and criminality (cf. Berg, Fearnly, Palerson, Pollock, & Vallance, 1967; Glueck & Glueck, 1950; Tuckman & Regan, 1967). This relationship remains even when social class and intelligence are controlled for. The data indicate that between one- and two-child families there is little or no difference in antisocial tendencies, but in family sizes of four and over, there seems to be a linear trend toward an increased frequency of antisocial behavior (cf. Terhune, 1974).

To summarize, demographic trends suggest that the size of the American family is getting smaller. Past research indicates that children from smaller families: (a) spent more time with their parents, (b) were more intelligent, (c) had higher self-esteem, (4) were more sociable, and (5) were less antisocial than children from larger families. Although the statistical relationship between family size and cognitive and personality traits of the child are small, they appear to be consistent across several studies. The results suggest that the quality and the nature of the parent–child interaction alter with family size, and that as family size decreases we may observe changes in parent–child interactions accompanied by an increase in the level of cognitive functioning in the children.

Increase in the Number of Working Mothers[2,3]

The increase in the number of working mothers has broad consequences in terms of the child's social, emotional, and cognitive development. First, it reduces the time that mothers have to interact with their children without necessarily increasing the time of the children with other family members, such as the fathers. Second, it increases the reliance on other nonfamily adults and also increases the number of the child's peer relationships, as the most common form of child care is group day care. These issues and their implications are discussed in this section.

Over the last three decades, the number of mothers in the labor force has been on the rise. As of 1980, 56.6% of women with children under 18 years were in the labor force, compared to 45% in 1975 and 35% in 1970. There has also been a steady rise of married women with children under 6 years old in the labor force, from 18.6% in 1960 to 30.3% in 1970 to 36.6% in 1975 to 45.0% in 1980. Obviously, a rapidly increasing proportion of young children is now being cared for during working hours by someone other than their mothers. One reason that younger mothers are entering the labor force is to supplement the relatively low earnings of the family if the husband's income is insufficient. In general, families in which the husband earns less than $5,000 per year have working wives. The mothers must work in order to support their families at an adequate economic level.

Women are working for self-fulfillment as well as for economic reasons. The advent of the feminist movement and the desire for self-actualization has moved many educated women into the labor force. Thus, whereas less-educated women may enter the labor force for economic reasons, more educated women may enter the labor force for personal-psychological reasons as well as financial ones. Women with more education spend more time (although not necessarily during the time of child rearing) in the labor force. However, surveys indicate that mothers and fathers with more education spend more time at child care than parents with less education (Leibowitz, 1974).

What can we conclude from these demographic trends? Certainly,

---

[2] The population statistics for this section are taken from U.S. Population Bulletins and Special Labor Force Reports e.g., *A Statistical Portrait of Women* in the U.S. Population Report series P-23, No. 58, 1975; *Statistical Abstract of the United States*, U.S. Department of Commerce, Bureau of the Census, 1980. The assistance of R. Green, Department of Labor, in compiling the information in this section is appreciated.

[3] By *working*, we mean to denote only that these are mothers who work out of the home rather than in the home.

an increase in the number of mothers who work should be related to a decrease in the amount of time that these mothers have to spend with their children. However, as indicated previously, the amount of time that the mother spends in the home is not necessarily indicative of the amount or quality of interaction time that the mother spends with her child (e.g., Robinson, 1977). The mother's time in the home can be spent in activities related to household tasks as well as child care. Unfortunately, there is no information on issues such as whether working mothers spend less time in child-oriented activities than nonworking mothers.

How maternal employment influences the quantity, the quality, and the nature of parent–child interactions may be related to the mother's reasons for working in the first place. Maternal attitudes toward employment may affect the sensitivity of a mother to her child's signals in several ways. If a mother chooses to work and feels fulfilled in pursuing a career, then her self-esteem is likely to be high, and this should have a positive effect on her relationship to her child. Indeed, Barret (1976) found that a mother's self-esteem is positively related to her responsiveness to her child. On the other hand, mothers who feel that they must work for economic reasons may feel guilty about abandoning their children. For example, they might overinterpret child signals and thus control interactions with their child to such an extent that a child's opportunity for initiating reciprocating interaction is diminished (Lamb, Chase-Lansdale, & Owen, 1979). Hock (1980) reported that working mothers who are ambivalent about working and being away from their babies tended to have infants who exhibited more negative reunion behavior. Similarly, nonworking mothers who did not value caretaking and parenting as their most important role (i.e., ambivalent mothers) tended also to have infants who displayed negative reunion behaviors. If negative reunion behavior reflects disturbances in the mother–infant relationship (Ainsworth et al., 1978), then regardless of work status *per se*, mothers who are ambivalent about either work or care of children may have poorer relationships with their children.

Paternal as well as adult peer-group attitudes toward maternal employment might also influence the mother's feelings about work and, indirectly, her relationship with her child. Thus, husband's and friends' positive attitudes toward maternal employment should be positively related to the mother's feelings about working and therefore result in a more positive family environment. Unfortunately, there is no information about the opinions of husbands-fathers and how they affect the mothers' attitudes and child-care practices.

Commitment to the child and knowledge about child development are other important factors that influence the impact of maternal employment on the child. A highly committed or knowledgeable mother is more likely to consider the probable influences that her time away from home will have on her child. Consequently, she may be more concerned about adequate alternative caretaking arrangements and the child's adjustment to the situation, as well as involving herself more in productive child activities when she is at home. Knowing that more mothers with younger children are now entering the work force does not supply us with enough information to be able to determine the direct consequence on mother–child relationships. In some cases, the husbands of working women tend to be more involved with, and closer to, their children than the husbands of nonworking women (e.g., Hoffman, 1972, 1979; Robinson, 1977). These data suggest that the presence of a working mother in the family enhances the quality of the father–child relationship. However, we suspect that these data may have been collected from households in which the mothers were well educated and chose to work and the husbands supported this decision. These fathers were also probably more educated, and fathers with more education tend to spend more time on child care than fathers with less education (Leibowitz, 1974). At any rate, it does not seem likely that an increase in the number of working mothers will precipitate an increase in the number of fathers who devote a substantial proportion of their day to child care (Weinraub, 1980).

According to Hofferth and Moore (1979), there has been only a very small increase in the hours of housework done by married men, who spend about one-sixth the time spent by working wives in housework. Even when the mother works, the division of labor in household chores seems to remain sexually segregated, although the gap between wives and husbands in total hours worked may be getting narrower. Nevertheless, the traditional roles of men and women are in transition. One eventual result of increased maternal employment should be the reallocation of work responsibilities within the household. A decrease in the chores and tasks allocated along traditional sexist lines may currently be under way. This means that the functions (e.g., activities such as caregiving, nurturance, and teaching) that characterize the role of different family members will change. Consequently, the child's socialization may be influenced differently, depending on to what extent the parents share caregiving and child-rearing activities. For example, the child whose mother and father devote an equal amount of time to caregiving will be likely to be *different* from the child who receives the majority of caregiving from the mother. How the quality of the father–child relationship is altered by the mother's working status will most likely be a function of such diverse factors as the nature of the wife–husband relationship, the economic status of the family and the father's attitudes and beliefs about his role in the family.

Maternal employment may also affect the child–sibling relationship in a variety of ways. For example, if children spend more time away from home in day-care arrangements as a consequence of their mother's working, and if the children are separated according to age, then they may have less time to interact with their siblings. Moreover, if maternal employment means that less time is available for each child, more sibling rivalry could result. Alternatively, children of working mothers may have extensive peer interaction in day-care centers and therefore may be more advanced in their peer relations and thus more cooperative with their siblings. School-age children of working mothers more often may be left alone together after school. As a consequence, increases in sibling contacts and in sibling caregiving and teaching may occur. Zajonc (1976) suggested that this situation may be beneficial for the older sibling's intellectual development through the mechanism of peer teaching.

As more mothers work, the child's care often becomes the task of a nonfamily member. Whether or not parents can compensate for the time away from their young children by increasing the quality and the nature of interaction when they are home, time away from the home does mean that alternative child-care needs arise and must be met. As more mothers of young children go to work, they are relegating the care of their children to other adults (Fein & Clarke-Stewart, 1973). Babysitters, women who look after small children in their homes, and persons in various institutional group settings have come to represent important figures in the child's social network.

Research on children with employed mothers and/or children in day-care settings (many of whose mothers work) suggests that maternal employment affects the child's cognitive, social, and emotional development. Schachter (1981) compared toddlers of employed mothers to a matched (middle-class, highly educated, predominantly firstborn) group of toddlers of nonemployed mothers. The toddlers of nonemployed mothers performed better on the Stanford-Binet IQ measure, although no difference was found in language development. Cohen (1978) also reported that toddlers of nonemployed mothers performed better on the Bayley test of infant development. However, no differences as a function of mother's employment status were found in cognitive development, as measured by the Bayley Mental Development Index and the Psychomotor Development Index, when Hock (1980) examined the performance of 8-month-olds. Hock's sample of infants of employed mothers were cared for in home settings, whereas the toddlers in the studies of Schachter and Cohen were cared for in a group day-care setting.

The confounding of factors such as the child's age and the type of substitute care should modulate the effects of maternal employment on the child's development. Studies that examine children in full day care versus home care suggest how maternal absence may influence the child's development. Studies have tended to show that, in middle-class samples, subsequent intellectual development reveals no differences between day-care–reared children and mother-home-care–matched controls (Caldwell, Wright, Honig, & Tannenbaum, 1970; Cochran, 1977; Moore, 1975). Children identified as *at risk* for poor cognitive development show more growth in day care versus home care on standardized tests of intelligence (Golden, Rosenbluth, Grossi, Policare, Freeman, & Brownlee, 1978; Ramey & Campbell, 1977; Ramey & Smith, 1976).

Research on socioemotional development supports the idea that the children of employed mothers may differ in personality but are not "better or worse" than the children of nonemployed mothers (Kagan, 1976; Moore, 1975). Schachter (1981) reported no difference between groups on any indexes of ego strength. Hock (1980) compared the attachment behavior of 12-month-olds with working and nonworking mothers in order to examine the socioemotional bond between mother and child as influenced by employment status. The infants of working and nonworking mothers did not differ in the social behaviors that they directed to their mothers. Specifically, in terms of the quality of the attachment relationship, no differences existed in the amount and the intensity of positive-approach or negative-avoidant-resistant behaviors.

Most of the studies on maternal absence do not look at whether the mothers are employed outside the home. Rather, they examine the effects of day care versus home care. It is assumed that day-care infants have mothers who work outside the home; however, among the middle class, at least, this may not be the case. Blehar (1974) found that children who were placed in day care before they were 3 years old tended to develop an anxious or ambivalent attachment to the mother. However, these findings are suspect because the observers were aware of the children's group status and the experimental hypothesis. Moskowitz, Schwarz, and Corsini (1977), conducting a study similar to Blehar's without experimenter bias, found no effect of attachment status on maternal employment. They also found that home-reared infants showed more distress on separation from their mothers than day-care children. With the exception of Blehar, the majority of researchers have found that day care does not have a negative influence on the child's tie to its mother (Caldwell *et al.*, 1970; Farran & Ramey, 1977; Kagan, 1976; Ricciuti, 1974). A significant disruption of the mother–child bond does not appear likely, especially if high-quality day care is the alternative (Belsky & Steinberg, 1978). The research that finds no deleterious effects of day care is based on comparisons of day-care centers with university-based, high-quality care to home rearing situations. Whether day-care settings of less quality or day-care home settings are also nondisruptive of the mother–child attachment quality is as yet undetermined.

As more mothers work and more children spend time in day-care centers or family day-care homes, the children may have a greater opportunity for interacting with peers at earlier ages. Thus, the child's social network may include an increasing number and variety of peer and adult friends. The development of peer relationships is an important aspect of social growth (cf. Lewis, Young, Brooks, & Michalson, 1975; Mueller, 1979; Suomi & Harlow, 1975). In the last decade, increased information has been gathered on peer relationships (Lewis & Rosenblum, 1975), and the data indicate that even the very young child can develop peer relations (Mueller & Vandell, 1978). These relations, although immature in some regards, can be considered operational at an early age. Peer interaction provides an excellent milieu for the development of some of the infant's capacities. The peer social context is quite different from the more pervasive caregiver milieu, where the adult guides, directs, controls, and plays. In contrast, peers are relatively equal in terms of their social skills, their capacity to formulate and carry out their goals, and the extent of their socialization. In the peer-oriented social context, the infants not only mutually socialize each other, but also have a better chance of successfully affecting one another. The development of social competence can be observed in this process (Lewis et al., 1975).

Thus, one direct consequence of increased maternal employment may be the increase in opportunities for peer-group interactions. Schachter (1981) found that toddlers with employed mothers were more peer-oriented and self-sufficient than toddlers with nonemployed mothers. Toddlers with nonemployed mothers were more adult-oriented and dependent. They tended to be more jealous about sharing adult attention and sought adult help and protection more frequently than toddlers with employed mothers. Other research indicates that day-care infants tend to be more peer-oriented and independent than home-reared children (Belsky & Steinberg, 1978; Hoffman & Nye, 1974; Moore, 1975). Children who enter day care early have also been reported to exhibit more physical and verbal aggression toward peers and adults than home-reared children (Schwartz, Stickland, & Krolick, 1974). Day-care children were also less cooperative with grown-ups, ran about more than sitting in one place, and interacted with agemates more than with adults. Children with mothers who work may be more likely to encounter a relatively higher number of peers to adults and may become more independent and peer-oriented and somewhat less amenable to the socializing influence of adults (hence showing more aggression and less cooperation, which may be the other side of the coin of independence).

To recapitulate, the structure and the nature of the family would appear to be one of the cultural variables in determining the effect of maternal employment on child care, as well as being directly affected itself by the employment. Thus, the causes of maternal employment, as well as paternal support (both in encouraging the mother's working as well as in spending time with the children), should be important determinants of potential outcomes. The role of the father as an indirect source of influence, such as supporting the mother and making her feel competent, may be a more important factor than is generally appreciated in determining the effect of maternal employment on the child's development (Lewis, Feiring, & Weinraub, 1981). In families where the husbands support maternal employment, disturbances of family relationships as a consequence of maternal employment should be minimal. Because of the limited data on whether husbands or working wives increase their child-care activity, it is difficult to pass judgment on this issue.

Increases in maternal employment may also affect the presence of nonrelative adults and peers in the social network of young children. When their mothers are working, both middle-class and poor children need to be cared for by some adult. Information on the influence of substitute care by adults other than the mother mostly concerns the effects of day care on the child's development. The data available on the day-care experience seem to indicate that alternative day care does not have a negative influence on the mother–child relationship or on the child's social development. Indeed, in the case of lower-class children, the day-care experience may have a positive impact on cognitive development. However, very little is known about the ways in which the more common types of substitute care, such as relatives, babysitters, and family day care, affect development (Belsky & Steinberg, 1978).

Similarly, how the child's increased exposure to peers may influence development has been explored mostly in the day-care setting. Early peer interaction may have a positive impact on the child's development of independence and social skills with agemates, although it may have less positive consequences for child–adult interactions. Finally, although it is known that the amount of maternal time available for child care decreases as a function of employment, the relationship between the total time spent at home and the time spent in child care is not known. Clearly, some child-care activities are more important than others for successful social, emotional, and cognitive development. It may be the case that only a small proportion of maternal time and activities is necessary for these tasks and that employment may not interfere with these activities. Still to be determined is the role of quality versus quantity of time in child care. Given the diverse factors that may have bearing on the nature of the social network (including the form of substitute care, the ratio of adults to children, and the nature of family relationships, such as the father's support of the mother and the mother's feelings about working), predicting the influence of maternal employment on child socialization cannot consist of simple evaluative statements.

It is clear that maternal employment, especially of the mothers of very young children, is on the increase. Maternal employment does result in less time and perhaps less maternal energy available for child care. Given these facts, it would seem possible to speak of the implications of maternal employment in children's development. However, because maternal activity takes place in the network of the family, there are too many other variables that may interact with maternal employment to allow us to do more than speculate on the impact of maternal employment on children.

INCREASE IN THE NUMBER OF SINGLE-PARENT FAMILIES[4]

In 1960, most children (88%) under 18 years of age lived with two parents, although these two parents were not necessarily their biological parents. By 1976, the percentage had dropped to 80%, and in 1979, it had dropped to 77.4%. In the last 30 years, the number of children under the age of 6 living in single-parent households has increased about fourfold, from 7% in 1948 to 15% in 1974 to 23% in 1978. In 1979, 29% of women maintaining single adult households were widowed, 33% were divorced, 17% had never been married, and 21% had husbands working away from home.

A major factor contributing to the number of single-parent families is divorce. Between 1965 and 1976, the U.S. divorce rate doubled from 2.5 to 5.0 per 1,000 marriages and was reported at 5.3 per thousand in 1979. Over the last three decades, there has been a large increase in the number of children affected by divorce; the number of children per 10,000 involved in divorce was 1,147 in 1978 compared with 347 in 1955. According to predictions made by Bane (1976), 28.8% of all children born in 1973 will be involved in divorce. Estimates made by Glick and Norton (1977) suggest that about 45% of all children born in 1977 are likely to live for a period of several months as a member of a one-parent family.

The number of single parents who are unwed mothers has also been on the rise. Children under 18 years of age were almost twice as likely to be living with never-married mothers in 1978 than in 1968. The proportion of out-of-wedlock births rose from 3.8% in 1940, 9.7% in 1968,

---

[4] The population statistics in this section are taken from the U.S. Population Bulletin, *Marrying, Divorcing and Living Together in the U.S. Today*, October 1977, and the Statistical Abstract of the United States, 1980.

and 13% in 1973, to 16.3% in 1978. The majority of unwed births are to women under the age of 20.

The data from a variety of sources make it clear that the number of families containing only one adult is increasing. It also appears that the number of children associated with one-parent families is increasing. The effects of one-parent families depend on several factors, one of which is whether it is the mother or the father who is the only parent. Although there are few data available and the numbers are quite small in relation to female adult one-parent families, there is an increase in the number of families with males as single parents. Nevertheless, the majority of single-parent families are headed by women, who must fulfill most of the parenting functions.

Many of the effects of father absence or the effects of a single-parent household may be explained by the differences in the mother's behavior toward her children as a result of lack of support rather than by the father's absence *per se* (Lewis *et al.*, 1981; Pedersen, 1976). Support can be viewed in many ways and includes economic support, emotional support (e.g., giving the mother attention, love, and care), interactive support (e.g., discussing child-rearing practices and problems), and finally energy support (actually helping with the care of the child). In many cases, mothers whose husbands are absent must provide economic support for the family whether or not they want to work. In 1980, 67% of mothers in one-parent families with children under 18 were in the labor force, compared with 54% of mothers in two-parent families. Divorced, single-parent mothers often experience downward economic mobility. In 1979, 48.6% of children under 18 in female-headed families lived on incomes below the poverty level. In 1978, the mean income for divorced female heads of household was $11,098, whereas for divorced male heads of household, the mean income was almost twice that amount: $19,402. Female household heads also tend to be concentrated in the low-paying occupations, such as clerical and household work. It is likely that these single-parent mothers, with less education and less income, are under a great deal of stress, which would both affect their own feelings about themselves and adversely affect the quality of the mother–child relationship.

Lynn (1974) described some of the difficulties experienced by mothers who were without husbands on a regular or temporary basis. Mothers without husbands felt psychologically worse and were not goal-oriented. They were more concerned about the children's educational achievements than mothers with husbands; they made more inappropriate efforts to help their children; they were more likely to be dissatisfied with their children's level of work; and they were less likely to be involved with the schools or to aspire to a college education for their children. Finally, mothers whose husbands were temporarily absent on a regular basis led less active social lives, worked less outside the home, were more overprotective of their children, and were more likely to be concerned with their child's obedience and manners than with his or her happiness and self-realization.

It seems evident that the father's emotional as well as economic support of the mother can affect the child in a variety of ways. For example, by allaying the mother's doubts, anxiety, and frustrations by discussion and mutual decision-making, and by making her feel more self-confident and secure, the father can enable the mother to be more responsive to the child. In addition, growing up in a loving, relatively consistent, and stable family atmosphere, the child may be more relaxed, more self-confident, and more likely to perceive social interaction as a pleasant, enjoyable activity.

In contrast, children who experience divorce are exposed to a family situation that can be extremely stressful (Hansen & Hill, 1964). Children who are involved in divorce may learn that becoming attached to a person requires the risk of loss, a risk that they may become fearful of taking. In fact, offspring from homes broken by divorce are more likely to experience divorce than those from homes broken by death (Pope & Mueller, 1976). This finding suggests an intergenerational transmission of divorce, perhaps a phenomenon related to the children's modeling the parents' intimate, interpersonal behavior.

Although there is relatively scant information, the data that do exist seem to indicate that mothers who are deprived of the support of a spouse or other adults are more likely to show deficits in their quality of maternal behavior. It is becoming increasingly obvious that the mother's sense of emotional support affects her maternal behavior, which, in turn, influences the child's development (Crockenberg, 1981; Lynn, 1974; Schaefer 1974). Feiring (1976) studied the relationship between maternal involvement and responsivity to the child and the mother's perception of how much support she received from the secondary parent (i.e., father or female relative). The results of this study indicated that a strong positive association existed between maternal ratings of support from the secondary parent and ratings of maternal involvement with and responsivity to the child.

The major weakness of a single-parent family may be the absence of two adults, each providing the other with mutual aid and support. The amount of support rather than the actual identity of the individuals in the adult–adult dyad may be of central importance in the functioning of the adult–child dyad. Indeed, one of the father's functions is that of being an additional adult (Lewis *et al.*, 1981). Although fathers may have specific sex-role functions because they are males, it may be that any additional adult may satisfy some support needs. For example, Aug and Bright (1970) studied different family structures and found that "normal" psychological development existed in those families with out-of-wedlock children that had the support of other family members and relatives. Such families functioned better than many two-parent families with in-wedlock children. However, in those out-of-wedlock families where additional support from family members or relatives was not evident, the functioning of the mother was impaired. The nature, as well as the mother's perception, of the support system in her family appears to be a crucial factor in determining the positive nature of the mother–child dyad (Colletta, 1981; Pedersen, 1976).

The absence of fathers in single-parent families can affect the child's development in ways other than depriving the mother of support. So far, we have focused on the mother–child relationship and how its quality might be affected by single parenthood with respect to divorced and unwed mothers. Of course, we must also mention that, as more and more children are affected by divorce, this will certainly have an impact on the father–child relationship. One would expect the amount of time that divorced fathers spend with their children to be even less than the amount of time that married fathers spend with their children. There is some anecdotal evidence that suggests that divorced fathers spend more time and more "important" time with their children once they no longer live in the same household. Although it is more difficult to discuss the quality of the father–child relationship and how this would be affected by divorce, there is considerable information on the effect of father absence that may be related to the father–child relationship itself.

As was the case for the effect of the father's absence on the mother–child relationship, it is the case that the cause of the father's absence should also affect the father–child dyad. For example, in the cases of the father's death or the father's divorce, it is clear that the former terminates interaction, whereas the latter need not affect the quality of the relationship between father and child. In fact, for some fathers, divorce and physical separation from the everyday activities of the house actually increase child contact and highlight the positive features of the relationship. Nevertheless, there is considerable evidence to indicate that father absence leads to significant differences in children's behavior. Whether these are directly due to a change in the father–child relationship or to changes in the entire family constellation is not known (Pedersen, 1976). The absence of a father in the home has been associated with characteristics such as lowered intellectual level (Deutch & Brown, 1964), impulsive cognitive style (Barclay & Cusumano, 1967), lower academic attainment (Blanchard & Biller, 1971), weaker sex-role identification (Biller, 1968; Hetherington, 1966), and more delinquent behavior (McCord, McCord, & Thurber, 1962).

Much of the literature points to the deleterious effects of single parenthood on children's development. Unfortunately, most of the work has confounded single parenting with a host of other variables; thus, it is almost impossible to separate out the effects of a single-parent family on the development of the child. Fiscal difficulties, the causes of the single-parent family, the social status of the family, the role of the absent parent, and the role of other family members are all important factors that must be considered in determining how single parenthood affects children's development. It is widely believed that divorce is deleterious for children. Some studies have shown that children from disrupted marriages are more likely to be delinquent, to be low achievers, and to be psychologically disturbed (Bane, 1976). However, recent reviews of the literature on marital disruption criticize these studies on the grounds that they lacked adequate controls (Bradwein, Brown, & Fox, 1974). Research adequately controlled for economic status does not provide support for the belief that divorce is disastrous for children. One study in particular found small or nonexistent differences between children from one- and two-parent homes of comparable economic status, on social adjustment, delinquent behavior, and school achievement (Murchinson, 1974).

Hetherington, Cox, and Cox (1979) conducted a longitudinal study of the effects of divorce on social interactions and play in middle-class preschool children two months, one year, and two years after divorce. The play patterns of children from divorced families, in comparison with a matched group of children in nondivorced families, were less socially and cognitively mature when measured shortly after divorce. Limitations and rigidity in fantasy were especially prevalent. In the first year following divorce, the children showed higher rates of dependent help-seeking behavior, acting out, and noncompliant behavior. Although these adverse effects had largely disappeared in the girls by two years following the divorce, they appeared to be more intense and enduring for the boys. Two years after parental divorce, even when the behavior of the divorced families had improved, the boys were perceived more negatively by peers and were responded to more negatively by peers and teachers than were children from nondivorced families or girls from divorced families. Consequently, the sex of the child may be an important factor in assessing the impact of single parenthood on child development.

The complexity of the problem of the single-parent family and its impact on child development is increased by the growth of new ideologies. Although there are few data on the subject, the interest in single-parent families is in small part due to some women who have decided to become mothers without having a husband. They differ from the more typical unwed mother in that they are usually more educated and have made a careful decision to have a child and raise it in a single-parent home. What assets or difficulties children from such families grow to have intellectually, emotionally, or socially remain to be seen.

In summary, more and more children are embedded in single-parent households, especially as a result of divorce and illegitimacy. Divorce would seem to reduce the amount of time that a parent spends with the child. Many single-parent mothers must work to support their children, and the fathers are no longer a part of the everyday household. Indeed, only a very small percentage of children under 18 live with their divorced fathers, although the percentage has increased somewhat from 1.1% in 1968 to 1.6% in 1979. In determining the effect of single-parent families on the development of the child, it is clear that "single-parentness" *per se* may not be the critical variable, and such factors as the economic condition of the family, the support system provided the single parent, and the reasons behind the single-parent status must be considered. Nevertheless, the reduction from at least two participating adults to one in the family must strain family resources, both financial and psychological. The reduction in the number and the availability of the remaining adults radically alters the social network of the children of single parents. The effects of enlarging this network through the involvement of other family members, such as the grandparents, may ameliorate many of the difficulties inherent in single-parent families. Nevertheless, the lack of an adult companion for the remaining parent affects that parent's life and consequently the child's. The rapid increase in the number of single-parent families probably necessitates new "family" structures and the participation of additional adults in the raising of the young.

Conclusion

The discussion of psychological effects on child development as a function of changing family structures is predicated on the belief that the child's development is a function of the nature of its social network (Lewis, 1982; Lewis & Feiring, 1978). It is insufficient to discuss developmental consequences without appreciating this fact. Although the mother–child relationship is important to the child's well-being and development (Ainsworth, 1969; Bowlby, 1969), it is clear that the effects of the entire social network need to be considered (Lewis, 1982; Lewis & Rosenblum, 1979). First, the social network itself affects the mother's relationship to the child. Thus, for example, if the father provides sufficient emotional and fiscal support, the mother is more able to carry out successful caregiving. We have labeled these effects *indirect* and have discussed their importance at length (Lewis & Feiring, 1981; Lewis &

Weinraub, 1976). Second, the social network directly affects the child's behavior. In general, it is now recognized that other social objects besides the mother are important to the development of the child. Peers (Lewis *et al.*, 1975; Mueller & Vandell, 1978), siblings (Dunn & Kendrick, 1979, 1980), and fathers (Lamb, 1976b; Lewis & Weinraub, 1976) are just a few of these others who exhibit strong, important, and varied influence on the course of the child's growth.

The understanding that development is related both directly and indirectly to the social network in which children are raised creates special problems when considering the effects of the three demographic variables that have shown large changes over the last 25–30 years. First, it compounds the complexity of understanding the consequences of these historical changes because we have not yet fully understood the nature of the social network itself. For example, the network is made up of many different people (e.g., parents, siblings, relatives, and adult and peer friends). Without understanding the relationships among these members, it becomes difficult to determine how the loss or change in role of one member will affect the child's growth. Thus, although the number of parents in the house may be decreasing, the increased involvement of grandparents could mean no net loss in the number of adults available and therefore in the amount of adult–child interaction. However, the activities of the parent may change, as well as the parents' role in socializing the child.

An important issue pertaining to the complexity and the diversity of the social network concerns the potential change in role relationships as they exist in the network itself. For example, although more mothers are working, it may be true that there is a growing increase in fathers' involvement in child care. Although there are no hard data to support fathers' increased involvement, there is some indication, at least among the better educated, that fathers are spending more time in child care. Thus, although maternal employment might be associated with less maternal care and even more peer involvement, fathers' increased involvement in the caregiving role itself might result in no net loss in regard to adult-to-child caregiving, although the distribution of the caregiving function by parent may alter.

What we do know is that the characteristics of the American family have been changing. There are more smaller families, more families with working mothers, and more single-parent families. These changes in the characteristic structure of families are surely having an impact on social networks, family relationships, and the development of the child. What we have offered here are conjectures about how changes in the demographic characteristics of the family might affect social networks and family relationships, as well as the child's growth. These speculations provide a bridge between issues of interest to demographers, sociologists, and psychologists. Rather than providing answers, our endeavor has given us the opportunity to consider what areas might be interesting to explore in research on the interface between the individual child, his or her family, and the larger social context.

REFERENCES

Ainsworth, M. D. S. Object relation, dependency and attachment: A theoretical review of the infant–mother relationship. *Child Development,* 1969, *40,* 969–1026.

Ainsworth, M. D. S., Blehar, M. C., Waters, E., & Wall, S. *Patterns of attachment: A psychological study of the strange situation.* Hillsdale, N.J.: Lawrence Erlbaum, 1978.

Aug, R. G., & Bright, T. A study of wed and unwed motherhood in adolescents and young adults. *Journal of the Academy of Child Psychiatry,* 1970, *9,* 577–592.

Ban, P., & Lewis, M. Mothers and fathers, girls and boys: Attachment behavior in the one-year-old. *Merrill-Palmer Quarterly,* 1974, *20*(3), 195–204.

Bane, M. J. Marital disruption and the lives of children. *Journal of Social Issues,* 1976, *32*(1), 103–117.

Barclay, A. G., & Cusumano, D. Father-absence, cross-sex identify, and field-dependent behavior in male adolescents. *Child Development,* 1967, *38,* 243–250.

Barret, R. Self-esteem and maternal behavior. Unpublished doctoral dissertation, University of Pittsburgh, 1976.

Belmont, L., & Marolla, F. A. Birth order, family size and intelligence. *Science,* 1973, *182,* 1096–1101.

Belsky, J. A family analysis of parental influence on infant exploratory competence. In F. Pedersen (Ed.), *The father–infant relationship: Observational studies in a family context.* New York: Praeger, 1980.

Belsky, J., & Steinberg, L. D. The effects of day care: A critical review. *Child development,* 1978, *49,* 929–949.

Berg, I., Fearnly, W., Palerson, M., Pollock, G., & Vallance, R. Birth order and family size of approved school boys. *British Journal of Psychiatry,* 1967, *113,* 793–800.

Biller, H. B. A note on father-absence and masculine development in young lower-class Negro and white boys. *Child Development,* 1968, *39,* 1003–1006.

Blanchard, R. W., & Biller, H. B. Father availability and academic performance among third-grade boys. *Developmental Psychology,* 1971, *4,* 301–305.

Blehar, M. C. Anxious attachment and defensive reactions associated with day care. *Child Development,* 1974, *45,* 683–692.

Borod, J., Grossman, J. C., & Eisenman, R. Extraversion, anxiety, creativity and grades. *Perceptual and Motor Skills,* 1971, *33,* 1106.

Bossard, J. H., & Boll, E. S. *The sociology of child development.* New York: Harper & Row, 1966.

Bowlby, J. *Attachment and loss, Vol. 1: Attachment.* New York: Basic Books, 1969.

Bradwein, R. A., Brown, C. A., & Fox, E. M. Women and children last: The social situation of divorced mothers and their families. *Journal of Marriage and the Family,* 1974, *36,* 498–514.

Bulatao, R. A., & Arnold, F. *Relationships between the value and cost of children and fertility: Cross-cultural evidence.* Paper prepared for the General Conference of the International Union for the Scientific Study of Population, Mexico City, August 8–13, 1977.

Caldwell, B. M., Wright, C. M., Honig, A. S., & Tannenbaum, J. Infant care and attachment. *American Journal of Orthopsychiatry*, 1970, *40*, 397–412.
Clarke-Stewart, A. Interactions between mothers and their young children: Characteristics and consequences. *Monographs of the Society for Research in Child Development*, 38/6–7, Serial No. 153 (1973).
Clarke-Stewart, K. A. And daddy makes three: The father's impact on mother and young child. *Child Development*, 1978, *49*(2), 466–478.
Clarke-Stewart, K. A. The father's contribution to children's cognitive and social development in early childhood. In F. A. Pedersen (Ed.), *The father–infant relationship: Observational studies in the family setting*. New York: Praeger, 1980.
Coates, D. L., & Lewis, M. *Relationships between cognitive behavior at six years and mother–infant interaction at three months*. Paper presented at the International Conference on Infant Studies, New Haven, Connecticut, April 1980.
Cochran, M. E. A comparison of group day and family child-rearing patterns in Sweden. *Child Development*, 1977, *48*, 702–707.
Cohen, S. E. Maternal employment and mother–child interaction. *Merrill-Palmer Quarterly*, 1978, *24*, 189–197.
Colletta, N. D. *The influence of support systems on the maternal behavior of young mothers*. Paper presented at the meeting of the Society for Research in Child Development, Boston, Massachusetts, April 1981.
Crockenberg, S. Infant irritability, mother responsiveness, and social support influences on the security of infant-mother attachment. *Child Development*, 1981, *52*, 857–865.
Deutch, M., & Brown, B. Social influences in Negro-white intelligence differences. *Journal of Social Issues*, 1964, *20*, 24–35.
Douglas, J. W. B., & Blomfield, J. M. *Children under five*. London: George Allen and Unwin, 1958.
Douglas, J. W. B., & Simpson, H. R. Height in relation to puberty, family size and social class. *Milbank Memorial Fund Quarterly*, 1964, *40*, 20–35.
Dunn, J., & Kendrick, C. Interaction between young siblings in the context of family relationships. In M. Lewis & L. Rosenblum, *The child and its family: The genesis of behavior* (Vol. 2). New York: Plenum Press, 1979.
Dunn, J., & Kendrick, C. The arrival of a sibling: Changes in patterns of interaction between mother and first-born child. *Journal of Child Psychology and Psychiatry*, 1980, *21*(2), 119–132.
Easterbrooks, M. A., & Lamb, M. E. The relationship between quality of infant-mother attachment and infant competence in initial encounters with peers. *Child Development*, 1979, *50*, 380–387.
Espenshade, T. J. *The value and cost of children*. Population Bulletin 32: 1 (1977).
Eysenck, H. J., & Cookson, D. Personality in primary school children: Family background. *British Journal of Educational Psychology*, 1970, *40*, 117–131.
Farran, D., & Ramey, C. Infant day care and attachment behaviors toward mothers and teachers. *Child Development*, 1977, *48*, 1112–1116.
Fein, G., & Clarke-Stewart, A. *Day care in context*. New York: Wiley, 1973.
Feiring, C. *The preliminary development of a social systems model of early infant–mother attachment*. Paper presented at the meetings of the Eastern Psychological Association, New York, April 1976.
Feiring, C., & Lewis, M. The child as a member of the family system. *Behavioral Science*, 1978, *23*, 225–233.
Feldman, S. S. *The impact of day care on one aspect of children's social emotional behavior*. Paper presented at the meetings of the American Association for the Advancement of Science, San Francisco, September 1974.

Glick, C. P., & Norton, A. J. Marrying, divorcing, and living together in the U. S. today. *Population Bulletin*, 1977, *32*, 5.

Glueck, S., & Glueck, E. *Unraveling juvenile delinquency*. Cambridge: Harvard University Press, 1950.

Golden, M., Rosenbluth, L., Grossi, M., Policare, H., Freeman, H., & Brownlee, E. *The New York City Infant Day Care Study*. New York: Medical and Health Research Association of New York City, 1978.

Hansen, D. A., & Hill, R. Families under stress. In *Christensen handbook of marriage and the family*. Chicago: Rand McNally, 1964.

Harlow, H., & Harlow, M. K. Age mate or peer affectional system. In D. S. Lehrman, R. Hinde, & E. Shaw, *Advances in the study of behavior* (Vol. 2). New York: Academic Press, 1969.

Hetherington, E. M. Effects of paternal absence in sex-typed behaviors in Negro and white preadolescent males. *Journal of Personality and Social Psychology*, 1966, *4*, 87–91.

Hetherington, E. M., Cox, M., & Cox, R. Play and social interaction in children following divorce. *Journal of Social Issues*, 1979, *35*, 26–42.

Hock, E. Working and nonworking mothers and their infants: A comparative study of maternal caregiving characteristics and infant social behaviors. In *Merrill-Palmer Quarterly—Behavior and Development*, April 1980, *26*(2), 79–101.

Hofferth, S. L., & Moore, K. A. Women's employment and marriage. In R. W. Smith (Ed.), *The subtle revolution*. Washington, D.C.: The Urban Institute, 1979.

Hoffman, L. W. Effects of maternal employment on the child—A review of the research. *Developmental Psychology*, 1972, *10*, 204–228.

Hoffman, L. W. Effects of the first child on the woman's role. In Miller & Newman (Eds.), *The first child and family formation*. North Carolina Press: Chapel Hill, 1977.

Hoffman, L. W. Maternal employment: 1979. *American Psychologist*, 1979, *34*, 859–865.

Hoffman, L. W., & Nye, F. I. *Working mothers*. San Francisco: Jossey-Bass, 1974.

Kagan, J. Emergent themes in human development. *American Scientist*, 1976, *64*, 186–196.

Kotelchuck, M. *The nature of the child's tie to his father*. Unpublished doctoral dissertation, Harvard University, 1972.

Lamb, M. E. Effects of stress and cohort on mother– and father–infant interaction. *Developmental Psychology*, 1976, *12*, 425–443.(a)

Lamb, M. E. *The role of the father in child development*. New York: Wiley, 1976.(b)

Lamb, M. E. Father–infant and mother–infant interaction in the first year of life. *Child Development*, 1977, *48*, 167–181.

Lamb, M. E., Chase-Lansdale, L., & Owen, M. T. The changing American family and its implications for infant social development: The sample case of maternal employment. In M. Lewis & L. Rosenblum (Eds.), *The child and its family: Genesis of behavior* (Vol. 2). New York: Plenum Press, 1979.

Landis, P. H. The families that produce adjusted adolescents. *Clearing House*, 1955, *29*, 537–540.

Leibowitz, A. Education and home production. *American Economic Review*, 1974, *64*(2), 243–250.

Lewis, M. The social network systems model: Toward a theory of social development. In T. Field (Ed.), *Review in human development*. New York: Wiley, 1982.

Lewis, M., & Coates, D. L. Mother-infant interactions and cognitive development in twelve-week-old infants. *Infant Behavior and Development*, 1980, *3*, 95–105.

Lewis, M., & Feiring, C. The child's social world. In R. M. Lerner & G. D. Spanier (Eds.), *Child influence in marital and family interaction: A life-span perspective*. New York: Academic Press, 1978.

Lewis, M., & Feiring, C. The child's social network: Social object, social functions, and their relationship. In M. Lewis & L. Rosenblum (Eds.), *The child and its family: The genesis of behavior* (Vol. 2). New York: Plenum Press, 1979.

Lewis, M., & Feiring, C. Direct and indirect interactions in social relationships. In L. Lipsitt (Ed.), *Advances in infancy research* (Vol. 1). New York: Ablex, 1981.

Lewis, M., & Feiring, C. Some American families at dinner. To appear in L. Laosa & I. Sigel (Eds.), *The family as a learning environment* (Vol. 1). New York: Plenum Press, 1982.

Lewis, M. & Goldberg, S. Perceptual-cognitive development in infancy: A generalized expectancy model as a function of the mother–infant interaction. *Merrill-Palmer Quarterly*, 1969, 51(1), 81–100.

Lewis, M., & Rosenblum, L. *Friendship and peer relations: The origins of behavior* (Vol. 4). New York: Wiley, 1975.

Lewis, M., & Rosenblum, L. *The child and its family: The genesis of behavior* (Vol. 2). New York: Plenum Press, 1979.

Lewis, M., & Schaeffer, S. Peer behavior and mother–infant interaction in maltreated children. In M. Lewis & L. Rosenblum (Eds.), *The uncommon child: The genesis of behavior* (Vol. 3). New York: Plenum Press, 1981.

Lewis, M., & Weinraub, M. The father's role in the infant's social network. In M. Lamb, *The role of the father in child development*. New York: Wiley, 1976.

Lewis, M., Young, G., Brooks, J., & Michalson, L. The beginning of friendship. In M. Lewis, M., Feiring, C., & Weinraub, M. The father as a member of the child's social network. In M. Lamb (Ed.), *The role of the father in child development* (2nd ed.). New York: Wiley, 1981.

Lynn, D. B. *The father: His role in child development*. Monterey, Calif.: Brooks/Cole, 1974.

Marjoribanks, K., & Walberg, H. J. Ordinal position, family environment and mental abilities. *Journal of Social Psychology*, 1975, 95, 77–84.

McCord, J., McCord, W., & Thurber, E. Some effects of paternal absence on male children. *Journal of Abnormal and Social Psychology*, 1962, 64, 361–369.

Monane, J. H. *A sociology of human systems*. New York: Appleton-Century-Crofts, 1967.

Moore, T. Exclusive early mothering and its alternatives: The outcome to adolescence. *Scandinavian Journal of Psychology*, 1975, 16, 255–272.

Moskowitz, D., Schwarz, J., & Corsini, D. Initiating day care at three years of age: Effects on attachment. *Child Development*, 1977, 48, 1271–1276.

Mueller, E. Toddlers + toys = An autonomous social system. In M. Lewis & L. Rosenblum (Eds.), *The child and its family: The genesis of behavior* (Vol. 2). New York: Plenum Press, 1979.

Mueller, E., & Vandell, D. Infant–infant interaction. In *Osofsky handbook of infant development*. New York: Wiley, 1978.

Murchinson, N. Illustration of the difficulties of some children in one-parent families. In *Finer Report of the Committee on One-Parent Families*. London: Her Majesty's Stationery Office, 1974.

Pedersen, F. A. *Mother, father, and infant as an interactive system*. Paper presented at the meetings of the American Psychological Association, Chicago, September, 1975.

Pedersen, F. A. Does research on children reared in father-absent families yield information on father influences? *The Family Coordinator*, 1976, 25, 459–464.

Pedersen, F. A. (Ed.). *The father–infant relationship: Observational studies in the family setting*. New York: Praeger, 1980.

Pedersen, F. A., & Robson, K. S. Father participation in infancy. *American Journal of Orthopsychiatry*, 1969, 39, 466–472.

Pedersen, F. A., Yarrow, L. J., Anderson, B. J., & Cain, R. L. Conceptualization of father influences in the infancy period. In M. Lewis & L. Rosenblum, (Eds.), *The child and its family: The genesis of behavior* (Vol. 2). New York: Plenum Press, 1979.

Pedersen, F. A., Anderson, B. J., & Cain, R. L., Jr. Parent–infant and husband–wife interactions observed at five months. In F. A. Pedersen (Ed.), *The father–infant relationship: Observational studies in the family setting.* New York: Praeger, 1980.

Pope, H., & Mueller, C. W. The intergenerational transmission of marital instability: Comparisons by race and sex. *Journal of Social Issues*, 1976, *32*(1), 49–66.

Ramey, C., & Campbell, F. The prevention of developmental retardation in high-risk children. In P. Mittler (Ed.), *Research to practice in mental retardation, Vol. 1: Care and intervention.* Baltimore: University Park Press, 1977.

Ramey, C., & Smith, B. Assessing the intellectual consequences of early intervention with high-risk infants. *American Journal of Mental Deficiency*, 1976, *81*, 318–324.

Reed, R. H., & McIntosh, S. Costs of children. In *Research Reports*, Vol. 2, Commission of Population Growth and the American Future. Washington, D. C.: Government Printing Office, 1972.

Rehberg, R. Q., & Westby, D. L. Parental encouragement, occupation, education, and family size: Artifactual or independent determinants of adolescent educational expectations? *Social Forces*, 1967, *45*, 362–374.

Ricciuti, H. Fear and development of social attachments in the first year of life. In M. Lewis & L. A. Rosenblum (Eds.), *The origins of human behavior: Fear.* New York: Wiley, 1974.

Robinson, J. P. *How Americans use time: A sociological perspective.* New York: Praeger, 1977.

Rosen, B. C. Family structure and value transmission. *Merrill-Palmer Quarterly*, 1964, *70*, 59–76.

Sander, L. The regulation of exchange in the infant–caretaker system and some aspects of the content–content relationship. In M. Lewis & L. Rosenblum (Eds.), *Interaction, conversation, and the development of language: The origins of behavior* (Vol. 5). New York: Wiley, 1977.

Schacter, F. Toddlers with employed mothers. *Child Development*, 1981, *52*, 958–964.

Schaefer, E. S. *The ecology of child development: Implications for research and the professions.* Paper presented at the meetings of the American Psychological Association, New Orleans, September 1974.

Schaffer, H. R., & Emerson, P. E. The development of social attachments in infancy. *Monographs of the Society for Research in Child Development*, 1964, *29*(4), Serial No. 94.

Schwartz, J. C., Stickland, R. G., & Krolick, G. Infant day care: Behavioral effects at preschool age. *Developmental Psychology*, 1974, *10*, 502–506.

Sears, R. K. Relations of early socialization experiences to self concepts and gender role in middle childhood. *Child Development*, 1970, *41*, 267–289.

Spanier, G. B., Lerner, R. M., & Aquilino, W. The study of child–family interactions: A perspective for the future. In R. M. Lerner & G. B. Spanier, *Child influences on marital and family interactions: A life-span perspective.* New York: Academic Press, 1978.

Sroufe, L. A., & Waters, E. Attachment as an organizational construct. *Child Development*, 1977, *48*, 1184–1199.

Statistical Abstract of the United States, U.S. Department of Commerce, Bureau of the Census, 1980.

A Statistical Portrait of Women, in the U.S. Population Report Series P-23, No. 58, 1975.

Suomi, S. J., & Harlow, H. F. The role and reason of peer relationships in rhesus monkeys. In M. Lewis & L. Rosenblum (Eds.), *Friendship and peer relations: Origins of behavior* (Vol. 4). New York: Wiley, 1975.

Svanum, S., & Bringle, R. Evaluation of confluence model variables on IQ and achievement test scores in a sample of 6 to 11 year old children. *Journal of Educational Psychology,* 1980, *72,* 427–436.

Taylor, J. *Systometrics.* Unpublished manuscript, University of Pittsburgh, 1975.

Terhune, K. W. *A review of the actual and expected consequences of family size.* U. S. Department of Health, Education and Welfare, Public Health Service, National Institutes of Health Publication No. (NIH) 75-779 (1974).

Tuckman, J., & Regan, R. A. Size of family and behavioral problems in children. *Journal of Genetic Psychology,* 1967, *3*(1), 151–160.

Turner, R. H. Some family determinants of ambition. *Sociology and Social Research,* 1962, *46,* 397–411.

U.S. Population Bulletin, *Marrying, Divorcing and Living Together in the U.S. Today,* October 1977.

Von Bertalanffy, L. *Robots, men, and minds.* New York: Braziller, 1967.

Weinraub, M. *The changing role of the father: Implication for sex role development in children.* Paper presented at the meetings of the American Psychological Association, Montreal, August 1980.

Zajonc, R. B. Family configuration and intelligence: Variations in aptitude scores parallel trends in family size and the spacing of children. *Science,* 1976, *192,* 227–236.

Zajonc, R. B., & Markus, G. B. Birth order and intellectual development. *Psychological Reviews,* 1975, *82,* 74–88.

Zajonc, R., & Markus, G. B. *The birth order puzzle.* Paper presented at the meeting of the Conference on the Social Network of the Developing Infant: The Origins of Behavior, Princeton, N.J., May 1977.

Zajonc, R. B., Markus, H., & Markus, G. B. The birth order puzzle. *Journal of Personality and Social Psychology,* 1979, *37,* 1325–1341.

5

Implications of Monogamy for Infant Social Development in Mammals

DEVRA G. KLEIMAN

INTRODUCTION

Monogamy is a relatively rare mating system among the mammals, probably because of the ability of the mammalian female to rear offspring in the absence of any parental investment by a male. Females both gestate and lactate, and they can typically ensure their offsprings' survivorship to the point of weaning and independence (Kleiman, 1977). Monogamy has evolved in diverse mammals, and the basis for its evolution appears to have differed in different groups. Some of the selective forces that may have influenced the evolution of monogamy in mammals include high rates of intrasexual aggression among females, limited polygyny potential in males, and nonshareable indispensable male parental care (Kleiman, 1977; Wittenberger & Tilson, 1980).

Monogamous mammals vary in the degree of sociality they exhibit (Kleiman, 1977, 1981). Among some species, there is a dispersed social system whereby the bonded pair are rarely seen together or with their offspring, a condition I have previously referred to as *facultative monogamy* (Kleiman, 1977, 1981). Other species exist as bonded pairs in close contact but are seen only seasonally with their young. In the more social

---

DEVRA G. KLEIMAN • Department of Zoological Research, National Zoological Park, Smithsonian Institution, Washington, D.C. 20008. The research in this paper was supported, in part, by NIMH 27241.

monogamous species, one sees nuclear families in which several generations of offspring cohabit with the mated pair, or with extended families, which include the mated pair, the offspring, other genetically related adults, and even unrelated individuals (Kleiman, 1981).

The selective forces influencing the evolution of the mating system, whether monogamy or polygamy, are independent of those factors selecting for the degree of sociality within a species, although they may act in concert during evolution. Thus, an analysis of the potential influence of monogamy on infant development must take into account the variability in sociality among species and, as a result, the variability in the numbers, ages, and sexes of conspecifics with which an infant might interact. For example, in the rufous elephant shrew *(Elephantulus rufescens)* (Rathbun, 1979) and the agouti *(Dasyprocta punctata)* (Smythe, 1978), the precocial young are hidden during early development and interact rarely with either parent. Because these species usually bear only singletons or twins, the potential for an infant to interact with more than one littermate is limited, as is the frequency of social interactions with the parents.

The extreme opposite condition prevails in some of the pack-hunting canid species (Frame, Malcolm, Frame, & Van Lawick, 1979; Malcolm, 1979) and in the marmosets and tamarins (Epple, 1975), whose young develop not only in close contact with the parental pair but with adult relatives, siblings of various ages, and littermates. Here, there is a rich opportunity for socialization, which occasionally extends beyond puberty.

Those selective factors promoting or retarding the evolution of sociality have been discussed by several authors (e.g., Alexander, 1974; Eisenberg, 1966; Wilson, 1975). For example, sociality may be negatively influenced by the adoption of an antipredator strategy involving crypsis, as in the small forest-dwelling elephant shrews, which rely on immobility and camouflage as protection from aerial and ground predators (Rathbun, 1979). By contrast, other species use sociality and group vigilance as an antipredator mechanism, as in the diurnal plains-dwelling African mongooses (Rood, 1983). In some species, sociality is promoted through the improved foraging potential of a larger social group; for example, canids (wild dogs and wolves) that hunt in packs may be able to kill larger prey through a group effort than with a solo hunting strategy (Kleiman & Eisenberg, 1973).

Regardless of the degree of sociality, there is a major difference between polygynous and monogamous mammals in infant development because the male, or father, almost always contributes to infant development, even if his investment is indirect (Kleiman & Malcolm, 1981). Among most polygynous species, the mother–young unit is the primary unit for infant socialization; highly social species usually incorporate a number of mother–young units without significant involvement by adult males. Yet, even in monogamous mammals that are asocial, the mother–father–young unit is the primary context in which infants are reared.

MONOGAMY, SOCIALITY, AND LIFE HISTORY CHARACTERISTICS

How certain life history characteristics are correlated in monogamous mammals depends on the degree of sociality exhibited, including the quality of the pair bond, the quality and quantity of parental care, the age of juvenile dispersal, and sex-role differences (Kleiman, 1981). These characteristics are summarized in Table 1. In asocial monogamous forms, the pair bond is weak, as expressed by a relative lack of affiliative interactions between the mated male and female. Although an elephant-shrew mated pair jointly occupies a single territory, the male and female rarely interact, do not synchronize their activity, and are rarely in the same location at the same time. Except at mating, a female is often

TABLE 1. Correlations among Life History Characteristics of Extremely Asocial (Facultative) and Extremely Social (Obligate) Monogamous Mammals[a]

	Asocial	Social
A. Pair bond	Weak	Strong
1. Interactions	a. Infrequent	Frequent
	b. Asynchronous	Synchronous
	c. Agonistic > affiliative	Affiliative > agonistic
B. Territorial behavior		
1. Long calls	—	♀ ≥ ♂
2. Scent marking	♂ > ♀	♀ ≥ ♂
3. Aggression to conspecific intruders	♂ > ♀ (Intrasex)	♀ ≥ ♂ (Intrasex)
C. Parental care		
1. Father	Indirect	Indirect and direct
2. Siblings and other "helpers"	—	Indirect and direct
D. Sexual dimorphism	Rare	Rare
E. Juvenile development		
1. Neonatal condition	Often precocial	Often altricial
2. Maturation rate	Rapid	Slow
3. Reproductive suppression	Rare	Typical

[a] From Kleiman (1981).

aggressive to the mate. However, territorial boundaries are rigid, and the territorial defense activity of each member of the pair provides little opportunity for mating outside the pair bond (Rathbun, 1979). At one level, monogamy is forced on elephant shrews (low polygyny potential) because of evolutionary and ecological restraints, such as a cryptic antipredator strategy, small size and mobility, the need for a familiar home range, and the disadvantages of social foraging.

The relatively weak pair bond is correlated with an absentee parental care system in elephant shrews. The precocial young are hidden, and the parents visit them singly, with the female nursing only infrequently. The young mature rapidly, and although they may remain on the parental home range after weaning, they eventually disperse to seek a territory and a mate. Because elephant shrews are small and not very long-lived, the turnover in pair mates is probably high, which may contribute to the weakness of the pair bond. This system is distinguished from an asocial polygynous condition by (1) the restriction of one pair to a single territory, implying an inability by the male to successfully encompass the territories of more than one female, and (2) an active role by the male and the female in eliminating intruders of their own sex, which serves to prevent adulterous matings.

Among the more social species, pair bonds are stronger and are characterized by pairs synchronizing activity, jointly exhibiting territorial behavior (mainly toward animals of the same sex), and exhibiting frequent affiliative social interactions. Sex roles with respect to maintaining the pair relationship are less dimorphic, with the female contributing equally (Kleiman, 1977, 1981). The close bonding of the male and the female permits the expression of direct parental care by the male from soon after the birth of the young.

Among monogamous species that are highly social (obligate monogamy), one rarely, if ever, sees an absentee parental care system with highly precocial young. Also, developmental time appears to be somewhat retarded, in that older juveniles and subadults may not become socially weaned from the parental family group until after the age of puberty. The degree to which the young disperse at sexual maturity (with the parents reverting to a pair condition), as opposed to the young becoming incorporated into the family, depends on the evolutionary forces selecting for a greater degree of sociality. As already mentioned, improved foraging efficiency and a group antipredator strategy may be two causes of the evolution of sociality among monogamous species. Without such selective forces, the young are likely to disperse at the earliest ages possible.

The retention of older juvenile offspring requires mechanisms for preventing postpubertal animals from independently reproducing within the family group. Kleiman (1980) discussed some of the inhibitory mechanisms that have evolved among different mammals exhibiting obligate monogamy, ranging from the physiological suppression of estrus and/or pregnancy in subdominant females in some marmosets and tamarins (Hearn, 1977; Lunn, 1978) to behavioral solutions, such as the killing of the offspring of subdominants or the inhibition of their lactation and nursing behavior in some canids (Altmann, 1974; van Lawick, 1973). Among males, physiological suppression seems to be rare; in most cases, it appears that subdominant males are simply prevented from breeding with the reproductively active female (tamarins—Epple, 1972; Kleiman, 1978, 1980) and/or breed with her during the least optimum period (dwarf mongooses, *Helogale parvula*—Rood, 1980). The different mechanisms used for the suppression of reproduction in subdominant individuals suggest that it is more important to inhibit female than male reproduction under conditions of obligate monogamy. Indeed, not only are the female suppressive mechanisms more highly evolved (e.g., physiological), but one also sees significantly more aggression among females, which may result in female emigration (Greenwood, 1980) and/or death (Kleiman, 1979; Frame et al., 1979).

Male Parental Care

An Overview

The influence of monogamy on infant social development can be evaluated only through an understanding of the quantity, the quality, and the timing of male parental care. Kleiman and Malcolm (1981) have separated indirect forms of male parental care from direct forms of investment. Among asocial monogamous species that exhibit facultative monogamy, it is more typical for males to provide care to their offspring in an indirect manner, by maintaining the integrity of the territory, providing safe refuges secure from predators, and ensuring, through the choice of the territory, that resources sufficient for the development of the offspring will be available. Such behaviors contribute not only to the survivorship of the offspring but to the survivorship of the father himself. Thus, the male's own survival and the survival of any offspring are both contingent on the same set of male behaviors. There is some slight evidence to indicate that the male's indirect investment increases with the appearance of young, which would clearly differentiate between the occurrence of the behavior in a parental versus an individual survivorship context. Classically, parental care by males of the indirect type has rarely been considered a form of parental investment. More attention has always been paid to the direct interactions of males with young that one sees only in the more social species. The differences in these behaviors have been described in Kleiman and Malcolm (1981).

There are several important points that should be considered with respect to the quality and the quantity of male direct parental care. First, other than nursing the young, the males and the females of the species tend to exhibit a high correlation in their parental care behaviors, in terms of both quantity and quality (Hartung & Dewsbury, 1979; Kleiman & Malcolm, 1981). Second, within any species, the form of parental care depends strongly on the life history strategy of the species. Because of morphological or ecological constraints, neither a male nor a female exhibits behaviors that are of high cost to them and of low benefit to the young. For example, neither an ungulate male nor an ungulate female carries food to its young because it would be impossible for a parent to provide sufficient food for survival without incurring a great cost. The parental feeding of weanlings occurs mainly in species where large food packets of high nutritional value can be provided to the offspring at a single time. Huddling with young, which may reduce thermoregulatory costs for the offspring, is a low-cost behavior that male and female parents can provide and that may have very positive benefits for the growth and development of the offspring (Dudley, 1974a).

Another aspect of parental care (which has great potential influence on the differential development of offspring) is the degree to which any behavior performed by the mother or the father can be shared among all the offspring (nondepreciable) or must be partitioned among several (depreciable) (Altmann, Wagner, & Lenington, 1977). A depreciable behavior is one that is reduced in availability to one individual once it has been expended on or used by another individual. Species differ in the degree to which even the same behavior pattern may be considered depreciable or nondepreciable. For example, if littermates typically rest separately from each other, any parent that slept with a single young would be said to be sharing a depreciable investment with that young, in contrast to a species whose parents can sleep together with an entire litter and thus not discriminate among the littermates. The importance of differentiating between depreciable and nondepreciable investment is that depreciable investment requires that the donor discriminate among the benefactors. In the case of parent–offspring interaction, the parent must thus make a choice among the available offspring of different ages and sexes when it performs parental care behaviors. For example, in a species that provides food to offspring at the time of weaning, the parents must make a choice with respect to which individual offspring will receive food at any given time if the food items are very small in relation to the size of each offspring and cannot be shared, as in marmosets and

TABLE 2. Categories of Indirect and Direct Parental Investment

Indirect	Direct
Resource acquisition	Carry or transport young
Resource maintenance	Huddle with young
Resource defense	Groom and clean young
Shelter construction and maintenance	Retrieve young
	Babysit
Sentinel and antipredator behavior	Play with or socialize young
	Active defense of young
Care of female	Provide food to young

tamarins (Hoage, 1982). In direct contrast, in species like the pack-hunting canids, large quantities of food may be stored in the stomach and regurgitated to the young in such a way that littermates may jointly share the food (Malcolm, 1979).

The six main categories of indirect investment and the eight main categories of direct investment by parents are detailed in Table 2 (see Kleiman & Malcolm, 1981). Variations in the correlation of the behavior of males and females (except for lactation) with respect to parental care often relate to the timing of the male's involvement and the retrictions imposed on the females, which promote different parental roles. For example, whereas females may retrieve infants that have been displaced from the nest at an early age, the male may not exhibit such behavior if he does not become involved in parental care activities until the young are sufficiently well developed so that retrieval is no longer important. Crab-eating fox *(Cerdocyon thous)* males do not appear to interact with their offspring until after the young emerge from the den (Brady, 1978).

Indirect Care

Indirect care by males is seen in almost all monogamous mammals, through their activities in territorial maintenance and defense. Indeed, it is rarely emphasized because it is so much less dramatic than direct care and more difficult to quantify. Koontz (1981, personal communication) has data indicating an increase in trail-cleaning behavior by captive male elephant shrews after the birth of young. The trails run throughout the territory and are maintained by the animal's sweeping debris (with the forefeet) off the substrate in selected locations (Rathbun, 1979). The trails are used for orientation and for escape from predators within the territory. A well-maintained trail system presumably benefits the young as they begin to explore within the parental home range. I. Porton (personal communication, 1982) has noted greater aggression toward keepers by male maned wolves *(Chrysocyon brachyurus)* after the mate has given birth—even when the pair are housed separately in adjacent cages.

Providing food to the mate may improve a female's condition during pregnancy and lactation. Anecdotal accounts of such behaviors abound, although quantitative documentation is weak (but see Brown & Mack, 1978, for the golden lion tamarin). In general, documentation showing changes in indirect care with the appearance of offspring is lacking.

Direct Care

The distribution of the categories of male parental care among selected mammalian species is shown in Table 3. The carrying of young is generally restricted to the primates (and bats that are little known). Among marmosets and tamarins, male carrying is initiated at any time between Day 1 and Day 30 (Christen, 1974; Epple, 1975; Hoage, 1977; Ingram, 1977). Among the gibbons (Hylobatidae), male carrying may or may not occur; it is common in the siamang *(Symphalangus syndactylus)* beginning when the infant is approximately 6 months old (Chivers, 1972, 1975). Carrying young is an energetically expensive behavior, which may explain its relative rarity among male primates.

Methods of providing food for young differ. Among canids that feed on relatively large prey items, food may be regurgitated to the young, beginning at weaning. Smaller prey items may be carried whole, or the young may be led to a kill. Prior to weaning, regurgitated food may be available to the female if the demands of lactation preclude her actively being involved in hunting (see Malcolm, 1979, for the African wild dog, *Lycaon pictus*).

In species feeding on smaller prey items, or high-energy foods that occur in small packets (e.g., fruits and insects), food items are usually carried directly to the young—again beginning at weaning (e.g., Hoage, 1982 for lion tamarins; Rasa, 1977, for dwarf mongooses). For species in which food provisioning is an important parental care activity, male involvement usually does not begin until the time of weaning, although the male may carry out other parental activities prior to that period.

Babysitting and huddling with the young are universal to most species in which male parental care is shown. Dudley (1974a,b) has shown how this behavior contributes to the growth and the survival of young, through improved temperature maintenance, in infant California mice *(Peromyscus californicus)*, which thermoregulate poorly. Even in species in which the infants have good thermoregulatory abilities, the heat conservation provided by the male may promote growth and develop-

TABLE 3. Categories of Direct Male Parental Care Seen in Selected Monogamous mammals[a]

	Huddle	Carry	Retrieve	Groom	Feed	Babysit	Defend	Play
Rodents								
(1) *Peromyscus californicus* (mouse)	X					X	?	?
(2) *Microtus ochrogaster* (prairie vole)	X		X	X		X	?	X
(3) *Castor fiber* (beaver)	X		X	X	X	X	X	X
Carnivores								
(4) *Helogale parvula* (dwarf mongoose)	X		X	X	X	X	X	X
(5) *Lycaon pictus* (hunting dog)	?		X	?	X	X	X	X
(6) *Speothos venaticus* (bush dog)	X		X	X	X	X	X	X
Primates								
(7) *Leontopithecus rosalia* (lion tamarin)	X	X	X	X	X	X	X	X
(8) *Symphalangus syndactylus* (siamang)	X	X	?	X	?	X	X	X

[a] From: (1) Dudley, 1974a,b; (2) Hartung and Dewsbury, 1979; Wilson, 1982b; Thomas and Birney, 1979; (3) Wilsson, 1971; Svendsen, 1980; (4) Rood, 1978; (5) Frame et al., 1979; van Lawick, 1973; Malcolm, 1979; (6) Jantschke, 1973; C.A. Brady, personal communication; I. Porton, personal communication; (7) Hoage, 1977, 1982; (8) Chivers, 1972, 1975.

ment by reducing the energy expended by the young on maintaining high body temperatures.

One of the thorniest contributions by the male is in the area of play and socialization. Wilson (1982a) has shown for the degu (*Octodon degus*, a nonmonogamous rodent) that certain forms of social interactions are reduced among siblings when the father is present (e.g., nosing of the body). This difference is significant because Wilson believes that nosing (or sniffing) interactions are of major consequence in the development of sibling relationships. Thus, the male's presence may have an inhibitory effect on his offspring.

The greater male social involvement with the offspring may inhibit the young, but in so doing, it may reduce parent–offspring conflict and permit the maintenance of longer term bonds among parents and young. Biben (in press) has examined infant development in three South American canid species (bush dogs, *Speothos venaticus*; crab-eating foxes; and maned wolves). Two findings relevant to this discussion are that (1) the young interact socially with the father more than with the mother in the two species with the stronger pair bonds and (2) the young "roll over" onto their backs in a submissive posture more toward the father than toward the mother or their littermates. This finding suggests a major role of the father in maintaining subordination behavior among offspring. It also indicates some degree of role differentiation in fathers and mothers during rearing, a finding that is also true for other monogamous mammals, both social and asocial.

To a large extent, the influence of paternal care on social and behavioral development in mammals may only be inferred because it requires the rarely tested assumption that the behavior of a father toward his young always has positive effects. Indeed, tests of the effects of "helpers" on the survivorship of young birds and mammals are only just appearing (jackals, *Canis aureus* and *C. mesomelas*—Moehlman, 1979, 1983), and in some cases, clear-cut benefits improving the survivorship of the young have not been obtained (Malcolm, 1979). Attempts to prove the necessity of paternal care are confounded by the fact that those species in which paternal care is most highly developed are also the species in which helpers are most common. There are few social monogamous mammals in which the pair alone care for their offspring and juveniles disperse prior to the subsequent litter's birth.

THE EFFECTS OF HELPERS

In most species of monogamous mammals in which the young are retained and act as "helpers," the influence of both the mother and the father is reduced because a greater percentage of parental care and social interactions directed toward the infants is performed by the helper class. In large groups of dwarf mongooses, Rood (1978) has found that both parents contribute less than the expected frequency of babysitting with and feeding the young. Similarly, Wolters (1978) has found that parental carrying frequencies are much lower in larger groups of cotton-top tamarins *(Saguinus oedipus)*. When more animals are available for interactions, the relationship between parents and offspring suffers some dilution, which may reduce the degree of parental control, to some extent, or result in the parents' having to assert themselves more dramatically when conflict arises.

COMPARISON OF NUCLEAR VERSUS EXTENDED FAMILIES

Definitions

As already discussd, the degree of sociality varies among monogamous species. As a result, some species live in permanent extended families with a single breeding pair and nonreproductives, including offspring of several ages, mature siblings of the breeding pair, and, occasionally, nonrelatives (e.g., dwarf mongoose—Rood, 1978, 1983). This form of extended family contrasts with that of species in which the breeding pair coexists only with their immature offspring. If reproduction occurs annually and each set of young disperses before the next breeding season, the pair may spend significant parts of the year alone (e.g., red foxes, *Vulpes vulpes;* crab-eating foxes). In other cases, there may always be some young together with the parents (e.g., most gibbon species). Obviously, if immature young provide assistance to the parents in parental care duties, then there is some overlap in the functioning of the two family systems. Yet, there are major differences in the evolutionary costs and benefits to infants and juveniles of developing in these two conditions. Table 4 details some of these advantages and disadvantages; however, it should be recalled that each family system is assumed to be an evolved strategy that is adaptive for any species in which it occurs.

Growth, Development, and Survivorship

In larger families, food acquisition for weanlings may be greater, in that larger families may collect more food, either as hunters or in locating rich food sources. The better foraging abilities of extended families may provide more—and more variable—foods for young animals as well as the opportunity to learn about appropriate and inappropriate food items. Food sharing by parents and helpers may reduce food competition among

TABLE 4. The Costs (C) and Benefits (B) to an Infant of Developing in a Nuclear (Parents Only) or Extended Family (Parents and Helpers)[a]

Nuclear family		Extended family
	Growth and development and survivorship	
C	Thermoregulation	B
C	Food acquisition	B
C	Food sharing	B
C	Food learning	B
C	Late dispersal	B
C	Protection	B
	Social integration	
C	Learn social roles	B
B	Behavioral suppression	C
B	Direct conflict	C
B	Aid relatives	C
	Reproduction	
C	Learn parental care	B
B	Reproductive suppression	C
B	Variance in reproductive success	C

[a] It should be remembered that these are not real costs and benefits since the social structure of a species is assumed to be an adaptive evolved strategy. However, the table points out the differences for infant mammals in developing in an extended vs. a nuclear family.

same-aged littermates. Thus, survivorship among young may be enhanced through greater access to food at all ages, including the preweaning period when the assistance of "helpers" in babysitting and bringing food to the mother may have positive effects on the mother's physical condition (which would positively affect milk production).

In extended families, the young are rarely left alone; thus, they do not need to expend as much energy on thermoregulation. The presence of older animals also protects them against predation. A negative correlation between the number of helpers available and the time left alone has been found by several authors (e.g., for jackals, see Moehlman, 1983). The protection afforded by a social group is significant even in those species in which the adults are still vulnerable to predation (e.g., dwarf mongooses—Rood, 1978). Indeed, because dwarf mongooses are individualistic foragers and feed on only small food items, the selective pressures favoring such large social groups derive mainly from their antipredator benefits.

The dispersal at or before sexual maturity of the young of monogamous species in nuclear families may be costly to those young in terms of their survivorship. They must forage and find protection in unfamiliar terrain and would certainly be the target of aggression by conspecific territorial pairs. Survivorship and successful reproduction ultimately depend on the acquisition as rapidly as possible, of both a mate and a territory.

Social Integration

An infant maturing in an extended family has the opportunity to gain much more complex social experiences than in a nuclear family because the age, sex, and reproductive experience of the group members may vary considerably. Thus, a youngster may learn a variety of social roles and be more socially flexible. Young from nuclear families typically interact only with parents and siblings of one age or reproductive class.

The social variety, however, has its costs. As the infant matures, it may experience behavioral suppression, both from its parents and from other older animals (Kleiman, 1979, 1980). The juvenile or pubertal individual may be cast frequently in a subordinate role that limits its potential for behavioral expression and may produce severe social stress. The same animal may also be the object of direct aggressive behavior, also by parents and other adults. Among wolves *(Canis lupus)* and New World marmosets and tamarins, survivorship may be jeopardized by the aggression of potential reproductive competitors (e.g., see Kleiman, 1979). Typically, this aggression occurs within each sex, and in some species, it is greater among females than among males (Frame *et al.*, 1979; Kleiman, 1979; Malcolm, 1979).

The maturing subadult or mature nonreproductive individual within an extended family typically aids its parents (or other relatives) in the care of their offspring. The cost of providing food and protection to such young is not clear; certainly, there may be some positive benefit if the young are close relatives, such as younger siblings, through an increase in inclusive fitness (Hamilton, 1964). Yet, as already discussed, there are few data available indicating a direct positive effect on infant survivorship due to the activities of helpers, and there are no known data on the costs to the helpers of helping.

Reproduction

Helpers may certainly benefit by gaining experience in parental care techniques. Such benefits through learning have been documented for some species of marmosets and tamarins (Hoage, 1977; Ingram, 1978).

The benefits of parental care experience in terms of the survivorship of the helper's offspring have also been shown in nonmonogamous species (e.g., chimpanzees, *Pan troglodytes*; see Nadler, this volume).

This benefit is, however, offset by two related costs. First, as long as an individual chooses to forgo a reproduction attempt by not risking emigration or not provoking a conflict with the reproductively dominant individual of its sex, it will not reproduce successfully. Methods of reproductive suppression have already been discussed (and see Kleiman, 1980).

Second, in all species in which extended families are the major reproductive unit, there is great variance in individual reproductive success. Indeed, it is likely that some adult individuals in groups may never reproduce during their lifetime—although survivorship may be jeopardized at any time that an animal attempts to improve its reproductive position, through emigration or direct conflict. Thus, although there are considerable benefits associated with development within an extended family, the ultimate cost of forfeiting reproduction is of major importance to an individual's fitness.

Some Concluding Speculative Comments

Parental Manipulation

Alexander (1974) discussed the retention of helpers in terms of the parental manipulation of offspring, as the reproductive success and fitness of the parents may be enhanced through the assistance and at the expense of nonreproductive offspring. Thus, the reproductive suppression of offspring may be seen as another example of parent–offspring conflict in which the parent dominates.

Incest Taboos

Although not quantitatively documented, there is some suggestion that parent–offspring and sibling matings are less common in species living in extended families. This may be an evolutionary response to the great variance in individual reproductive success among males and females in extended families and may thus ensure that inbreeding, with its negative consequences, will be rare (Ralls, Brugger, & Ballou, 1979). The mechanisms whereby matings are inhibited among relatives appear to be behavioral; thus, individuals that grow up together from an early age avoid sexual relationships (for an interesting example in the prairie vole, *Microtus ochrogaster*, see Carter, Getz, Gavish, McDermott, & Arnold, 1983). Exact genetic relations need not be known; the major variable would be in the degree of familiarity from an early age.

Sex Ratios, Polyandry, and Female Conflict

I have already indicated that among some species existing in extended families, female competition and aggression appear to be greater than in males. Such species (e.g., African wild dogs and some marmosets and tamarins) tend to exhibit sex ratios biased toward males among adults in nature and, in some cases, to have sex ratios significantly biased toward males at birth (Kleiman, 1979; Kleiman & Eisenberg, 1973; Malcolm, 1979). Female survivorship may be lower than that of males *in utero*, and again at and after puberty, when females are more often forced to emigrate from their natal group. It is of interest that there have been reports of multiple males mating with the dominant reproductive female in these or related species (Malcolm, 1979), a finding suggesting a polyandrous condition (greater reproductive variance among females than among males). Because of the typically greater parental effort of mammalian females relative to males, polyandry would not be expected to evolve among mammals. Its occurrence may indicate that male assistance in rearing young is sufficiently crucial to a female's reproductive success so that, for both sexes, fitness is enhanced by the increased uncertainty of paternity and a near-polyandrous condition.

Role Differentiation in Fathers and Mothers

Despite the high correlation within the males and the females of a species in the quality and the quantity of parental care behaviors (Hartung & Dewsbury, 1979), fathers and mothers do differ in their interactions with their offspring, as already mentioned. Thus, different parental roles exist. The direction of the difference is likely to be based on evolutionary and ecological constraints, and it needs to be further investigated in light of each species' natural history. For example, whereas marmoset and tamarin males may assist in carrying infants, permitting the female the possibility of unimpeded foraging, a wolf or a wild dog male may take a major responsibility for leading a hunt while the mother remains at the den site and guards the young. Because the male can regurgitate food to the female and the offspring, both the carrying and the hunting strategies of the males serve to maintain the nutritional status of the mate. They do, however, result in major differences in the father's form and frequency of interaction with his offspring.

ACKNOWLEDGMENTS

The author would like to thank J. Malcolm, J. Rood, M. Biben, F. Koontz, H. J. Wolters, I. Porton, and C. Brady for access to unpublished data and stimulating discussions.

REFERENCES

Alexander, R. D. The evolution of social behavior. *Annual Review of Ecology and Systematics,* 1974, *5,* 325–383.

Altmann, D. Beziehungen zwischen sozialer Rangordnung ung Jungenaufzucht bei *Canis lupus* L. *Zoologische Garten Jena,* 1974, *44,* 235–236.

Altmann, S. A., Wagner, S. S., & Lenington, S. Two models for the evolution of polygyny. *Behavioral Ecology and Sociobiology,* 1977, *2,* 397–410.

Biben, M. Comparative ontogeny of social behaviors in three South American canids, the maned wolf, crab-eating fox, and bush dog: Implications for sociality. *Animal Behaviour,* in press.

Brady, C. A. Reproduction, growth and parental care in crab-eating foxes *Cerdocyon thous* at the National Zoological Park, Washington. *International Zoo Yearbook,* 1978, *18,* 130–134.

Brown, K., & Mack, D. S. Food sharing among captive *Leontopithecus rosalia. Folia Primatologica,* 1978, *29,* 268–290.

Carter, C. S., Getz, L. L., Gavish, L., McDermott, J. L., & Arnold, P. Male-related pheromones and the activation of female reproduction in the prairie vole *(Microtus ochrogaster). Biology of Reproduction,* 1980, *23,* 1038–1045.

Chivers, D. J. The siamang and the gibbon in the Malay Peninsula. *Gibbon and Siamang,* 1972, *1,* 103–135.

Chivers, D. J. The siamang in Malaya: A field study of a primate in tropical rain forest. *Contributions to Primatology,* 1975, *4,* 1–335.

Christen, A. Fortpflanzungsbiologie und Verhalten bei *Cebuella pygmaea* and *Tamarin tamarin. Zeitschrift für Tierpsychologie Beiheft,* 1974, *14,* 1–78.

Dudley, D. Contributions of paternal care to the growth and development of the young in *Peromyscus californicus. Behavioral Biology,* 1974, *11,* 155–166.(a)

Dudley, D. Paternal behavior in the California mouse, *Peromyscus californicus. Behavioral Biology,* 1974, *11,* 247–252.(b)

Eisenberg, J. F. The social organizations of mammals. *Handbuch der Zoologie,* VIII (10/7). Berlin: De Gruyter, 1966. (Leiferung 39)

Epple, G. Social behavior of laboratory groups of *Saguinus fuscicollis.* In D. D. Bridgwater, (Ed.), *Saving the lion marmoset.* Wheeling, W. V.: Wild Animal Propagation Trust, 1972.

Epple, G. The behavior of marmoset monkeys (Callithricidae). In L. A. Rosenblum, (Ed.), *Primate behavior* (Vol. 4). New York: Academic Press, 1975.

Frame, L. H., Malcolm, J. R., Frame, G. W., & Van Lawick, H. Social organization of African wild dogs *(Lycaon pictus)* on the Serengeti Plains, Tanzania, 1967–1978. *Zeitschrift für Tierpsychologie,* 1979, *50,* 225–249.

Greenwood, P. J. Mating systems, philopatry, and dispersal in birds and mammals. *Animal Behavior,* 1980, *28,* 1140–1162.

Hamilton, W. D. The genetical theory of social behaviour, I, II. *Journal of Theoretical Biology,* 1964, *7,* 1–16, 17–52.

Hartung, T. G., & Dewsbury, D. A. Paternal behavior in six species of muroid rodents. *Behavioral and Neural Biology*, 1979, 26, 466–478.

Hearn, J. P. The endocrinology of reproduction in the common marmoset, *Callithrix jacchus*. In D. G. Kleiman (Ed.), *The biology and conservation of the Callitrichidae*. Washington, D.C.: Smithsonian Institution Press, 1977.

Hoage, R. J. Parental care in *Leontopithecus rosalia rosalia*: Sex and age differences in carrying behavior and the role of prior experience. In D. G. Kleiman, (Ed.), *The biology and conservation of the Callitrichidae.*Washington, D.C.: Smithsonian Institution Press, 1977.

Hoage, R. J. Social and physical maturation in captive lion tamarins, *Leontopithecus rosalia rosalia* (Primates: Callitrichidae). *Smithsonian Contributions to Zoology*, 1982, 354, 1–56.

Ingram, J. C. Interactions between parents and infants, and the development of independence in the common marmoset *(Callithrix jacchus)*. *Animal Behavior*, 1977, 25, 811–827.

Ingram, J. C. Preliminary comparisons of parental care of wild-caught and captive-born common marmosets. In H. Rothe, H. J. Wolters, & J. P. Hearn (Eds.), *Biology and behaviour of marmosets*. Göttingen, West Germany: Eigenverlag H. Rothe, 1978.

Kleiman, D. G. Monogamy in mammals. *Quarterly Review of Biology*, 1977, 52, 39–69.

Kleiman, D. G. The development of pair preferences in the lion tamarin *(Leontopithecus rosalia)*: Male competition or female choice. In H. Rothe, H. J. Wolters, & J. P. Hearn (Eds.), *Biology and behaviour of marmosets*. Göttingen, West Germany: Eigenverlag H. Rothe, 1978.

Kleiman, D. G. Parent–offspring conflict and sibling competition in a monogamous primate. *American Naturalist*, 1979, 114, 753–760.

Kleiman, D. G. The sociobiology of captive propagation in mammals. In M. Soulé & B. Wilcox (Eds.), *Conservation biology*. Sunderland, Mass.: Sinauer Associates, 1980.

Kleiman, D. G. Correlations among life history characteristics of mammalian species exhibiting two extreme forms of monogamy. In R. D. Alexander & D. W. Tinkle (Eds.), *Natural selection and social behavior*. New York: Chiron Press, 1981.

Kleiman, D. G., & Eisenberg, J. F. Comparisons of canid and felid social systems from an evolutionary perspective. *Animal Behavior*, 1973, 21, 637–659.

Kleiman, D. G., & Malcolm, J. The evolution of male parental investment in mammals. In D. J. Gubernick & P. H. Klopfer (Eds.), *Parental care in mammals*. New York: Plenum Press, 1981.

Lunn, S. F. Urinary oestrogen excretion in the common marmoset, *Callithrix jacchus*. In H. J. Wolters & J. P. Hearn (Eds.), *The biology and behaviour of marmosets*. Göttingen, West Germany: Eigenverlag H. Rothe, 1978.

Malcolm, J. R. *Social organization and communal rearing in African wild dogs (Lycaon pictus)*. Ph.D. dissertation, Harvard University, 1979.

Moehlman, P. D. Jackal helpers and pup survival. *Nature* (London), 1979, 277, 382–383.

Moehlman, P. D. Socioecology of silverbacked and golden jackals (*Canis mesomelas* and *Canis aureus*). In J. F. Eisenberg & D. G. Kleiman (Eds.), *Advances in the Study of Mammalian Behavior*. The American Society of Mammalogists, Special publication no. 7, 1983, pp. 423–453. (Available from Secretary-Treasurer Gordon L. Kirkland, Jr., Vertebrate Museum, Shippensburg State College, Shippensburg, Pennsylvania 17257.)

Ralls, K., Brugger, K., & Ballou, J. Inbreeding and juvenile mortality in small populations of ungulates. *Science,* 1979, 206, 1101–1103.

Rasa, O. A. E. The ethology and sociology of the dwarf mongoose, *Helogale undulata rufula*. *Zeitschrift für Tierpsychologie*, 1977, 43, 337–406.

Rathbun, G. The social structure and ecology of elephant shrews. *Zeitschrift für Tierpsychologie Supplement*, 1979, 20, 1–76.

Rood, J. P. Dwarf mongoose helpers at the den. *Zeitschrift für Tierpsychologie*, 1978, 48, 277–287.

Rood, J. P. Mating relationships and breeding suppression in the dwarf mongoose. *Animal Behavior,* 1980, *28,* 143–150.

Rood, J. P. The social system of the dwarf mongoose. In J. F. Eisenberg & D. G. Kleiman (Eds.), *Advances in the Study of Mammalian Behavior.* The American Society of Mammalogists, Special publication no. 7, 1983, pp. 454–488. (Available from Secretary-Treasurer Gordon L. Kirkland, jr., Vertebrate Museum, Shippensburg State College, Shippensburg, Pennsylvania 17257.)

Smythe, N. The natural history of the Central American agouti *(Dasyprocta punctata). Smithsonian Contributions to Zoology,* 1978, *257,* 1–52.

Svendsen, G. E. Population parameters and colony composition of beaver *(Castor canadensis)* in Southeast Ohio. *American Midland Naturalist,* 1980, *104,* 47–56.

Thomas, J. A., & E. C. Birney. Parental care and mating system of the prairie vole, *Microtus ochrogaster. Behavioral Ecology and Sociobiology,* 1979, *5,* 171–186.

van Lawick, H. *Solo: The story of an African wild dog.* Boston: Houghton Mifflin, 1973.

Wilson, E. O. *Sociobiology: The new synthesis.* Cambridge: Harvard University Press, 1975.

Wilson, S. C. Contact-promoting behavior, social development, and relationship with parents in sibling juvenile degus *(Octodon degus). Developmental Psychobiology,* 1982, *15,* 257–268.(a)

Wilson, S. C. Parent-young contact in prairie and meadow voles. *Journal of Mammalogy,* 1982, *63,* 300–305.(b)

Wilsson, L. Observations and experiments on the ethology of the European beaver *(Castor fiber L.).* Stockholm, Viltrevy. 1971, *8,* 117–266.

Wittenberger, J. F., & Tilson, R. L. The evolution of monogamy: Hypotheses and evidence. *Annual Review of Ecology and Systematics,* 1980, *11,* 197–232.

Wolters, H. J. Some aspects of role taking behaviour in captive family groups of the cottontip tamarin *Saquinus oedipus oedipus.* In H. Rothe, H. J. Wolters, & J. P. Hearn (Eds.), *Biology and behaviour of marmosets.* Göttingen, West Germany: Eigenverlag H. Rothe, 1978.

6

Biological Contributions to the Maternal Behavior of the Great Apes

RONALD D. NADLER

INTRODUCTION

My research on biological contributions to maternal behavior of the great apes, while not recognized as such at the time, was initiated early in 1972. I was studying the sexual behavior of gorillas under laboratory conditions and learned that one of the females I had tested was pregnant. Because I saw this as an opportunity to extend my studies of mother–infant relations to a third species of great ape, research on chimpanzees and orangutans already being under way, I conducted a literature search to determine what was known about this subject in gorillas. I learned that little was known about mother–infant relations in gorillas, in part because there had been relatively few gorillas born in captivity in comparison to chimpanzees and orangutans, and because most of the infants that were born were separated from their mothers within the first week of life (Kirchshofer, 1970). In fact, of the first 29 live-born gorillas in captivity, only 1 was reared by its mother beyond a year of age. Even

---

RONALD D. NADLER • Yerkes Regional Primate Research Center, Emory University, Atlanta, Georgia 30322. Preparation of this manuscript and research by the author described herein were supported by National Science Foundation Grants BNS 75-06287 and BNS 79-23015 to the author and U.S. Public Health Service Grant RR-00165 (Division of Research Resources, National Institutes of Health) to the Yerkes Regional Primate Research Center of Emory University.

more ominous for my planned research were reports that several of the infants that had initially been allowed to remain with their mothers had been neglected or abused, and a few had actually been killed by the mother. Kirchshofer (1970) proposed that the inadequate and abusive maternal behavior of the gorillas was related to their early rearing experience in captivity. Essentially, all the gorillas in captivity had been separated from their own mothers within a year or two, as a result of being captured. Most had been reared at zoos with one or more peers, but no adult conspecifics. Kirchshofer (1970) concluded that the absence of adult conspecifics on which to model their behavior and a lack of experience with infants accounted for the failure of captive gorilla mothers to learn the appropriate care of infants during their formative years, prior to the time that they delivered their own offspring.

Kirchshofer's conclusion may have been based on the results of laboratory research conducted on the maternal behavior of rhesus monkeys (Harlow, Harlow, Dodsworth, & Arling, 1966) and chimpanzees (Davenport, 1979; Davenport & Rogers, 1970). These studies showed that early separation from the mother and isolation from conspecifics during the early stages of development seriously incapacitated these animals in terms of their ability to interact socially with other members of their species when they were eventually introduced to them. Most of the females reared under such restricted conditions in the laboratory eventually mated with wild-born males and delivered offspring, but few displayed maternal behavior toward their infants. It should be noted, however, that although the gorilla mothers superficially resembled these other primates in terms of their early restricted experience and later maternal behavior, there were some potentially significant differences between them and the other species. Whereas the rhesus monkeys and chimpanzees were separated from their mothers on the day of birth, the gorillas lived with their mothers for one to two years prior to capture. Moreover, although the monkeys and chimpanzees were generally inadequate in their sexual behavior, the gorillas mated in a species-typical manner (e.g., Hess, 1973; Nadler, 1975c, 1976). These and other differences among these species are considered further below.

Maternal Behavior of Gorillas in Captivity

In preparation for the birth of our gorilla, the firt gorilla to be born at the Yerkes Regional Primate Research Center of Emory University, the female was treated similarly to the many chimpanzees and orangutans that had previously given birth at the center (Nadler, 1974). The female was separated from its companions and placed in a "maternity cage," where its behavior could be carefully monitored, where vitamin and food supplements could be administered, and from which an infant could be more easily retrieved if the circumstances warranted.

Based on the early indications following the birth, my concerns regarding the maternal behavior of our female appeared to have been overly pessimistic. Although the mother was somewhat hesitant in her handling of the infant initially (Figure 1) and awkward in positioning the infant on its ventrum (Figure 2), its maternal behavior improved steadily over the first hour postpartum (Figure 3). The mother and infant then slept intermittently for a considerable period of time, and suckling was eventually observed 27 hours postpartum. The first bout of suckling appeared to result from a series of interactions between the infant and the mother. For example, coordinated vocalizing between the infant and the mother developed in the form of initial vocalizations made by the infant and responding growls made by the mother. Associated with the infant's vocalizations were rooting behavior and suckling sounds. As time progressed, the mother responded to the infant's vocalizations less by vocalizing itself and more frequently by readjusting the position of the infant. Finally, after 27 hours, these readjustments led to nipple contact and a period of prolonged (13 minutes) nursing (Figure 4). It was clear, therefore, that the basic components of adequate maternal

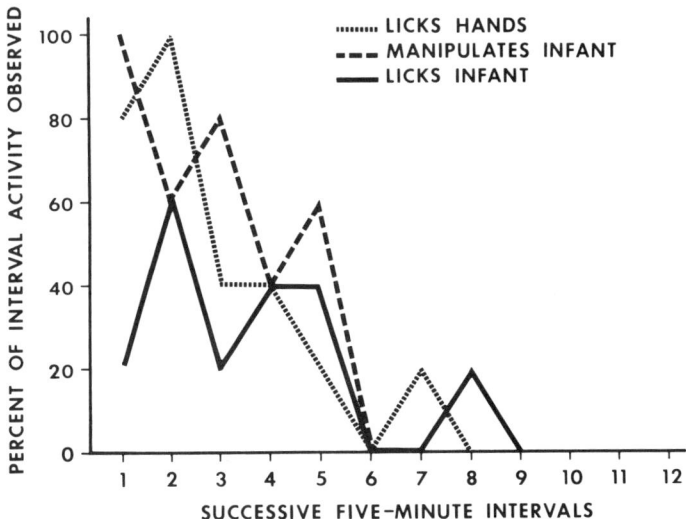

FIG. 1. Behavior of a primiparous gorilla during initial contact with its newborn. Curves reflect percentage of five-minute intervals activity was observed during first hour postpartum (from Nadler, 1974).

FIG. 2. Inappropriate dorsal-ventral positioning of an infant gorilla (from Nadler, 1974).

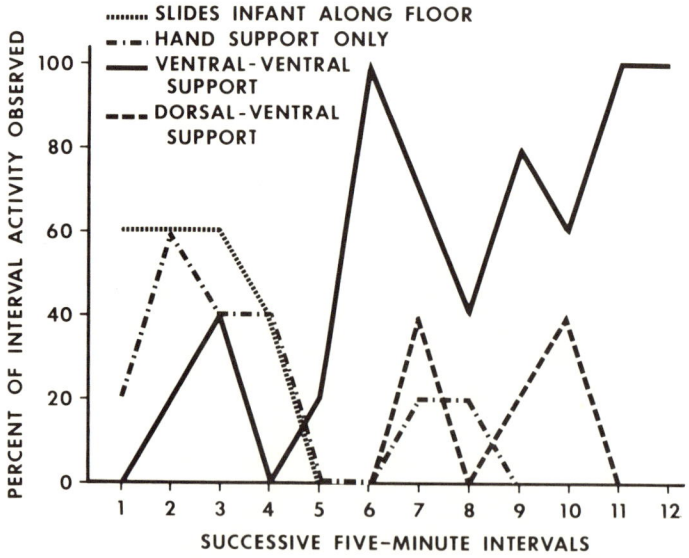

FIG. 3. Pattern of contact and support exhibited by a primiparous gorilla during the first hour postpartum. Curves reflect percentage of five-minute intervals activity was observed (from Nadler, 1974).

FIG. 4. Infant gorilla nurses while supported in a ventral-ventral position.

behavior could develop in a gorilla despite the limitations of its early rearing experience. It was proposed that the maternal behavior of this gorilla mother developed as a result of responses reinforced and learned during the early postnatal period (Nadler, 1974), consistent with earlier proposals regarding the development of maternal behavior in chimpanzees (Rogers & Davenport, 1970; Yerkes & Tomilin, 1935).

Despite this auspicious beginning of the mother–infant relationship, however, a situation developed one week after the birth that resulted in a decision to separate the infant from its mother. One evening, after most of the staff had left for the night, the mother was observed pushing the infant out of contact on the floor. During the week prior to this, two similar crises had occurred and had been ameliorated, first by moving the prior female cage-mate of the mother into the cage adjoining the maternity cage, and then by transferring the mother from the maternity cage to her original cage in the gorilla housing area. On both these occasions, the actions taken had brought about a temporary cessation of the mother's mistreatment of her infant. With the final recurrence, however, concern for the infant's well-being prompted its removal for rearing in the center's nursery.

The study described above, although terminating with the removal of the infant from the mother, did not appear to support Kirchshofer (1970) in her conclusion regarding the etiology of infant neglect and abuse in gorillas. Kirchshofer proposed that experience with adult conspecifics and infants was necessary for a female gorilla to learn adequate maternal behavior. This first gorilla birth at the Yerkes Center, as well as several others, suggests otherwise (see Nadler, 1974). In one mother–infant dyad at the Central Park Zoo, New York, an initially adequate mother began acting in an agitated manner within a few days after the birth, paced the cage, and relinquished support of its clinging, screaming infant. The adult male was reintroduced into the cage with the mother and the infant, and the mother resumed solicitous care of the infant. A somewhat similar situation developed at the Cincinnati Zoo with a mother that had been separated from an earlier infant after three months because she had picked wounds on the skin of her offspring. When the same behavior was initiated with a subsequent infant, the adult male was placed with the mother and the infant. Again, the addition of a familiar conspecific to the dyad terminated the mistreatment of the infant. A final example is that of the second gorilla born at the Yerkes Center to the same female described above (Nadler, 1975b). The female cared adequately for her second infant for approximately three months before showing signs of agitation and mistreatment of the infant (Figure 5). Following the return of the mother's female cage-mate to the cage with the dyad, the mother retrieved her infant to the ventral position and

FIG. 5. Gorilla mother in social isolation restrains infant from ventral-ventral contact.

successfully reared it beyond 1 year of age. These several examples suggest that inadequacies in early experience, to the extent that they influence the maternal behavior of captive gorillas, are confounded and perhaps compounded by conditions of captivity at about the time of birth and thereafter. The mothers described above all demonstrated their abilities to care adequately for their infants prior to the time that they began mistreating them, in one case for fully three months. The problem of infant mistreatment in captive gorillas, therefore, did not appear to reflect a deficiency in learning *per se*.

In order to gain additional information on the maternal behavior of captive gorillas, I conducted a survey of gorilla births in captivity similar to that of Kirchshofer, but I included data on more than 90 mother–infant dyads (Nadler, 1975a). The results of that survey show that multiparous females had a slightly higher percentage of live-born infants than primiparous ones, but that the majority of infants from both classes of female were separated from their mothers within one week of birth (Table 1). Most of the infants of primipara were separated, in fact, on the day of birth for reasons of neglect or abuse. For the multipara, infants were frequently separated within an hour of birth, but for reasons related to past experience with other infants (e.g., neglect or abuse by the mother, injury or illness of the previous infants). Because they were separated

TABLE 1. Percentage of Premature, Stillborn, and Live-Born Gorilla Infants of Primiparous and Multiparous Mothers and Disposition of the Live-Born Infants[a]

	Primipara ($n = 41$)	Multipara ($n = 53$)
Premature	22 ($n = 9$)	11 ($n = 6$)
Stillborn	7 ($n = 3$)	4 ($n = 2$)
Live	71 ($n = 29$)	85 ($n = 45$)
Separated	83 ($n = 24$)	80 ($n = 36$)
Less than one week	71 ($n = 17$)	69 ($n = 25$)
Abuse/neglect	59 ($n = 4/6$)	28 ($n = 1/6$)
Failure to suckle	23 ($n = 4$)	16 ($n = 4$)
Other	18 ($n = 3$)	56 ($n = 14$)
One week or later	29 ($n = 7$)	31 ($n = 11$)
Abuse/neglect	28 ($n = 2/0$)	27 ($n = 1/2$)
Injury or threat of	43 ($n = 3$)	9 ($n = 1$)
Illness or threat of	28 ($n = 2$)	64 ($n = 7$)
Retained by mother	10 ($n = 3$)	20 ($n = 9$)
Insufficient information	7 ($n = 2$)	0 ($n = 0$)

[a] From Nadler (1975a).

from their infants so quickly following the births, many gorilla mothers were probably not given an adequate opportunity to exhibit maternal behavior. On the positive side, the data indicate that 10% of the liveborn infants of primipara and 20% from multipara were retained by the mothers for prolonged periods of time. The latter finding of the apparently adequate maternal behavior of some captive females suggests that factors other than those related to their own early experience contributed to the superior performance of some females in comparison to the majority of others.

An analysis of the maternal competency of only the females that were given an adequate opportunity to exhibit maternal behavior provided additional insights. Among this more restricted group, fully 33% of the primiparous females and almost 60% of the multipara retained their infants for at least one week. Because the survival of an infant for one week requires at least the rudiments of maternal competency, this analysis suggests that separation of an infant *per se* is not sufficient evidence for a conclusion that the mother lacked the ability to rear her infant. Other factors, exclusive of behavioral inadequacy, could account for the failure of captive gorillas to nurture their young, including a failure to lactate, weakness of the infant or the mother, and conservatism on the part of the attending authorities.

The finding that multiparous female gorillas were superior to primiparous ones with respect to maternal behavior supported the earlier data on this relationship reported by Kirchshofer (1970). The data are consistent with the interpretation that experience with her own first delivery and/or infant contributes in a positive way to a female gorilla's responsiveness and care of subsequent infants. The data also suggest that factors that influence the maternal behavior of captive female gorillas other than experience with their first offspring are better studied by analysis of data on primiparous females. The examples of improved maternal behavior exhibited by previously abusive gorilla mothers after a conspecific was introduced into the cage with the mother–infant dyad suggested the basis for such further analysis (Nadler, 1980). These latter cases, in other words, raised the possibility that the social isolation of the mother and the infant before and following the birth might be the critical factor differentiating competent from abusive gorilla mothers.

Analysis of the social environment associated with the competent and the inappropriate maternal behavior of primiparous female gorillas was revealing. When only mothers for which the data were available regarding maternal behavior were considered, it was found that seven of the eight mothers that provided adequate care to their infants were living with a male or as part of a larger social group at the time of delivery and afterward. In contrast, eight of nine mothers that abused or neglected their infants were socially isolated with their infants during the postnatal period. This difference is statistically significant ($p < .01$, Fisher Exact Probability Test, two-tailed) and suggests that the abusive behavior exhibited by so many primiparous gorillas in captivity was related to their social isolation from conspecifics at the time the abuse occurred. The data are compatible with the hypothesis that a congenial social environment facilitates the expression of competent maternal behavior by captive gorillas.

This hypothesis was subsequently tested in a prospective study of gorilla maternal behavior conducted at the Field Station of the Yerkes Center. A group of gorillas was established in an outdoor compound that provided them with the social environment thought to be important for maternal behavior. Approximately nine months after the group was formed, three females gave birth to their first offspring. All three mothers showed some hesitancy and ineptitude in their initial handling of their infants (Figure 6), but all quickly improved maternally and have since reared their offspring beyond 4 years of age. The results of this study, therefore, further support the hypothesis that congenial social companionship facilitates the maternal behavior of captive gorillas (Figure 7), whereas social isolation contributes to neglect and abuse of the offspring.

FIG. 6. Primiparous gorilla mother carries infant inappropriately on her back.

FIG. 7. Group of gorilla mothers with infants in an outdoor compound.

The foregoing analysis of the maternal behavior of captive gorillas does not support the hypothesis of Kirchshofer (1970) that the inadequacies of gorilla mothers described above are related to deficiencies in early experience that preclude them from learning the appropriate treatment of an infant. That early experience plays a role in the development of the maternal behavior of female chimpanzees was suggested many years ago (Yerkes & Tomilin, 1935) and clearly demonstrated more recently (Rogers & Davenport, 1970). Examination of the research on this close biological relative of the gorilla facilitates further clarification of the regulation of maternal behavior in gorillas.

Experiential Influences on Maternal Behavior of Chimpanzees and Orangutans

Yerkes and Tomilin (1935) reported on the maternal behavior of five female chimpanzees and stated that "structurally determined patterns of activity are neither adequate nor dependable, but instead require facilitation, modification, and supplementation through experience." They further reported that the primiparous female "acts as if frightened, bewildered, or uncertain what to do" (p. 343). Despite differences in the maternal experience of their subjects, however, all the mothers reared their infants with reasonable competency. It should be noted that most of these mothers were living alone with their infants from their birth until the first year of life.

Rogers and Davenport (1970) compared the maternal behavior of two groups of female chimpanzees that differed in terms of their early experience with their own mothers. One group, consisting primarily of captive-born females, had remained with their mothers for less than 18 months, whereas the second group, of mostly wild-born females, had remained with their mothers for 18 months or longer. Several differences between the groups were reported. In general, the females that had remained with their mothers for more than 18 months were superior maternally to those that had had less mothering. Both groups, however, showed improvements in maternal behavior with successive infants, the greatest improvement occurring between the first and second infants. Essentially, all the longer mothered mothers performed at the two highest ratings of maternal behavior for all their infants beyond the first. Again, it should be noted that these animals were confined alone in cages with their infants (Figure 8).

In a subsequent study on chimpanzees, infants were separated from

FIG. 8. Chimpanzee mother cares adequately for its infant in social isolation.

their mothers on the day of birth and were reared under conditions of social deprivation until they were 2 years old (Davenport & Rogers, 1970). Although the conditions of rearing varied somewhat, all the infants were confined in individual cubicles for the entire two years. No differences between rearing conditions were apparent in the later maternal behavior of the subjects (Davenport, 1979). These early-restricted socially deprived females were all socially debilitated in general and were essentially devoid of solicitous behavior toward their offspring (Figure 9). Little, if any, improvement was observed with successive infants. These results suggest that early social restriction from birth through the first year or so of life severely compromises a female chimpanzee's ability to rear offspring. Whereas females that have some early contact with their mothers show improvement in the adequacy of their maternal behavior as a result of experience with their offspring (Rogers & Davenport, 1970), the severely deprived females do not appear to learn from their past experiences with infants (Davenport, 1979).

The research on chimpanzees illustrates two dimensions of experience related to maternal behavior: early experience with the mother during the first year or so of life, and later experience following the birth of a first infant. Given some, as yet incompletely defined, period of early experience with her mother, a female chimpanzee is more-or-less pre-

FIG. 9. Chimpanzee mother, socially deprived as an infant, provides inappropriate support for its own infant.

pared to profit from maternal experience in adulthood and to develop competent maternal behavior despite her rearing without adults or infants following separation from her mother. Without such early experience, later learning of maternal behavior is apparently unlikely.

To what extent is it possible to extrapolate results from the chimpanzee research to the gorillas? In terms of their early experience with their mothers, the captive gorillas described above closely resemble the longer-mothered group of chimpanzees from the Rogers and Davenport (1970) study. Both the chimpanzees and the gorillas were captured at 1–3 years of age, and both were reared in captivity with peers, but no adult conspecifics. The female chimpanzees were quite competent in their subsequent rearing of offspring, as were the female gorillas that lived in a social setting. These data, therefore, further support the interpretation that the deficiencies in maternal behavior of most captive female gorillas are related to the conditions prevailing at the time they give birth, rather than to the conditions of their early rearing. If this interpretation is accurate, what factor accounts for the finding that gorilla mothers appear to require a social environment in order to care adequately for their offspring, whereas chimpanzee mothers do not?

Before suggesting an answer to the above question, it is appropriate first to consider some related information on a third species of great ape,

FIG. 10. Orangutan mother cares adequately for its infant in social isolation.

the orangutan. Although many fewer data are available on the maternal behavior of orangutans in comparison to the other apes, they appear to resemble chimpanzee females, rather than gorillas, in terms of the issue under consideration. In other words, like chimpanzee mothers, orangutan mothers do quite well rearing their infants alone in individual cages (Figure 10). As noted above, the individual caging of chimpanzee and orangutan mother–infant dyads has been used successfully for years at the Yerkes Center, and the results of a recent study confirm that mother–infant relations under such conditions are congenial for these species (Miller & Nadler, 1981).

Maternal Behavior of Orangutans and Chimpanzees under Natural Conditions

In order to suggest a biological basis for these differences in the maternal behavior of the great apes, I now consider the species-typical pattern of infant rearing by these species in the natural habitat. Such analysis is consistent with the view that ultimate causation is reflected in the adaptation of species-typical behavior to the conditions of the environment in which the species evolved. Fortunately, a considerable number of field studies have been conducted on the great apes in recent years and have provided data relevant to the conditions of infant rearing.

Although relatively few data are available on the details of mother–infant relations in orangutans, the (social) conditions under which mothers rear their offspring are probably clearest for this species, among the great apes. Early field research (Davenport, 1967; MacKinnon, 1971) revealed that orangutans live in a semisolitary, spatially dispersed form of social organization, a finding corroborated by later studies (Galdikas, 1979; Horr, 1975; Rijksen, 1978; Rodman, 1973). Male orangutans live alone for most of their adult lives, and females socialize primarily with their own offspring. Given their spatially dispersed social structure and rather limited social interactions, it follows that orangutan mothers rear their offspring under natural conditions with a minimum of input from other conspecifics. Certainly, individual differences exist, and subspecific differences in sociality are reported (Galdikas, 1979), but the data indicate that female orangutans generally rear their offspring in the absence of the complex social relationships so characteristic of many group-living species of primates.

Based on a superficial evaluation of the social organization of chimpanzees, one might conclude that infant rearing in this species is dramatically different from that of orangutans. Chimpanzees live in large communities of individuals, encompassing all possible age and sex classifications (Goodall, 1965; Kortlandt, 1962; Reynolds & Reynolds, 1965; Sugiyama, 1968). Such communities do not consist of a single aggregation of animals, however, but are fragmented into small groups or bands that frequently change in composition. The most stable groupings of chimpanzees consist of an adult female and its younger offspring (van Lawick-Goodall, 1968). Evidence has been presented that female chimpanzees, in fact, are rather similar to female orangutans in that they are rather evenly dispersed throughout the community in separate core areas (Wrangham, 1979) and, especially when their infants are young, interact little with other animals in the community (Sugiyama, 1969). The field research on chimpanzees, therefore, suggests that infant rearing in this species, as in the orangutans, entails little social interaction with individuals outside the mother–infant dyad and no permanent associations between the mother and other members of the community other than her own offspring.

MATERNAL BEHAVIOR OF GORILLAS UNDER NATURAL CONDITIONS

Given the basic similarity ascribed to chimpanzees and orangutans with respect to the absence of permanent associations between mothers and other unrelated conspecifics, it remains to be demonstrated that gorilla mothers differ in this regard when studied under natural conditions. Early research demonstrated (Schaller, 1963) and subsequent studies confirmed (Harcourt, Stewart, & Fossey, 1976) that gorillas live in a relatively stable, harem form of social organization. Gorilla groups typically consist of one adult leader or silver-backed male, several females with offspring, and, sometimes, one or more younger black-backed males, thought to be offspring of the silver-back. Males, as well as females, may leave the natal group, but only females transfer between groups. Such transfers by females are more common during early adulthood, however, and in general, females remain and bear several offspring in the same group. Gorilla mothers, therefore, differ from their chimpanzee and orangutan counterparts in that, under natural conditions, they rear offspring in the company of other conspecifics with which they maintain a relatively long period of close association. It is interesting to note that although gorilla harems are relatively enduring over time, they are not characterized by very much social interaction among adult females (Harcourt, 1979). In the Harcourt study, those females that spent considerable time together were either related (mother-daughter), known to each other since immaturity, or in association as a result of their common proximity to the leader male. Because the adult females in a harem are unrelated, for the most part, it appears that their associations, in general, are based on their common attraction to the leader male.

Interpretations and Implications

The field studies on the great apes suggest that female chimpanzees and orangutans are similar in that, under natural conditions, they rear their offspring relatively independently in the absence of any long-term social relationships with unrelated conspecifics. These same species, under laboratory conditions of caging, rear their offspring quite competently when confined alone with them in individual cages. Female gorillas living in their natural habitat differ from the other female apes in that they rear their infants in the company of other unrelated conspecifics with which they have a relatively long history of association. The basis for the social facilitation of maternal behavior in gorillas, if such, in fact, is the case, is not immediately apparent. The relatively little social interaction among adult females in nature suggests that they do not support each other in a positive way. Their relationship appears to be indirect, based on their joining the same harem (male). Females with relatively young infants spend more time in proximity to the leader male than do females with older offspring. This latter point may be relevant to interpreting the inappropriate maternal behavior of captive mothers that are socially isolated with their offspring. In other words, neglect and abuse of infants in captivity may reflect the inability of female gorillas to pursue species-typical modes of infant rearing, that is, establishing spatial proximity to a familiar conspecific. The privileged position of mothers within the primate group, reflected, perhaps, in the proximity of gorilla mothers to the leader male, has been suggested as a social reinforcer of adequate maternal behavior (Weisbard & Goy, 1976). Close proximity to the leader could convey some benefits to the mother, such as protection from inter- or intraspecific attacks. As noted above, it is the mothers with relatively young infants that are spatially closest to the leader. Because infant rearing among chimpanzees and orangutans does not include such relationships under natural conditions, their species-typical behavior in this regard is not disrupted by the conditions of social isolation in captivity to the same degree that it is in the gorillas.

There is also evidence from field research on gorillas that group members, especially juveniles and relatively young adult females, stimulate protective and possessive responses on the part of mothers toward their offspring by showing considerable curiosity toward the infants (Harcourt, 1979; Stewart, 1977). Stewart (1977) proposed that such responses by the mother might strengthen her attachment to her infant. Captive mothers in social isolation, therefore—especially primiparous ones—may not develop as strong an attachment to their infants because of the absence of such stimulation from conspecifics.

Although it is not possible to assert specifically which characteristics of the social group (harem) support the expression of adequate maternal behavior by gorillas, the data discussed above are consistent with the hypothesis that maternal behavior in this species is facilitated by social contacts under natural conditions. Data on captive gorillas also support this hypothesis in that adequate maternal behavior was displayed when mothers reared their infants in the company of other conspecifics, abused or neglected their infants in the absence of such companionship, and terminated abuse or neglect when provided conspecific companions. This analysis suggests that the maternal behavior of the great apes, a species-typical aspect of behavior, is adapted to the conditions under which the species live in the natural habitat. Alteration of the environment in which mothers rear their infants, such as maintaining the species in captivity, appears to influence maternal behavior adversely in relation to the degree that the captive environment prevents or distorts species-typical modes of infant rearing. Expressed in positive terms, adequate maternal behavior is displayed by great apes under (captive) conditions that permit an approximation of the social environment that normally exists during infant rearing under natural conditions.

It is useful, when appropriate, to extrapolate research results on nonhuman primates to issues of interest related to our own species. In the present context, two possibilities for extrapolation of research results exist, one general and one specific. The general hypothesis derived from this research is that certain aspects of the maternal behavior of the great apes, species closely related biologically to humans, are differentially adapted to species-typical forms of social organization. The possible implication of this hypothesis for humans is that the maternal behavior of our species is also adapted to a species-typical form of social organization or society. In order to explore the further implications of such a possibility, one would examine evidence on human social systems—not modern systems, but those of more primitive societies that, perhaps, better reflect the roles of natural and sexual selection that were operative as our species evolved. The more specific hypothesis regarding the maternal behavior of the great apes is that the mother's responsiveness and treatment of its offspring are compromised when the mother gives birth and rears an infant under social conditions that deviate in certain aspects from conditions in the natural habitat. The possible implication of this latter hypothesis is that human maternal (or parental) behavior is comparably compromised when humans rear offspring under conditions that deviate from a species-typical form of social organization. It is not at all clear, of course, that there is a "species-typical form of social organization" for the human species, and the varied patterns of child rearing among societies today indicate, certainly, that widely different child-rearing practices are more-or-less successful.

Despite these caveats, several authors have attempted to apply the lessons derived from nonhuman primate research on maternal behavior to the issue of child abuse in humans (e.g., Horenstein, 1977; Nadler, 1980; Suomi, 1978). In relation to the topic considered above, one can ask the question, "Does the abuse of offspring by socially isolated gorilla mothers bear any relationship to the abuse of children by mothers in our society?" Having posed the question, I must state that it is not my purpose to attempt an answer in this chapter. It is appropriate to point out, however, that early social stress in childhood, such as abuse and neglect by parents, is a common characteristic of human parents who abuse their children (Kempe, Silverman, Steele, Droegemueller, & Silver, 1962; Parke & Collmer, 1975). Social isolation of the parent is another characteristic frequently found among human child abusers (Elmer, 1967; Young, 1964). Whether these relationships in the human reflect homologies or merely analogies to the nonhuman data is a provocative question for the scientist interested in biological contributions to human behavior.

REFERENCES

Davenport, R. K., Jr. The orang-utan in Sabah. *Folia Primatologica*, 1967, 5, 247–263.

Davenport, R. K. Some behavioral disturbances of great apes in captivity. In D. A. Hamburg & E. R. McCown (Eds.), *The great apes*. Menlo Park, Calif.: Benjamin/Cummings, 1979.

Davenport, R. K., & Rogers, C. M. Differential rearing of the chimpanzee: A project survey. In G. H. Bourne (Ed.), *The chimpanzee* (Vol. 3). Basel: Karger, 1970.

Elmer, E. *Children in jeopardy: A study of abused minors and their families*. Pittsburgh: University of Pittsburgh Press, 1967.

Galdikas, B. M. F. Orangutan adaptation at Tanjung Puting Reserve: Mating and ecology. In D. A. Hamburg & E. R. McCown (Eds.), *The great apes*. Menlo Park, Calif.: Benjamin/Cummings, 1979.

Goodall, J. Chimpanzees of the Gombe Stream Reserve. In I. DeVore (Ed.), *Primate behavior*. New York: Holt, Rinehart and Winston, 1965.

Harcourt, A. H. Social relationships among adult female mountain gorillas. *Animal Behavior*, 1979, 27, 251–264.

Harcourt, A. H., Stewart, K. J., & Fossey, D. Male emigration and female transfer in wild mountain gorilla. *Nature*, 1976, 263, 226–227.

Harlow, H. F., Harlow, M. K., Dodsworth, R. O., & Arling, G. L. Maternal behavior of rhesus monkeys deprived of mothering and peer association in infancy. *Proceedings of the American Philosophical Society*, 1966, 110, 58–66.

Hess, J. P. Some observations on the sexual behaviour of captive lowland gorillas, *Gorilla g. gorilla* (Savage and Wyman). In R. P. Michael & J. H. Crook (Eds.), *Comparative ecology and behaviour of primates*. London: Academic Press, 1973.

Horenstein, D. The dynamics and treatment of child abuse: Can primate research provide the answers? *Journal of Clinical Psychology*, 1977, *33*, 563–565.

Horr, D. A. The Borneo orang-utan: Population structure and dynamics in relation to ecology and reproductive strategy. In L. A. Rosenblum (Ed.), *Primate behavior: Developments in field and laboratory research*. New York: Academic Press, 1975.

Kempe, C. H., Silverman, F. N., Steele, B. F., Droegemueller, W., & Silver, H. K. The battered child syndrome. *Journal of American Medical Association*, 1962, *181*, 17–24.

Kirchshofer, R. Gorillazucht in zoologischen Garten und Forschungsstationen. *Zoologische Gärten*, 1970, *38*, 73–96.

Kortlandt, A. Chimpanzees in the wild. *Scientific American*, 1962, *206*, 128–138.

MacKinnon, J. R. The orang-utan in Sabah today. *Oryx*, 1971, *11*, 141–191.

Miller, L. C., & Nadler, R. D. Mother–infant relations and infant development in captive chimpanzees and orang-utans. *International Journal of Primatology*, 1981, *2*, 247–261.

Nadler, R. D. Periparturitional behavior of a primiparous lowland gorilla. *Primates*, 1974, *15*, 55–73.

Nadler, R. D. Determinants of variability in maternal behavior of captive female gorillas. In S. Kondo, M. Kawai, A. Ehara, & S. Kawamura (Eds.), *Symposia of the Fifth Congress of the International Primatological Society*. Tokyo: Japan Science Press, 1975 (a).

Nadler, R. D. Second gorilla birth at the Yerkes Regional Primate Research Center. *International Zoo Yearbook*, 1975, *15*, 134–137 (b).

Nadler, R. D. Child abuse: Evidence from nonhuman primates. *Development of Psychobiology*, 1980, *13*, 507–512.

Parke, R. D., & Collmer, C. W. Child abuse: An interdisciplinary analysis. In E. M. Hetherington (Ed.), *Review of child development research* (Vol. 5). Chicago: University of Chicago Press, 1975.

Reynolds, V., & Reynolds, F. Chimpanzees of the Budongo Forest. In I. DeVore (Ed.), *Primate behavior*. New York: Holt, Rinehart and Winston, 1965.

Rijksen, H. D. *A field study on Sumatran orang utans (Pongo pygmaeus abelii Lesson 1827)*. In H. Veeneman & B. V. Zonen (Eds.), *Ecology, behavior and conservation*. (Monograph). Wageningen, the Netherlands, 1978.

Rodman, P. S. Population composition and adaptive organisation among orang-utans of the Kutai Reserve. In R. P. Michael & J. H. Crook (Eds.), *Comparative ecology and behaviour of primates*. London: Academic Press, 1973.

Rogers, C. M., & Davenport, R. K. Chimpanzee maternal behavior. In G. H. Bourne (Ed.), *The chimpanzee* (Vol. 3). Basel: Karger, 1970.

Schaller, G. B. *The mountain gorilla*. Chicago: University of Chicago Press, 1963.

Stewart, K. J. The birth of a wild mountain gorilla *(Gorilla gorilla beringei)*. *Primates*, 1977, *18*, 965–976.

Sugiyama, Y. Social organization of chimpanzees in the Budongo Forest, Uganda. *Primates*, 1968, *9*, 225–258.

Sugiyama, Y. Social behavior of chimpanzees in the Budongo Forest, Uganda. *Primates*, 1969, *10*, 197–225.

Suomi, S. J. Maternal behavior by socially incompetent monkeys: Neglect and abuse of offspring. *Journal of Pediatric Psychology*, 1978, *3*, 28–34.

van Lawick-Goodall, J. The behaviour of free-living chimpanzees in the Gombe Stream Reserve. *Animal Behavior Monographs*, 1968, *1* (3),161–311.

Weisbard, C., & Goy, R. W. Effect of parturition and group composition on competitive drinking order in stumptail macaques *(Macaca arctoides)*. *Folia Primatologica*, 1976, *25*, 95–121.

Wrangham, R. W. Sex differences in chimpanzee dispersion. In D. A. Hamburg & E. R. McCown (Eds.), *The great apes*. Menlo Park, Calif.: Benjamin/Cummings, 1979.
Yerkes, R. M, & Tomilin, M. I. Mother–infant relations in chimpanzee. *Journal of Comparative Psychology*, 1935, *20*, 321–359.
Young, L. *Wednesday's Children: A Study of Child Neglect and Abuse*. New York: McGraw-Hill, 1964.

7

The Social Network of the Young Child
A Developmental Perspective

MICHAEL LEWIS, CANDICE FEIRING, AND
MIRIAM KOTSONIS

Imagine that one could invisibly watch a washday in a typical middle-class American home and watch a washday in a pretechnological Indian society in the Amazon basin.

In the first scene, a mother and her 3-year-old child are alone in their home. The 3-year-old wants attention and would like to play with someone. The mother is busy collecting the laundry, putting it in the machine, and trying to straighten the house. The child is upset: there is no one to play with because the mother is busy. The mother is upset because she has work to do and because she would like to spend some time talking to her friend, Jane, with whom she has not spoken in several days.

Contrast this with a second scene. Picture an entire village down by the river. There, mothers and children are congregated, all doing their individual family tasks together. Mothers sit by the side of the river pounding their wash against large rocks and stumps of trees. The 3-year-olds play with their peers of the same age, and younger and older

---

MICHAEL LEWIS AND CANDICE FEIRING • Department of Pediatrics, Rutgers Medical School—University of Medicine and Dentistry of New Jersey, New Brunswick, New Jersey 08903. MIRIAM KOTSONIS • Bell Laboratories, Holmdel, New Jersey 07733.

children play as their mothers work. The mothers work and at the same time are able to interact with other mothers, who also vary in age, some being older kin (their mothers or grandmothers), others being peers of varying age.

If we focus on the first scene, it becomes obvious that the social network of children is nearly restricted to their mothers. All functions that constitute good caregiving must be satisfied by a single person. The child's mother must be the nurturer, the teacher, the game player, the protector, the socializer, and the provider. At the same time, this adult female needs to derive most of her daily satisfaction from her young child. The network of the adult (mother) is as restricted as that of the child.

Not so for the Indian family. By the river, the 3-year-old has other children—kin, nonkin, older, and younger—with whom to play. Moreover, there are other adults, kin and nonkin, friends of the mother, and unfamiliar people. Likewise, the mother has a large array of other adults with whom to talk, to share, and even to play. For the most part, these are women; however, in this particular group, preadolescent boys and old men also wander down to the river. The network of the child, like the network of the mother, is a rich nexus, an interconnection of relationships. Within this nexus, the needs of both can be served, at times by each other, at times by others.

These two scenes serve to contrast the different social networks that are possible. In one, the American family, the child's network consists primarily of the mother–child dyad. In the Indian family, the child's network consists of many others, and the possible dyads are quite numerous. These two possible networks serve to contrast two opposite social milieus of which a young child may be a part. The differences in the social milieus point out the need to explore the nature of the child's network and point to the fact that even in the American family the mother–infant relationship needs to be considered only one of the possible dyads in a social network. Until recently, it was the relationship between mother and child that received the most attention, both in theory and in empirical study. Although the child in the American family is embedded in a social network that often includes mother, father, siblings, other kin (grandparents, aunts/uncles, cousins), peers, adult friends of parents, and teachers, relatively little attention has been paid to these people and the role that they play in the child's social development. In the discussion to follow, we explore the nature of the young child's network: first, theoretically, by considering the possible models of social development; second, by examining the mechanisms that allow for such development; and third, empirically, by examining the network of a group of 3-year-old children who differ in social class, gender, and birth order. We also examine some preliminary data on the developmental changes in the social network as these children become older.

Epigenetic versus Social Network Models of Development

The two theories that are most relevant to social development rest on different sets of assumptions concerning the role of others besides the mother in the child's social development. A vignette of each model can be seen in the data presented by Rosenblum and Kaufman (1968) on macaque monkeys. Bonnet macaques form physically close, large groups of which mother and infant are only two members. Loss of the mother results in some distress for the infant, but other members of the group are able to alleviate this distress. This type of organization lends itself clearly to social network explanations.

In contrast, pigtail macaques form simple groups consisting of only the mother and the infant. Loss of the mother here is more catastrophic: there are no significant others to mitigate the infant's distress, and the young pigtail macaque fails to thrive. In this type of social organization, explanations of future development, as determined by a first relationship, seem appropriate.

For humans, however, there is no single pattern of social organization. We do know that most children are exposed to intensive caregiving relationships and a variety of other relationships. Widening our view to encompass both aspects broadens the explanatory power of models of development.

Epigenetic Model

The epigenetic model is the most accepted and most utilized theory of the development of social relationships. Its central thesis is the assumption that one set of social experiences is directly connected to the next. More specifically, this model argues for a linear relationship of such a kind that the infant first adapts to one relationship, and then all subsequent ones follow from this primary relationship. This model is characterized by three features: (1) its fixed sequence; (2) its determinism; and (3) its trait or structural quality.

Fixed Sequence

The epigenetic theory argues for a linear progression in which the infant first adapts to one person within the family network, usually the mother, and then all subsequent social relationships follow from this basic adaptation (Bowlby, 1969). More recently, the epigenetic model has been expanded so that the child's adaptation is viewed as first involving the development of a relationship with the mother, then with the father and siblings, and later with others, for example, peers.

Determinism

Not only do the epigenetic theories of Freud and Bowlby postulate a fixed sequence in the development of relationships, but they also assume that later social experiences are determined by early ones, particularly the relationship between mother and child. This view is still widely held by others who see the mother–infant attachment as determining the child's later peer relationships (Arend, Gove, & Sroufe, 1979; Matas, Arend, & Sroufe, 1978). Although much has been made of the seminal influence of early relationships on subsequent social relationships, the data are rather sparse. Although Bowlby (1951) pointed out that without a proper mother–infant relationship the child is "at risk" for a wide set of developmental dysfunctions, including death, the longitudinal nature of the proof necessary to back up this claim has made the demonstration of this theory difficult.

Harlow's (1969) demonstration that motherless monkeys are at risk (1) for dysfunctional peer relationships (i.e., they have difficulty mating) and (2) for mothering (i.e., they make poor mothers who maltreat and even kill their young) is impressive evidence of the deterministic nature of early relationships. However, such a deterministic view has been challenged by a variety of other data, most of which pertain to mother–child and child–child relationships.

As has been pointed out repeatedly, the effects of maternal deprivation on monkeys as reported by Harlow and Harlow (1965) were due not to the lack of a mother–child relationship but rather to the *complete* social isolation in which the monkeys were raised. Baby monkeys raised with other babies showed few of the effects originally reported (Harlow, 1969). This finding not only addresses the deterministic property of relationships but also demonstrates the multiple—as opposed to the fixed—aspects of social development. Lewis, Young, Brooks, and Michalson (1975), in discussing peer development, pointed out that rather than being determined in a linear fashion, peer relationships may evolve in parallel with other relationships. This view of a multiple-affect system was also offered by Harlow and Harlow (1965) and has found support in a variety of other studies. Thus, rather than viewing peer relationships as determined by the mother- or father–child attachment relationship, we can regard them as developing in tandem with the parent–child relationship. For example, a study by Cowen, Pedersen, Balijian, Isso, and Trost (1973) suggested that there is a greater risk for adult dysfunctional social behavior in individuals with poor peer interaction in childhood. Harlow (1969) also looked at infant monkeys raised by mothers without peer contact and another group raised with peers without maternal contact. Although there were some problems with the design (isolating mother and infant may have disturbed the mother's normal behavior), the general findings indicate that peer attachment in the absence of maternal attachment does not lead to subsequent peer dysfunction, whereas the absence of peer attachment results in serious disturbances.

Evidence of the importance of peers can also be found in Furman, Rake, and Hartup (1979), who treated 24 socially withdrawn preschoolers with peer therapy. The results of this intervention increased the social activity of the isolate group. Although it is not known for certain, it can be assumed that the maternal attachment in these isolates was inadequate; nevertheless, successful peer experiences subsequently produced adequate peer interactions. This study and ones like it argue for a strictly deterministic interconnection between relationships of the kind depicted by the epigenetic model. Lewis and Schaeffer (1981) studied a group of abused and neglected infants and were concerned with the question of whether such children—who have insecure attachment relations with their mothers—would have good peer relations if they were placed in a day-care setting for eight hours a day and were given peer experience. After four months in the day-care setting, gross observation could not distinguish the abused children from the nonabused group. The children's behavior toward peers, their free play behavior, and their social interactions with caregivers indicated no obvious and discernible differences between the two groups. The results of this study, along with those of Furman *et al.* (1979), confirm the belief that poor infant–mother relationships do not have as a necessary consequence poor peer relationships.

Although evidence of the deterministic effect of one relationship on another exists (Arend *et al.*, 1979; Matas *et al.*, 1978; Waters, Wippman, & Sroufe, 1979), most of the data leave this question open. This is not to imply that one relationship cannot exert an effect on another. Indeed, in our social systems model, just such an effect is envisioned. The difference between the positions concerns the deterministic nature of the effect. This question will be explored in the section on trait or character effect.

Traits

The issue and controversy surrounding the nature of a trait or an enduring aspect of personality have dominated much contemporary thinking (Pervin, 1978). The issue is raised again when we consider the effect of one relationship on another, and it accounts for the major differences between the epigenetic and the social network systems models.

In the epigenetic view, the mother–infant or earliest attachment relationship endows the infant with a trait or characteristic that is located within the organism, sometimes referred to as *social competence* (Sroufe, 1979). This trait or its absence then determines subsequent relationships. Sroufe (1979) and Block and Block (1979) have associated it with ego skills; however, it could be a trait such as self-esteem, self-efficacy, or some combination. Whatever its nature, it is the presence or absence of the trait that influences other relationships. An often-used metaphor is that the child is like an empty vessel that needs filling. Once filled, the child can move on to new relationships. If the child is not filled, its movement will be inhibited, or the new relationships will differ in nature or degree from what they would be if the child were filled. The task of the earliest attachment relationship is to fill the vessel. Thus, the explanation for the establishment of relationships rests with a mechanism that resides in the organism.

Also related to this trait issue is the notion of a critical period. Although this point is seldom stated explicitly, most theories assume that the attachment relationship in the opening months of life is critical and that, if this relationship is inadequate, then subsequent relationships, no matter how adequate, are not sufficient to alter the destructive effect (Klaus & Kennell, 1976).

The notion of a trait provides a mechanism for the deterministic nature of the epigenetic model. One relationship can affect another through the creation of a trait in the child. The child then brings this trait to bear on its next relationship. Moreover, this trait or its absence, based on the outcome of the first relationship, is not easily affected by experience.

There are data that do not fit this model. Interestingly, the motherless-monkey literature again provides important information. Recall that the poor mothering behavior of motherless monkeys is used as an example of the consequences deriving from the lack of some trait associated with having had a good mothering experience. In other words, these monkeys, because of lack of mothering, become poor mothers. What is not often attended to is the fact that, after having their first children, these motherless monkeys appear to be quite normal in their mothering of their second children (Rupenthal, Arling, Harlow, Sackett, & Suomi, 1976; Suomi, 1978). The more time a mother spends with the first child, even though she maltreats it, the better mother she appears to be with the second child.

A similar analysis can be applied to the data on the peer behavior of maltreated or abused human infants (Lewis & Schaeffer, 1981). If such infants have poor attachment relationships with their mothers, and if this leads to a deficiency in a trait, then one would expect this missing trait to result in poor peer relationships. That this effect is eliminated by giving the infants peer experience again suggests that the concept of a trait as the deterministic factor may be overstated.

Social Network Systems Model

The epigenetic model relies on features and mechanisms that are developmental in nature. As an example, the trait characteristic provides for a change in the internal structure of the child, a change brought about by a particular set of experiences. The social network systems model is less an individual and more a social model of development. The causes of social behavior are found in the structure of the social system. The forces influencing social relationships and behaviors exist, at least in part, in the structure of the system. When the system is altered, some behaviors and relationships change.

Systems in general and social systems in particular can be characterized by a number of features: (1) they have elements; (2) the elements are related; (3) the elements are nonadditive (i.e., the sum of the elements does not equal the total system); (4) the elements operate under a steady-state principle so that the elements have the ability to change and yet maintain the system; and (5) the systems are goal-oriented (Feiring & Lewis, 1978; Lewis, 1982; Lewis & Feiring, 1979, 1981).

Elements

Systems are composed of sets of elements. The influences of the social system elements in the young infant's life are not necessarily limited to the mother. The research literature of the past 5–10 years indicates that the infant is embedded in a complex system of elements including mother, father, siblings, other relatives (especially grandparents, uncles, and aunts), and peers. Moreover, the infant is capable of interacting with these other people and of forming relationships with them. Some of these relationships may be attachments, some love without attachment, and some friendships. The constraints on these relationships probably rest on cultural factors. Thus, the first requirement of a social network systems analysis is met: there is a multitude of possible elements.

Interconnection of Elements

Systems are characterized not only by sets of elements but by a set of interrelated elements, that is, by elements that are influenced by each other (Monane, 1967). Within the family, the interaction of elements can be at several levels. At the simplest level, the infant affects its parents (Lewis & Rosenblum, 1974), the parents affect the infant, and the parents affect each other. Such effects come about through direct interactions among family members. Recall that elements are not restricted to individuals but may be dyads or even larger units. When larger elements are considered, the study of the interrelation of elements becomes more complex. For example, a child can affect not only each parent separately but also the parental interaction. The research on family size (number of children) and the age of the child shows that children affect marital satisfaction (Rollins & Galligan, 1978). Likewise, the father can influence the mother and the child individually, as well as the mother–child interaction. Many different effects of this complex nature have been observed. Pedersen (1975) and Feiring (1975) observed the parental relationship as it affects mother–child interactions, and several investigators have studied the effect of the father on the mother–child relationship (Clarke-Stewart, 1978; Lamb, 1978; Lewis & Weinraub, 1976; Lewis, Feiring, & Weinraub; 1981, Pedersen, Anderson, & Cain, 1977; Parke, 1978). Cicirelli (1975) looked at the influence of siblings on the mother–child relationship, and Dunn and Kendrick (1979) studied the effect of a newborn child (sibling) on the older-child–mother relationship.

Nonadditivity

Social systems also possess the quality of nonadditivity; that is, knowing everything about the elements that comprise a system does not reveal everything about the operation of the whole. Any set of elements behaves quite differently within the system from the way individual elements do in isolation. This rule holds for simple elements as well as for more complex ones. Within the family, how an individual person behaves alone can be quite different from how that person behaves in the presence of another. Clarke-Stewart (1978) observed mothers, fathers, and children in dyadic (parent–child) and triadic (mother–father–child) interactions. She showed that the quantity and the quality of behavior in the isolated mother–child subsystem is changed when this dyad is embedded in the mother–father–child subsystem.

Pedersen *et al.* (1977) reported that whereas the father-husband frequently divides his behavior between the child and spouse in a three-person subsystem, the mother-wife spends much more time in dyadic interaction with the child than with her husband. The child also exerts influence on the parental system. Rosenblatt (1974) found that the presence of one or more children reduced adult–adult touching, talking, and smiling in selected public places such as the zoo, the park, and shopping centers. Lewis and Feiring (1982), looking at interactions at the dinner table, found that the number of children at the table affects the amount of mother–father verbal interaction and the amount of positive affect exhibited between the two. As the number of children in the family increases, the amount of positive affect between the parents decreases.

Steady State

Social systems are characterized by steady states. The term *steady state* describes the process whereby a system maintains itself while always changing to some degree. A steady state is characterized by the interplay of flexibility and stability by which a system endeavors to maintain a viable relationship among its elements and its environment. Social systems are defined as goal-oriented, and steady-state processes are directed toward goal achievement. However, the same general goal may be served by different patterns of behavior as the system changes to adapt to its environment. Within development, such processes appear to be essential because behavioral changes occur in the child as a function of age. Lewis and Ban (1971) have shown that the nature of the interactions between child and parent changes over the first two years of the child's life, although the function of maintaining a relationship, which these interactional behaviors serve, remains the same.

Goals

Social systems are also characterized by their purposeful quality. The family system is generally thought to exist in order to perform certain functions (or goal-oriented activities) that are necessary both to the survival of its members and to the perpetuation of the specific culture and society. The family's functions are often stated to be procreation and child rearing; thus, it is suggested that the family is the principal agent of these societal goals. Beyond this level of generality, there are numerous other ways of describing and defining family functions (Lewis & Feiring, 1978; Parson & Bales, 1955).

The epigenetic model of social development not only has restricted its conceptualization of the nature of the family relationships available to the young child by focusing more-or-less exclusively on the mother but also has limited its delineation of the types of activities or goals engaged in by the other family members. If only caregiving functions or goals are considered, then it makes some sense to study the mother as the most important (and only) element. However, other functions in the child's life—including, for example, play and teaching—may involve family members other than the mother. The contrast between the two theories makes it clear that the number and the nature of the people in the child's network are important to study.

The Role of Direct and Indirect Effects

Lewis (Feinman & Lewis, 1981; Lewis & Feiring, 1981; Lewis & Weinraub, 1976), Bronfenbrenner (1977), Lamb (1978), and Parke, Power, and Gottman (1979) have discussed some of the various ways that elements within a network may influence each other. Direct and indirect effects have been identified as two major classes of influence processes (Lewis & Feiring, 1981; Lewis & Weinraub, 1976). Direct effects are those interactions that represent the influence of one person on the behavior of another when both are engaged in mutual interaction. Direct effects are usually observed in dyadic interactions but could involve more than two people, such as when a teacher instructs a class of students. Such effects involve information gathered from participation in an interaction with another person or object and always involve the target person as one of the active participants in the interaction. Direct effects have been studied for each member of the infant's family: mother–infant (Brazelton, Koslowski, & Main, 1974; Lewis, 1972; Stern, 1974); father–infant (Brazelton, Yogman, Als, & Tronick, 1979); and sibling–infant (Greenbaum & Landau, 1977).

The indirect effects are two classes of interactions or influence. In the first, indirect effects are those sets of interactions that affect the target person but that occur in the absence of that person. Although these sets of interactions affect the target person, they may be best described as influences that play their role indirectly in development by affecting direct effects or interactions. For example, the father affects the mother–infant relationship by supporting the mother (both emotionally and financially). Several studies have demonstrated this effect (Barry, 1970; Feiring & Taylor, in press; Heath, 1976).

Another cateogry of indirect effects includes those that occur in the presence of the target person, even though the interaction neither is directed toward nor involves that person. These effects are based on information that is gathered from sources other than direct interaction with another person and may be the result of (1) observation of another's interaction with persons or objects or (2) information gathered from another about the attitudes, behaviors, traits, or actions of a third person. Such indirect influences have been studied under a wide number of labels, including *identification, modeling, imitation,* and *incidental, vicarious,* and *observational learning* (see Lewis & Feiring, 1981). In the case of young infants and their interactions with others, several recent studies by Clarke-

Stewart (1978), Feinman and Lewis (1981), and Feiring, Lewis, and Starr (in press) have pointed out how the mother's interaction with others can affect the child's behavior toward others, particularly strangers. Thus, for example, if the mother acts positively toward a stranger, the child's behavior toward that stranger will be more positive than if the mother shows no interest at all (Feiring, Lewis, & Starr, in press).

In short, the infant must adapt to a social network containing elements that vary in number and complexity. These elements influence the infant both directly through their interactions with the child and indirectly through their interactions with each other. The child establishes interactions within a network of already-existing interactions.

The study of indirect effects becomes essential once one considers the entire social network. When considering only a single dyadic interaction, these effects go unnoticed. It is only when multiple relationships are considered that these indirect effects become obvious: the study of the social network increases the complexity of our theories and analyses because (1) more than a single person's effect on the child needs to be considered, and (2) indirect effects become obvious.

THE SAMPLE CHARACTERISTICS OF CHILDREN AND THEIR FAMILIES

The sample of children on whom we have social network data at 3 years of age belonged to 117 families. All the children and families were participants in a longitudinal research study that was conducted from the time the children were 3 months old. The breakdown of the sample by the socioeconomic status (SES), sex, and birth order of the children under study is given in Table 1.

As Table 1 indicates, there was a fairly even split by the categories of socioeconomic status and sex of child. Specifically, there were 59 males

TABLE 1. Distribution of 3-Year-Old Sample by Socioeconomic Status, Birth Order, and Sex of Child

	Total	Upper-middle SES	Middle SES	Only	First-born	Second-born	Third-born	Fourth- and later-born
Males	59	31	28	6	13	19	17	4
Females	58	27	31	9	10	18	15	6
Total	117	58	59	15	23	37	32	10

and 58 females, and there were 58 upper-middle-SES and 59 middle-SES families.[1] The distribution of the sample by birth order, it should be noted, approximated the proportions in cross-section U.S. families, as reported in a 1977 Census Bureau publication (Lewis & Feiring, 1981), having first-, second-, third-, and fourth-borns.

Description and Measurement of the Social Network

As an area of research, the study of social networks is almost 30 years old (Barnes, 1954). The seminal work of Bott (1957, 1971), *Family and Social Network*, is widely known to both sociologists and psychologists. Bott reported that, contrary to popular belief, middle-class couples did not live in isolation but formed coalitions with other middle-class couples in a network of social kinship. Rather than being isolated nuclear families separated from their kin, middle-class, as compared to working-class, families had more contact with nonrelative friends on a daily basis, although they maintained contact with the extended family on a less frequent basis. In interviews with couples, Bott discussed with both husbands and wives the type and frequency of contact with people in the categories of relatives, friends, neighbors, and organizations (e.g., school, clubs, unions, and religious groups). The couples also kept diaries, which helped the researchers determine the nature of the social networks. One interesting result of Bott's work involved the relationship between social networks and family roles. Although networks have been examined in regard to adults, relatively little work has been done on networks, roles, and the child's development (for exceptions, see Cochran & Brassard, 1979; Feiring & Lewis, 1981b; Lewis & Feiring, 1979, 1981).

Since Bott's work, social network theory and research have received the most study and attention from the field of sociology. Attributes of the social network—including size, variety of membership, density, connectedness, reciprocity, frequency, and function of contact—have been studied in regard to the nature of the marital relationship and the social

---

[1] The families were divided into two instead of the five Hollingshead categories, with the division at 27.5. The lower numbers were assigned to the upper-middle-SES group and the higher numbers to the middle-class group (see Feiring & Lewis, 1981a, for a full description of the SES levels of the parents). In education (the component of SES that seems most influential in child development outcomes), the majority of upper-middle-class mothers had completed college, and the majority of the middle-class mothers had finished high school. The majority of upper-middle-SES fathers had college degrees with additional graduate work, and the majority of middle-SES fathers had finished high school and had at least two years of college education.

and geographic mobility patterns of families (e.g., Lee, 1979). An extensive literature exists that employs quantitative techniques for describing and specifying the characteristics of social networks (e.g., Holland & Leinhardt, 1979). Social networks have also been utilized as the unit of intervention in conducting therapy (e.g., Attneave, 1976; Pattison, 1975). However, there have been few attempts to measure and describe the broad network of people and contacts in a child's social system and to understand the developmental influence that they might have, especially on the young child.

In general, the examination of the social network, as in Bott's work (1957, 1971), has attempted to delineate through interviews and questionnaires the kind and frequency of contact with kin and nonkin groups. In addition, information about the connectedness of network members (i.e., who knows whom) and the network density (the number of people who know each other relative to those who do not) is often of interest. Once the point of anchorage (e.g., a person, a conjugal pair, or a nuclear family) is determined, information is gathered on kin and nonkin contact. In the data collected in this study, the point of anchorage was the 3-year-old child. In order to clarify some basic attributes of the young child's extended social network, data were collected on the composition of the children's networks as perceived by their mothers. The mothers were asked to complete an adapted version of the Pattison Psychosocial Network Inventory (Pattison, 1975). In questionnaire form, the mother was asked to list the persons in the child's social network in the categories of family, relatives, friends of the parents, and friends of the child. The mother was asked to specify the relationship of each person listed to the 3-year-old child (e.g., for the relatives category: cousin, grandparent, and so on), and to indicate the amount of contact the person had with the child. Contact could be on a daily, weekly, monthly, biyearly, or yearly basis (*contact* was defined as including face-to-face or by phone or letter). From the mother's report, we were thus able to get an idea of (1) the number of people and the kinds of people who comprised the 3-year-old's network as well as (2) the frequency of contact with these people.

Self-report data are full of problems. Obviously, the mother's report reflects her perception of the child's network and may not yield the most accurate picture of whom the child sees and how often. However, it is not possible to interview the 3-year-olds for this data, and practically speaking, it was not possible to observe the children over a long period to determine whom they saw and how often. Because the initial purpose was to get an idea of the child's social network, we decided to use the traditional questionnaire method, recognizing its problems but utilizing this procedure as the most efficient means of data collection to obtain a general mapping of the child's network. For future analyses, we have thought of using a phone check technique and diary keeping by the parents in order to get multiple measures of the child's network. However, for an initial exploratory analysis, the mother is probably the person most likely to be aware of the people with whom her child comes in contact (especially as, in this sample, most of the mothers did not work full time outside the home).

Once we had collected data on the kinds of people and contact the 3-year-old children experienced in their social networks, these people were grouped by categories that would make sense in depicting the young child's social world. Lewis and Feiring (1978, 1979) had previously described a possible construction of the child's social world that may be present in the infant's conceptualization of people. In particular, the three categories of age, gender, and familiarity are attributes of the social world that the child acquires early. Whether the infant uses these categories as a means of ordering the social array that comprises its social network is undetermined; however, the child's ability to use these categories to make sense of the social network suggested that they might provide a meaningful way of analyzing the social network data. Consequently, in the analysis to follow, people were divided into three groups: *adults* versus *peers* (i.e., age); *males* versus *females* (i.e., gender); and *relatives* versus *nonrelatives*. Although the nuclear family component of *relatives* may be strongly related to degree of familiarity, a comparison of *relatives* and *nonrelatives* was made primarily because sociologists have stressed the importance of kin in defining social structure. Both sociologists and sociobiologists have argued that kin have a special role to play, as compared to nonkin, in defining the group membership of the child (Wilson, 1975).

Describing the Child's Social Network

Characteristics of the Social Network: Total Sample

Table 2 presents the social network of 117 three-year-olds as reported by their mothers. Each of the terms designating frequency of contact (e.g., weekly or monthly) indicates that the child had contact with the person(s) in question at least as often as the given quantity of time, but not as often as the next smaller quantity of time. For example, *weekly* means contact at least weekly, but not daily. The data represent the number of people the child saw in a year and the number of people seen by frequency of contact. Examination of Table 2 indicates that the children were in contact with significant others besides their parents

TABLE 2. Different Types of People in 3-Year-Olds' Social Networks: Mean Number of People Contacted Daily through Yearly for the Total Sample ($n = 117$)

	Mean N people	N seen daily	N seen weekly	N seen monthly	N seen 6-monthly	N seen yearly
Mother's parents and grandparents	1.77	.19	.86	.53	.16	.03
Father's parents and grandparents	1.57	.08	.58	.63	.21	.07
Mother's siblings	1.46	.08	.50	.44	.33	.11
Other aunts/uncles of child	1.83	.03	.41	.72	.50	.17
Cousins of child	1.47	0.0	.31	.56	.51	.09
Other relatives	1.24	.03	.20	.59	.27	.15
Adults known by child	6.04	1.15	2.50	1.61	.67	.12
Adult school personnel/teachers	.68	.30	.38	—	—	—
Babysitters	.32	.03	.19	.10	—	—
Male friends (peer)	2.97	.81	1.57	.40	.15	.03
Female friends (peer)	2.81	.78	1.33	.52	.15	.03
Nuclear family (parents and siblings)	3.47	3.47	0.0	0.0	0.0	0.0

and siblings. Peers, relatives, and friends of the parents were involved in the child's life and therefore played an important role. Not shown in the table is the obvious fact that nuclear family members were seen on a daily basis, whereas others were seen less frequently. The data in Table 2 suggest that, for this sample of 3-year-old children, contact with peers and adults was more likely to occur on a weekly basis. Contact with grandparents, cousins, aunts, and uncles did not occur very often on a daily or weekly basis; however, note that the mother's parents were seen on a weekly basis more often than other relatives. It has often been noted by sociologists that mothers keep in contact with their parents more than do fathers (Bott, 1971; Reiss, 1962; Rosser & Harris, 1965; Willmont & Young, 1960). This finding suggests that the young child has more exposure to the mother's family and its behavior, attitudes, and style than to the father's. Although there is little doubt that the impact on the child of significant others increases with age, there is evidence to indicate that, even in the initial stages of the lifespan, people other than the mother are involved with the infant. The involvement of others besides the nuclear family in the child's network varies by the culture as well as the age of the child (Lewis & Feiring, 1979). Konner (1975), for one, has shown that young !Kung San infants are touched significantly more by others besides their mothers than are Cambridge, Massachusetts, infants.

Table 3 presents the social network by gender, age, and relatedness

TABLE 3. The Social Network by Age, Kin, and Gender Categories[a]

	Total N	Daily	Weekly	Monthly	6 Months–1 Year
Age					
Adults	14.91	1.88	5.61	4.62	2.79
	(16.89)	(3.86)			
Peers	7.25	1.59	3.21	1.49	.96
	(8.74)	(3.07)			
Kin					
Relatives	9.34	.40	2.85	3.48	2.61
	(12.81)	(3.87)			
Nonrelatives	12.81	3.07	5.97	2.63	1.15
Gender					
Males[b]	(10.92)				
Females[b]	(15.06)				

[a] Numbers in parentheses include the nuclear family. Inclusion of nuclear family members alters only the total number and the daily contact means. Note that the frequency-of-contact categories are exclusive; that is, those seen daily are not included in the weekly-through-yearly frequency categories.

[b] Our data have been coded only for total number of males and females, including nuclear family members.

(the three categories of interest in describing the social world of the child) as a function of the total number of people and the frequency of contact.[2] The analyses presented here were conducted twice, with the use of social network scores that both included and excluded nuclear family members. The inclusion of nuclear family members increased the size of the network and the frequency of contact for the categories of *relatives* (parents and siblings), *adults* (parents), *peers* (siblings), *males* (father and brothers), and *females* (mother, sisters), whereas it did not change the number or frequency of contact for *nonrelatives*. The daily contact of *adults*, *peers*, *relatives*, *males*, and *females* was increased by inclusion of the nuclear family, whereas inclusion did not change contact on a weekly, monthly, and semiannual/annual basis in any category (as the nuclear family was always recorded exclusively in the daily category). Overall, the analysis did not yield different significant findings depending on whether the nuclear family members were included or not. However, both sets of means were included in the tables because the data convey a somewhat different picture of the child's network depending on whether the nuclear family members are included for comparison of categories. The data are discussed here with the nuclear family excluded.

For the sample as a whole, twice as many adults were seen as children ($F_{1,109} = 20.37$, $p < .001$) and, except for a daily basis, more adults were seen more frequently (weekly, $F = 39.89$, $p < .01$; monthly, $F = 96.19$, $p < .0004$; semiannual/annual, $F = 33.20$, $p < .0004$). Although there appear to be no differences for the number of relatives versus nonrelatives in the child's network, this breakdown of network members varies as a function of frequency of contact. On a weekly basis, more nonrelatives were seen than relatives ($F = 39.11$, $p < .04$), whereas on a monthly and semiannual/annual basis, there were more relatives than nonrelatives seen ($F = 2.62$, $p < .10$ for monthly; $F = 29.64$, $p < .004$ for semiannual/annual).

Data for the gender division are limited to the number of people. There appeared to be more females than males in these 3-year-olds' lives, ($F_{1,109} = 87.01$, $p < .0001$). Given that females remain the primary caregivers in our society, it is no wonder that they appear more frequently in the network. Even when it is not the mother who cares for the child during the day, child care usually remains the task of other females (Feiring & Lewis, this volume). Mothers also filled out the questionnaire and might have biased the report of the number of males and females seen. However, it is more likely that there were more females in the child's network because mothers are the directors of the young

---

[2] It should be noted that, for all ANOVAS completed, the degrees of freedom for the *F* test were 1,109 unless otherwise stated.

child's social experience and most often mediate the child's contact with others. In fact, sociologists have called mothers the "kin keepers of the family" (Bott, 1971; Lee, 1979).

Characteristics of the Social Network by Socioeconomic Status

Table 4 presents the breakdown of the social network by the socioeconomic status of the child's family.

Adults versus Peers

Examine first the difference in the number and the frequency of contact of adults versus peers as a function of socioeconomic status. As mentioned before, there was an overall effect for adults versus peers, collapsing across socioeconomic status (with adults seen more than peers, even when nuclear family members are not considered). There is also an SES effect so that members of the upper-middle-SES group saw more people (both adults and peers) than did members of the middle-SES group ($F = 7.39$, $p < .01$). This finding suggests that upper-middle-SES families expose their children to more people than do middle-SES families. There appears to have been not only a difference in the total number of people between SES groups, but also differences in the frequency of contact, which are significant for the daily ($F = 3.79$, $p < .05$) and the semiannual/annual ($F = 8.82$, $p < .004$) contact categories.

Relatives versus Nonrelatives

Observation of the mean number of people seen reveals that upper-middle in comparison to middle SES saw relatively more nonkin than kin ($F = 6.62$, $p < .01$). When comparisons are made that include the nuclear family, the difference between the upper-middle and the middle groups in the kin–nonkin distinction is even clearer. This difference also holds for the number of daily ($F = 5.70$, $p < .02$) and weekly ($F = 7.02$, $p < .01$) contacts. Interestingly, the upper-middle-SES groups saw proportionally more relatives than nonrelatives semiannually/annually than did the middle-SES groups ($F = 5.52$, $p < .02$). The data indicate that contact with kin was experienced more on a daily basis by the middle-SES child, whereas kin contact was less frequent for the upper-middle-SES child.

In general, the upper-middle-SES children came in contact with a higher proportion of nonrelatives to relatives than did the middle-SES children. These findings are consistent with the work of sociologists who report a negative relationship for proximity and contact with kin

TABLE 4. SES Differences[a]

	Total number		Daily contact		Weekly contact		Monthly contact		Semiannual to annual contact	
	Upper middle	Middle	Upper middle	Middle	Upper middle	Middle	Upper middle	Middle	Upper middle	Middle
Adults	16.71 (18.71)	13.14 (15.10)	2.29 (4.29)	1.47 (3.44)	5.57	5.64	5.24	4.02	3.60	2.00
Peers	8.05 (9.59)	6.46 (7.90)	2.05 (3.59)	1.14 (2.58)	3.26	3.17	1.53	1.44	1.21	0.71
Relatives	9.64 (13.17)	9.05 (12.46)	0.22 (3.76)	0.58 (3.98)	2.22	3.47	3.74	3.22	3.45	1.78
Nonrelatives	15.12	10.54	4.12	2.03	6.60	5.34	3.03	2.24	1.36	0.93
Males	(11.81)	(10.05)								
Females	(16.40)	(13.75)								

[a] Numbers in parentheses include the nuclear family. Inclusion of nuclear family members alters only the total number and the daily contact means. Note that the frequency-of-contact categories are exclusive; that is, those seen daily are not included in the weekly-through-yearly frequency categories.

and higher levels of SES (Adams, 1970; Berardo, 1967; Leslie & Richardson, 1961).

Males versus Females

Because of the coding system used, the data for males and females are limited to the total number of males and females, including nuclear family members. As indicated previously, there were more females than males for the sample as a whole, and this finding did not alter as a function of socioeconomic status. This result suggests that a more predominant female presence exists in the young child's life, a finding that transcends particular beliefs about child rearing and reflects the general cultural attitude of the role of women in child care (cf. Lewis et al., 1981).

The Characteristics of the Social Network by Sex of Child

Table 5 presents the social network categories as a function of the child's gender.

Relatives versus Nonrelatives

As a function of the target child's sex, there were no differences in the number and the frequency of contact with relatives and nonrelatives or with adults versus peers.

Males versus Females

Although both male and female children saw more females than males, there was an interaction between the sex of the child and the ratio of males to females ($F = 4.78$, $p < .03$). The male children saw relatively fewer females than did the female children (male difference score: female peers − male peers = 3.22; female difference score: female peers − male peers = 5.06).

In a separate analysis, the sex of nonrelated peers as a function of the gender of the target child was examined. For the total number of peers, as well as frequency, there was a gender-of-child by gender-of-peer interaction. For the total number of peers in the male child's network, the means were 3.64 versus 2.49 for male and female peers, respectively. For the number of peers in the female child's network, the means were 2.28 versus 3.14 for males and females, respectively. The data show that the male children had more males than females in their network, whereas for the females the reverse was true ($F = 6.55$, $p < .03$). This finding is supported by observational data on the social behavior

TABLE 5. Social Networks by Gender of Target Child[a]

	Total number		Daily contact		Weekly contact		Monthly contact		Semiannual to annual contact	
	Male	Female	Male	Female	Male	Female	Male	Female	Male	Female
Adults	15.00	14.81	1.42	2.34	5.90	5.31	4.85	4.40	2.83	2.76
	(16.98)	(16.79)	(3.41)	(4.31)						
Peers	7.32	7.17	1.68	1.50	3.22	3.21	1.46	1.52	0.97	0.95
	(8.83)	(8.64)	(3.19)	(2.97)						
Relatives	9.03	9.66	0.37	0.43	2.71	3.00	3.42	3.53	2.53	2.69
	(12.53)	(13.10)	(3.86)	(3.88)						
Nonrelatives	13.29	12.33	2.73	3.41	6.41	5.52	2.88	2.38	1.27	1.02
Males	(11.56)	(10.28)								
Females	(14.78)	(15.34)								

[a] Those numbers in parentheses include the nuclear family. Inclusion of nuclear family members alters only the total number and the daily contact means. Note that the frequency-of-contact categories are exclusive; that is, those seen daily are not included in the weekly-through-yearly-frequency categories.

preference of preschoolers for same-sex as compared to mixed-sex dyadic interaction (Jacklin & Maccoby, 1978).

In general, the major difference in network by gender of the child appeared to be the interaction between the sex of the child and the sex of the network member. Even by the age of 3 years the male children appeared to be relatively more oriented to males in the network than to females, whereas the female children were more oriented to females than to males.

Although there were no significant effects, there was some tendency for the male children to be more oriented toward peers than toward adults (daily contacts) and more oriented toward nonrelatives than toward relatives (total number). The reverse appeared to be the case for the female children, who tended to show relatively more adult contact than contact with peers (daily contact) and more kin than nonkin contact (total number, daily). Such findings lend support to the view that males may be directed out of the home and away from adults and kin, whereas females are oriented toward adults and kin (Lee, 1979; Scanzoni, 1979). Such socialization practices, if true, should result in adult-oriented behavior (Maccoby & Jacklin, 1974) and a kin-oriented (kin-keeper) behavior on the part of females.

The Characteristics of the Social Network by Birth Order of the Child

Table 6 presents the social network categories as a function of the child's birth order.

Adults versus Peers

The data indicate that the firstborns saw more people, both adults and peers ($F = 9.34$, $p < .003$) and made more frequent contacts with others (daily, $F = 3.63$, $p < .05$) than did later-borns. Moreover, there was a birth-order by age-of-network-member interaction that was significant when the nuclear family was included ($F = 6.53$, $p < .012$). When the nuclear family was not included, this interaction was only a trend indicating that later-borns' interaction with younger members was in part influenced by their play with siblings. Examination of daily contact shows this more clearly. The firstborns had more adult than peer contact, whereas the later-borns had equal contact with adults and peers, regardless of the inclusion of the nuclear family.

The interaction between birth order and age of people (adult versus peer) is of particular theoretical interest. The data suggest how the predominance of adults to peers in the social network could be related to

TABLE 6. Birth Order[a]

	Total number		Daily contact		Weekly contact		Monthly contact		Semiannual to annual contact	
	First	Later	First	Later	First	Later	First	Later	First	Later
Adult	17.76	13.53	2.89	1.39	6.00	5.42	5.55	4.18	3.32	2.54
	(19.74)	(15.52)	(4.87)	(3.38)						
Peer	8.53	6.63	1.63	1.57	3.45	3.10	1.92	1.28	1.53	0.68
	(8.92)	(8.65)	(2.03)	(3.58)						
Relative	10.89	8.59	.66	.28	3.18	2.70	3.74	3.35	3.32	2.27
	(13.26)	(12.59)	(3.03)	(4.28)						
Nonrelative	15.39	11.57	3.87	2.68	6.26	5.82	3.74	2.10	1.53	0.96
Males	(12.11)	(10.35)								
Females	(16.95)	(14.15)								

[a] Numbers in parentheses include the nuclear family. Inclusion of nuclear family members alters only the total number and the daily contact means. Note that the frequency-of-contact categories are exclusive; that is, those seen daily are not included in the weekly-through-yearly contact categories.

the child's cognitive development. It is often reported that firstborn children perform better on IQ measures than later-born children (Lewis & Jaskir, 1983; Terhune, 1974). Because adults may provide a more intellectually stimulating environment, a larger ratio of adults to peers should facilitate cognitive development (Zajonc, Markus, & Markus, 1979). Perhaps the fact that firstborn children have contact with a predominance of adults over peers (i.e., a higher average intellectual environment in the social network) may be an important factor, in addition to parental education and verbal ability, in enhancing cognitive performance.

Relatives versus Nonrelatives

The firstborns saw more nonrelatives than relatives, whereas the later-borns saw more relatives than nonrelatives, when nuclear family members were included ($F = 3.91, p < .05$). When nuclear family members were excluded, both birth-order groups saw more nonrelatives, although the difference between nonrelatives and relatives was still greater for the firstborns (4.50 vs. 2.99 for firstborns and later-borns, respectively). When nuclear family members were excluded, daily contacts also showed the birth-order by relatedness interaction ($F = 5.30, p < .02$): the firstborns saw relatively more nonkin than kin on a daily basis than did the later-borns.

Males versus Females

There were no differences as a function of birth order for the number of males and females with whom the child had contact.

In summary, the social network was somewhat different for the firstborns than for the later-borns. The firstborns saw more people more frequently than did the later-borns. The firstborns saw relatively more adults than peers and saw relatively more nonkin than kin. Such findings suggest that the structure of the social network may contribute to both cognitive and social interaction differences that have been attributed to parent–child relations (Feiring & Lewis, 1981b; Lewis & Kreitzberg, 1979). Although the parent–child relationship is undoubtedly responsible for some of the reported birth-order differences, the fact that the social network also differs by birth order raises the possibility that the observed differences between groups stem from multiple sources.

The Effect of Working Mothers

Because the fact that its mother works might affect the nature of the child's social network and because the number of working mothers might differ as a function of the variables investigated, we conducted an analysis of the differences in children's social networks as a function of employment outside the home.

The number of mothers working for some time outside the home during the week was 48 (41%), a finding consistent with the number of working mothers found in the population at large (see Feiring & Lewis, this volume). Comparisons of general and specific categories (see Table 2) showed no significant differences as a function of employment. There were more peer friends (6.00 vs. 5.62) and more babysitters (.44 vs. .23) for those whose mothers worked outside the home, although the differences were small and insignificant. These results may best be explained by the fact that most of the employed mothers were working outside the home on a part-time basis. In addition, the use of preschool and peer groups was a fairly frequent phenomenon for the total sample; thus, contact with others outside the home was likely for the children of both employed and nonemployed mothers.

Age Changes in the Social Network

Preliminary data on these children at 6 years of age are available. The data at 6 years were collected in the same way as before. The data presented here give a first outline of how children's networks change as they begin to spend relatively more time outside the home (at school) and relatively less time in the home. In the analysis, the networks of 110 six-year-old children were considered.

Adults versus Peers

From 3 to 6 years, the number of contacts with different adults increased (3 years, $\overline{X} = 14.89$; 6 years $\overline{X} = 16.56$), whereas the number of adults seen daily (3 years, $\overline{X} = 1.88$; 6 years, $\overline{X} = 1.43$) and weekly (3 years, $\overline{X} = 5.61$; 6 years, $\overline{X} = 2.95$) decreased. The number of peers known at 3 years was somewhat greater than at 6 years (3 years, $\overline{X} = 7.25$; 6 years, $\overline{X} = 6.66$). Whereas the 3-year-olds saw peers more often on a weekly basis (3 years, $\overline{X} = 3.21$; 6 years, $\overline{X} = 1.83$), daily contact with peers was greater for the 6-year-olds ($\overline{X} = 2.83$) than for the 3-year-olds ($\overline{X} = 1.59$). Taken together, these data suggest that school brings children more into contact with friends on a daily basis. As children enter school and are away from home more often, they can make contact with friends independently from parents and family members. Also, when the child enters school, the mother may become less informed about the child's contact with others (adults and peers), as she does not have as much opportunity as formerly to see the child outside the home. This may explain the fact that mothers reported fewer peers known at 6 than at 3 years. It appears that the mother reported those friends of her child who played with the child at home (whom she knew and saw more frequently) and omitted the peers and friends whom the child saw primarily at school. At any rate, the data indicate that frequent contact with peers became more prevalent as the child got older. Also, the number of adults contacted on a daily or weekly basis dropped with age. It is possible that whereas the child at 6 may have more contact with teachers at school, contact with the mother's friends drops.

Females and Males

At 6 years, as had been the case at 3 years, there were more females (3 years $\overline{X} = 15.06$; 6 years $\overline{X} = 15.64$) than males (3 years $\overline{X} = 10.92$; 6 years $\overline{X} = 10.68$) in the child's social network. As indicated previously, this ratio may be due, in part, to the mother's report as well as to the role of women as the caretakers of children.

Relatives and Nonrelatives

There were more nonrelatives than relatives in the 6-year-old's network (relatives $\overline{X} = 9.12$; nonrelatives $\overline{X} = 12.81$), than as had been the case for the 3-year-olds (relatives $\overline{X} = 9.34$; nonrelatives $\overline{X} = 13.13$), and this finding held even if the nuclear family was included. Excluding the nuclear family, daily contact with relatives decreased from 3 years (.40) to 6 years (.22), whereas daily contact with nonrelatives increased from 3 years (3.07) to 6 years (4.26).

In summary, these preliminary data on the change in the social network from 3 to 6 years suggest that peers become more a daily part of the child's network and that contact with nonrelatives increases, while the number of females (especially female adults) remains the same.

Social Networks and the Young Child's Experience of the Social World

The data that we have presented demonstrate several points. First, the child has a considerable amount of contact (as judged by number and frequency) with significant others besides the mother, and these influences on the child's development must be taken into account when studying social development. Second, the characteristics of these others, considered here in terms of three dimensions (age, relatedness or familiarity, and gender) are not equally distributed in terms of the number and the frequency of contacts. Thus, for example, regardless of child or family characteristics, females make more contact than males with young children. Third, the characteristics of the child interact in affecting the characteristics of the social network. Social class, the sex of the child, and birth order are only three possible characteristics that have been examined. The data indicate that, viewed separately, each of the three characteristics exerts some influence on the nature of the social network. Our sample was not large enough for us to consider these three characteristics in one analysis because the individual cell size was too small. These child characteristics have been thought to engender the differences in parent–child interaction that have been considered responsible for subsequent development. Although parent–child relations are affected by birth order (Lewis & Kreitzberg, 1979), gender (Goldberg & Lewis, 1969; Feiring & Lewis, 1979), and SES (Feiring & Lewis, 1981a), the fact that the social network is also affected suggests that two sources of influence may be at work in affecting later developmental outcomes. Thus, social network effects as well as parental effects need to be taken into account when predicting subsequent behavior. For example, subsequent peer interactions may be a consequence both of early parent–child interactions and of early peer experience. Without an appreciation of the whole social network, the effects of the influence of others besides the parents may be lost. For example, the finding that there are marked differences in adult and peer contact as a function of SES, birth order, and gender suggests that the opportunity to engage in early peer contact may play an important role in later peer relationships.

Differences in the structure of the social network should be influenced by other factors, such as the developmental status of the child (normal vs. handicapped or sick); the age of the child; and the culture (Lewis & Feiring, 1979). The handicapped child may have a more restricted social network because of the child's limited ability to initiate contact, the lack of opportunity for peer interaction, and parental discomfort and depression. These factors, in turn, may have adverse developmental effects. The child thus both influences and is influenced by its network.

Cultural values clearly affect the structure of the social network. As indicated in the earlier example, some cultures have an elaborate network of kin, whereas others are more nonkin-based (cf. Bott, 1971). Cultures with social institutions such as day care or social patterns such as communal living obviously engender networks different from those that do not. The impact of these structures on individuals and relationships has yet to be explored.

Finally, the nature of the social network is influenced by the age of the child. When children are infants, unable to move about and make contact in their environments, they are dependent on others to initiate contact and to expose them to additional social contacts. As children become older and more mobile, the degree of contact and the ability to make one's own contacts increase. Nevertheless, the degree of contact across age remains a function of other variables, such as culture, SES, and sex.

In the present sample, the social network was populated by more adults than children as a function of the age of the child: 3-year-olds, who must rely on their mothers to make peer contact, see fewer children than 6-year-olds, who go to school and can make peer contact without parental assistance. By 6, children start to see peers on a daily basis and thus have the opportunity to develop friendships more independently from parents.

Kin versus nonkin contact does not change as much in the early stages of development, in part because kin contact is more controlled by parents. The sex of network members indicates that, in the early stages of the life cycle, females are more prevalent than males. As we suggested, this may be influenced by the societal female role as child caretaker. Even so, there is some indication of a sex-of-child by sex-of-peer interaction, demonstrating the effect of gender identity at an early age.

Taken together, these data on the changing social network of the child indicate that any theory of social development needs to consider the child's total social experience rather than simply the child's contact with a single element (i.e., a parent). In fact, a more accurate account of social development requires going beyond an understanding of the influences of the nuclear family and should encompass the study of adults and peers, kin and nonkin alike. Finally, like all systems, the social network is in dynamic relationship to the child: the child's characteristics are a product of and a contributor to the social network.

REFERENCES

Adams, B. N. Isolation, function and beyond: American kinship in the 1960s. *Journal of Marriage and the Family,* 1970, *3R,* 575–597.

Arend, R., Gove, F., & Sroufe, L. Continuity in early adaptation: From attachment theory in infancy to resiliency and curiosity at age five. *Child Development,* 1979, *50,* 950–959.

Attneave, C. L. Social networks as the unit of intervention. In P. J. Guerin, (Ed.), *Family therapy: Theory and practice.* New York: Gardner Press, 1976.

Barnes, J. A. Class and committees in a Norwegian island parish. *Human Relations,* 1954, *7,* 39–58.

Barry, W. A. Marriage research and conflict: An integrative review. *Psychological Bulletin,* 1970, *73*(1), 41–54.

Berardo, F. M. Kinship interaction and communications among space-age migrants. *Journal of Marriage and the Family,* 1967, *29,* 541–554.

Block, J., & Block, J. The role of ego control and ego-resiliency in the organization of behavior. In W. A. Collins (Ed.), *Minnesota symposia on child psychology* (Vol. 13). New York: Erlbaum, 1979.

Bott, E. *Family and social network.* London: Tavistock Institute of Human Relations, 1957.
Bott, E. *Family and social network* (2nd ed.). New York: Free Press, 1971.
Bowlby, J. *Maternal care and mental health.* Geneva: WHO, 1951.
Bowlby, J. *Attachment and loss: Attachment* (Vol. 1). New York: Basic Books, 1969.
Brazelton, T. B., Koslowski, B., & Main, M. The origins of reciprocity: The early mother–infant interaction. In M. Lewis & L. Rosenblum (Eds.), *The effect of the infant on its caregiver: The origins of behavior* (Vol. 1). New York: Wiley, 1974.
Brazelton, T. B., Yogman, M., Als, H., & Tronick, E. The infant as a focus for family reciprocity. In M. Lewis & L. Rosenblum (Eds.), *The child and its family: The genesis of behavior* (Vol. 2). New York: Plenum Press, 1979.
Bronfenbrenner, U. Toward an experimental ecology of human development. *American Psychologist,* 1977, *32,* 513–531.
Cicirelli, V. G. Effects of mother and older sibling on the problem solving behavior of the younger child. *Developmental Psychology,* 1975, *11,* 749–756.
Clarke-Stewart, K. A. And daddy makes three: The father's impact on mother and young child. *Child Development,* 1978, *49*(2), 466–478.
Cochran, M. M., & Brassard, J. A. Child development and personal social networks. *Child Development,* 1979, *50,* 601–616.
Cowen, E., Pedersen, A., Balijian, H., Isso, L. D., & Trost, M. A. Long term follow-up of early detected vulnerable children. *Journal of Consulting and Clinical Psychology,* 1973, *41,* 438–446.
Dunn, J., & Kendrick, C. Interaction between young siblings in the context of family relationships. In M. Lewis & L. Rosenblum (Eds.), *The child and its family: The genesis of behavior* (Vol. 2). New York: Plenum Press, 1979.
Feinman, S., & Lewis, M. *Maternal effects on infants' responses to strangers.* Paper presented at the meeting of the Society for Research in Child Development, Boston, April 1981.
Feiring, C. *The influence of the child and secondary parent on maternal behavior: Toward a social systems view of early infant–mother attachment.* Doctoral dissertation, University of Pittsburgh, 1975.
Feiring, C., & Lewis, M. The child as a member of the family system. *Behavioral Science,* 1978, *23,* 225–233.
Feiring, C., & Lewis, M. Sex and age differences in young children's reaction to frustration: A further look at Goldberg and Lewis (1969) subjects. *Child Development,* 1979, *50,* 848–853.
Feiring, C., & Lewis, M. Middle class differences in the mother–child interaction and the child's cognitive development. In T. Field (Ed.), *Culture and early interactions.* Hillsdale, N.J.: Erlbaum, 1981.(a)
Feiring, C., & Lewis, M. *The social network of three-year-old children.* Paper presented at the Society for Research in Child Development Convention, Boston, April 1981.(b)
Feiring, C., & Taylor, J. The influence of the infant and secondary parent on maternal behavior. In N. D. Colletta & D. Belle (Eds.), *Support systems and family functions.* Sage Publications, in press.
Feiring, C., Lewis, M., & Starr, M. Indirect effects and infants' reactions to strangers. *Developmental Psychology,* in press.
Furman, W., Rake, D. F., & Hartup, W. W. Rehabilitation of socially-withdrawn children through mixed-age and same-age socialization. *Child Development,* 1979, *50,* 915–922.
Goldberg, S., & Lewis, M. Play behavior in the year-old infant: Early sex differences. *Child Development,* 1969, *40,* 21–31.
Greenbaum, C. W., & Landau, R. Mother's speech and the early development of vocal behavior: Findings from a cross-cultural observation study in Israel. In P. H. Leiderman, S. R. Tulkin, & A. Rosenthal (Eds.), *Culture and infancy: Variations in the human experience.* New York: Academic Press, 1977.

Harlow, H. F. Age-mate or peer affectional system. In D. S. Lehrman, R. A. Hende, & E. Shaw (Eds.), *Advances in the study of behavior* (Vol. 2). New York: Academic Press, 1969.

Harlow, H. F., & Harlow, M. D. The affectionate systems. In A. M. Schrier, H. F. Harlow, & F. Stollnitz (Eds.), *Behavior of nonhuman primates* (Vol. 2). New York: Academic Press, 1965.

Heath, D. H. Competent fathers: Their personality and marriage. *Human Development*, 1976, *19*, 26–39.

Holland, P. W., & Leinhardt, S. (Eds.). *Perspectives on social network research*. New York: Academic Press, 1979.

Jacklin, C. N., & Maccoby, E. E. Social behavior at thirty-three months in same-sex and mixed-sex dyads. *Child Development*, 1978, *49*, 557–569.

Klaus, M., & Kennell, J. *Mother–infant bonding*. St. Louis: C. V. Mosby, 1976.

Konner, M. Relations among infants and juveniles in comparative perspective. In M. Lewis & L. Rosenblum (Eds.), *Friendship and peer relations: The origins of behavior* (Vol. 4). New York: Wiley, 1975.

Lamb, M. E. The development of sibling relationships in infancy: A short-term longitudinal study. *Child Development*, 1978, *49*, 1189–1196.

Lee, G. R. Effects of social networks on the family. In W. R. Burr, R. Hill, F. Nye, & I. Neiss (Eds.), *Contemporary theories about the family Vol. 1: Research-based theories*. New York: Free Press, 1979.

Leslie, G. R., & Richardson, A. H. Life cycle, career patterns and the decision to move. *American Sociological Reviews*, 1961, *26*, 894–902.

Lewis, M. State as an infant-environment interaction: An analysis of mother–infant interaction as a function of sex. *Merrill-Palmer Quarterly*, 1972, *18*, 95–121.

Lewis, M. The social network systems: Toward a general theory of social development. In T. Field (Ed.), *Review of human development* (Vol. 1). New York: Wiley Interscience, 1982.

Lewis, M., & Ban, P. *Stability of attachment behavior: A transformational analysis*. Paper presented at a symposium on Attachment: Studies in Stability and Change, at the meeting of the Society for Research in Child Development, Minneapolis, April 1971.

Lewis, M., & Feiring, C. The child's social world. In R. M. Lerner & G. D. Spanier (Eds.), *Child influences on marital and family interaction: A life-span perspective*. New York: Academic Press, 1978.

Lewis, M., & Feiring, C. The child's social network: Social object, social functions and their relationship. In M. Lewis & L. Rosenblum (Eds.), *The child and its family: The genesis of behavior* (Vol. 2). New York: Plenum Press, 1979.

Lewis, M., & Feiring, C. Direct and indirect interactions in social relationships. In L. Lipsitt (Ed.), *Advances in infancy research* (Vol. 1). New York: Ablex, 1981(a).

Lewis, M., & Feiring, C. First-born and only children. Final report to the Population Research Bureau. Contract No. 1-HD 82849, Washington, D.C., 1981(b).

Lewis, M., & Feiring, C. Some American families at dinner. In L. M. Laosa & I. E. Sigel (Eds.), *Families as learning environments for children*. New York: Plenum Press, 1982.

Lewis, M., & Jaskir, J. Infant intelligence and its relationship to birth order and birth spacing. *Infant Behavior and Development*, 1983, 117–120.

Lewis, M., & Kreitzberg, V. The effects of birth order and spacing on mother–infant interactions. *Developmental Psychology*, 1979, *15*(6), 617–625.

Lewis, M., & Rosenblum, L. (Eds.). *The origins of fear: The origins of behavior* (Vol. 2). New York: Wiley, 1974.

Lewis, M., & Rosenblum, L. (Eds.). *Interaction, conversation, and the development of language: The origins of behavior* (Vol. 5). New York: Wiley, 1977.

Lewis, M., & Schaeffer, S. Peer behavior and mother–infant interaction in maltreated children. In M. Lewis & L. Rosenblum (Eds.), *The uncommon child: The genesis of behavior* (Vol. 3). New York: Plenum Press, 1981.

Lewis, M., & Weinraub, M. The father's role in the child's social network. In M. E. Lamb (Ed.), *Role of the father in child development*. New York: Wiley, 1976.

Lewis, M., Young, G., Brooks, J., & Michalson, L. The beginning of friendship. In M. Lewis & L. Rosenblum (Eds.), *Friendship and peer relation: The origins of behavior* (Vol. 4). New York: Wiley, 1975.

Lewis, M., Feiring, C., & Weinraub, M. The father as a member of the child's social network. In M. Lamb (Ed.), *The role of the father in child development* (2nd ed.). New York: Wiley, 1981.

Maccoby, E. E., & Jacklin, C. N. *The psychology of sex differences*. Stanford, Calif.: Stanford University Press, 1974.

Matas, L., Ahrend, R. A., & Sroufe, L. A. Continuity of adaptation in the second year: The relationship between quality of attachment and later competence. *Child Development*, 1978, *49*, 547–556.

Monane, J. H. *A sociology of human systems*. New York: Appleton-Century-Crofts, 1967.

Parke, R. Perspectives in father–infant interaction. In J. Osofsky (Ed.), *Handbook of infancy*. New York: Wiley, 1978.

Parke, R. D., Power, T. G., & Gottman, J. M. Conceptualizing and qualifying influence patterns in the family triad. In M. E. Lamb, S. S. Suomi, & G. R. Stephenson (Eds.), *Social interaction analysis: Methodological issues*. Madison: University of Wisconsin Press, 1979.

Parsons, T., & Bales, R. F. *Family socialization and interaction process*. Glencoe, Ill.: Free Press, 1955.

Pattison, M. A psychosocial kinship model for family therapy. *American Journal of Psychiatry*, 1975, *132*, 1246–1251.

Pedersen, F. A. *Mother, father and infant as an interactive system*. Paper presented at the meetings of the American Psychological Association, Chicago, September 1975.

Pedersen, F. A., Anderson, B. J., & Cain, R. L. *An approach to understanding link-ups between the parent–infant and spouse relationship*. Paper presented at the Society for Research in Child Development meetings, New Orleans, 1977.

Pedersen, F. A., Yarrow, L. J., Anderson, B. J., & Cain, R. L. Conceptualization of father influences in the infancy period. In M. Lewis & L. Rosenblum (Eds.), *The child and its family: The genesis of behavior* (Vol. 2). New York: Plenum Press, 1979.

Pervin. L. A. *Current controversies and issues in personality*. New York: Wiley, 1978.

Reiss, P. J. The extended kinship system: Correlates of attitudes on frequency of interaction. *Marriage and Family Living*, 1962, *24*, 333–339.

Rollins, B. C., & Galligan, R. The developing child and marital satisfaction of parents. In R. Lerner & G. Spanier (Eds.), *Child influences on marital and family interaction*. New York: Academic Press, 1978.

Rosenblatt, P. C. Behavior in public places: Comparisons of couples accompanied and unaccompanied by children. *Journal of Marriage and the Family*, 1974, *36*, 750–755.

Rosenblum, L. A., & Kaufman, I. C. Variations in infant development and response to maternal loss in monkeys. *American Journal of Orthopsychiatry*, 1968, *38*, 418–426.

Rosser, C., & Harris, G., *The family and social change*. London: Routledge and Kegan Paul, 1965.

Rupenthal, G. C., Arling, G. L., Harlow, H. F., Sackett, G. P., & Suomi, S. J. A 10-year perspective of motherless-mother monkey behavior. *Journal of Abnormal Psychology*, 1976, *85*(4), 341–349.

Scanzoni, J. *Sex roles, women's work, and marital conflict*. Lexington, Mass: D. C. Heath, 1979.

Sroufe, L. The coherence of individual development. *American Psychologist,* 1979, *34,* 834–841.
Stern, D. N. The goal and structure of mother–infant play. *Journal of the American Academy of Child Psychiatry,* 1974, *13,* 402–421.
Suomi, S. J., & Harlow, H. F. Early experience and social development in rhesus monkeys. In M. E. Lamb (Ed.), *Social and personality development.* New York: Holt, Rinehart and Winston, 1978.
Terhune, K. W. *A review of the actual and expected consequences of family size.* U.S. Department of Health, Education & Welfare, Public Health Service, National Institute of Health Pub. No. (NIH) 75-779, 1974.
Waters, E., Wippman, A., & Sroufe, L. A. Social competence in preschool children as a function of the security of earlier attachment to the mother. *Child Development,* 1979, *50,* 821–829.
Willmont, P., & Young, M. *Family and class in a London suburb.* London: Routledge and Kegan Paul, 1960.
Wilson, E. O. *Sociobiology.* Cambridge: Belknap Press of Harvard University Press, 1975.
Zajonc, R. B., Markus, H., & Markus, G. The birth order puzzle. *Journal of Personality and Social Psychology,* 1979, *37,* 1325–1341.

8

Grandparents as Support and Socialization Agents

BARBARA R. TINSLEY AND ROSS D. PARKE

INTRODUCTION

What is a grandparent?

> Everybody should try to have one, especially if you don't have television, because grandmas are the only grown-ups who have got time.—Patsy Gray, age 9
>
> —Ruth Goode, *A Book for Grandmothers*
>
> Happy or sad I think that grandmothers are the wisest, most understanding people in the world (excluding, of course, grandfathers).—Jane Moore, age 10
>
> My Grandad is a safety shield against an angry mom.—Rebecca Smith, age 11.
>
> She's the person who tells me all the things about my parents they would rather not have me know.—Sarah Scott, age 15
>
> My grandpa used to rough and tumble with my brother but now both of them are past it.—Richard Thompson, age 12
>
> Grandad grows lovely raspberries and always pretends not to notice us eating them.—Tracey Knight, age 8
>
> She has a part of her own and a future which belongs to everyone. She leads an empty life of her own which is filled by the lives of others. Most of

---

BARBARA R. TINSLEY AND ROSS D. PARKE • Department of Psychology, University of Illinois at Urbana-Champaign, Champaign, Illinois 61820. Preparation of this chapter was supported by NICHD Grant HEW Ph5 05951, NICHD Training Grant HDO 7205-01, and the National Foundation March of Dimes.

all she is a person who will always have time to see you when the rest of
the world is busy.—Gill Webb, age 14

Grandpas are delightful things that date back to the last century.—Simon
Welch, age 10

—Richard and Helen Exley (Eds.), *To Grandma and Grandpa*

In recent years, there has been an expansion of our views of the cast of individuals who influence the socialization of infants and children. Over the past decade, we have watched a shift away from an exclusive focus on the mother–child dyad as the primary unit of analysis. During the 1970s, the father came to be portrayed as an active and competent contributor to infant caregiving and development (Lamb, 1976; Parke & O'Leary, 1976; Pedersen & Robson, 1969), and by the end of the decade, the father–mother–child triad was beginning to receive the attention that it deserved—both theoretically (Lewis, 1979; Lewis & Weinraub, 1976; Parke, Power, & Gottman, 1979) and empirically (Belsky, 1980; Pedersen, 1980). Families, in turn, were increasingly recognized as functioning within a variety of other social systems, including both formal and informal support systems, as well as cultural systems (Bronfenbrenner, 1979; Cochran & Brassard, 1979; Parke & Lewis, 1981). Currently, we are attempting to move beyond the recognition of this embeddedness of families and to begin to delineate (1) the links between these social systems; (2) the functions that different systems play; and (3) the impact of variations in the amount and the quality of contact on family functioning and child development.

Our purpose is to present a closer examination of one aspect of this complex set of relationships——namely, the role played by grandparents—in modifying family functioning and child development. The issue of grandparents can be approached from a number of different perspectives. Our focus is a developmental one, with a major concern being the impact of grandparents on children's developmental progress. In contrast, a gerontological perspective would include a focus on the impact of being a grandparent on the aging individual (Neugarten & Weinstein, 1964), and a sociological perspective would include a concentration on intergenerational relationships in the three-generation family (Troll & Bengston, 1979).

Reasons for Child Developmentalists' Neglect of Grandparents

Grandparents have received scant attention from child development researchers for both conceptual and methodological reasons. However, it is argued here that this neglect is unjustified and demonstrates an overly simplistic characterization of the function of grandparents as a significant influence on child development.

One major reason for our historical neglect of grandparents stems from our characterization of the American family as an isolated nuclear unit. Both contemporary (Bronfenbrenner, 1977) and historical analyses (Laslett, 1976–1977) suggest that, over the last three centuries, in response to urbanization and industrialization, Western families have increasingly consisted of nuclear units of parents and young children. This view has increasingly come under attack on both conceptual and empirical grounds. Earlier theorists often failed to distinguish between structure and function in their discussions of the "isolated nuclear family" (Elder, 1978; Hare-Mustin, 1981). Although *structural* nuclear households have predominated in Western cultures since industrialization, families do not *function* in this manner (Adams, 1970; Sussman, 1959; Uzoka, 1979). Even Parsons (1943, 1954), who originated the concept of "isolated nuclear families," suggested that nuclear families function within complex kin networks (Hare-Mustin, 1981). Moreover, some of the literature on family configuration offers a broader typology of family structures: extended, modified extended, modified nuclear, and nuclear, varying with respect to psychological and economic dependence on kin, decision-making processes, contact frequency, and geographical proximity (Yorburg, 1975). There now appears to be a consensus with regard to characterizations of contemporary Western family structure; although *structural* isolation is common, *functional* isolation of nuclear families from kin is not the rule. Research has shifted from a debate over whether the nuclear family is isolated to a search for the conditions that affect the degree of social connectedness of nuclear families with their extended kin and the consequences of these variations (Lee, 1980). This conceptual shift toward a recognition that the nuclear family is embedded in a network of not only extended kin but nonkin figures, such as friends and neighbors, as well as formal support systems, is evident across a number of disciplines, including anthropology (Whiting & Whiting, 1975); sociology (Troll & Bengston, 1979); psychiatry (Reiss, 1981); and child development (Cochran & Brassard, 1979).

This shift toward a new conceptual view of the relationships between nuclear families and wider social networks also reflects a closer examination of the impact of recent demographic changes on family contact patterns. The potential influence of grandparents has increased over the past century. As Hagestad (1982) noted:

> Increased general life expectancy combined with smaller more closely spaced families (Glick, 1979; Neugarten & Moore, 1968) has produced greater "life overlaps" between parents and children. These factors have also given us

multigenerational families, in which three or four generations of parents and children coexist within a given lineage at the same time (Troll, 1971; Troll, Miller, & Atchley, 1979). (p. 486)

In our culture, women become grandparents at approximately 49–51 years of age and men at 51–53 years old; thus, there is a 20 to 30-year period of grandparenthood. In spite of upward shifts in the age of marriage and first birth, there still remains a lengthy period in which the generations overlap (Troll, 1981). Although there is a general acceptance of the myth that intergenerational kin ties were more prevalent prior to industrialization, Hareven (1977) has reminded us the "the opportunity for a meaningful period of overlap in the lives of grandparents and grandchildren is a twentieth-century phenomenon" (p. 62). In fact, it is only recently that most children have reached adulthood with one or both parents still living (Glick, 1979; Uhlenberg, 1980); thus, more grandchildren have the opportunity to develop a relationship with their grandparents than in 1900 (Nimkoff, 1961).

At present, approximately 70% of older people in the United States have grandchildren (Atchley, 1972; Kahana & Kahana, 1971). Not only are grandparents more available for interaction than in the past, but evidence from contact patterns indicate that there is a significant amount of cross-generational contact. In spite of the geographic mobility of the U.S. population, the proportion of aged parents having contact with an *adult child* at least once a week is extremely high, varying from 77% to 83% over the years 1957–1975 (Shanas, 1979). Although it is not clear how many of these adult children are parents themselves and therefore provide grandchildren for grandparent–grandchild contact, it is likely that these figures reflect a high degree of grandparent–grandchild contact as well. In an informal survey of 300 grandparents and grandchildren, Kornhaber and Woodward (1981) found a full continuum of involvement of grandparents with their grandchildren, ranging from close and regular contact (25%) to minimal or no contact (5%). Most grandparents fell in the middle range, with 70% reporting that they had only intermittent or irregular contact. Even though a significant percentage of grandparents do not have the opportunity for regular face-to-face contact because of such factors as geographic separation, Litwak and Szeleniji (1969) have suggested that kinship systems maintain their viability in the face of American family mobility because of progress in modern communication aids (e.g., telephone, rapid mail delivery, and transportation); the form that grandparent–family relationships assume is at least partially dependent on contiguity and ease of access (Gilford & Black, 1972).

These figures lead to two conclusions. First, there is a high degree of variability in the amount of grandparent–nuclear-family contact across families. Second, it is clear that grandparents have the potential to affect how parents and grandchildren function and develop.

A second conceptual reason for our neglect of grandparental influences is a long-standing theoretical assumption concerning the primacy of the mother–child dyad as the most important relationship in a child's social and cognitive development. This assumption led to an exclusion of other influence agents within the family, such as fathers and siblings (Lewis, 1982), as well as more peripheral agents such as peers and extended-family members, including grandparents. This view has been successfully challenged in recent years for fathers (Lamb, 1981; Parke, 1981a); siblings (Dunn & Kendrick, 1982); and peers (Asher & Gottman, 1981); but it has not been addressed with respect to grandparents. However, a revised emphasis on the embeddedness of the family in a larger social environment has led to a reconsideration of extrafamilial agents, including grandparents.

A further reason for our neglect of grandparents was the lack of a life-span perspective, with its explicit recognition of the plasticity inherent in development, a lack that has led to a static view of adults in general and of grandparents in particular. A life-span perspective alerts us to the changes that may ensue across adult development, which, in turn, may alter the potential quality of the contribution that grandparents may make to the child's development. Consistent with this perspective are recent findings that suggest that different phases of the family life cycle, such as pregnancy, parenthood, and grandparenthood, are associated with variations in adult sex-role attitudes and behavior; specifically, the evidence suggests that the onset of grandparenthood increases interest in and responsivity to infants and children. Evidence from a variety of sources offers support for this hypothesis. In an interview study of over 300 grandparents, Kornhaber and Woodward (1981) found that the birth of a first grandchild elicited nurturant reactions in adults, as indexed by such reactions as their desire to see the grandchild and a heightened positive affect. Although the study is suggestive, the often-reported lack of correspondence between self-report and observed behavior underscores the necessity of supplementing these interview data with direct behavior observations. Feldman and Nash, in an extensive series of studies, have offered a more convincing evaluation of this hypothesis by examining how male and female adults in different stages of the life cycle react to unfamiliar infants (Abrahams, Feldman, & Nash, 1978; Feldman & Nash, 1978, 1979; Feldman, Birigen, & Nash, 1981; Feldman, Nash, & Cutrona, 1977).

These studies indicate that both men's and women's behavioral responses to unfamiliar infants in a waiting room reflected their level of involvement with infants in their current life situation. Whereas mothers of infants were more responsive to unfamiliar infants than pregnant women, cohabiting women, married childless women, mothers of adolescent children, "empty-nest" women, or grandmothers of infants (Abraham et al., 1978; Feldman & Nash, 1978, 1979), grandmothers of infants were more responsive to unfamiliar infants than mothers of adolescent children or "empty-nest" mothers (Feldman & Nash, 1979). Moreover, grandfathers were more responsive overall behaviorally than men at any other stage of life, except fathers of young children; specifically, grandfathers approached and touched unfamiliar infants more often than other men (Abrahams et al., 1979; Feldman & Nash, 1978, 1979). In the Feldman and Nash picture-preference measure, grandfathers chose more pictures of babies as favorites than fathers of adolescents or males with "empty nests" (Feldman & Nash, 1979). A third measure, the Bem Sex Role Inventory (BSRI) was also administered to the subjects in the Feldman and Nash studies: the grandfathers described themselves as having more feminine characteristics than "empty-nest" fathers, and grandmothers showed higher masculinity ratings than "empty-nest" mothers (Feldman & Nash, 1979). Furthermore, cross-sex-typed attitudes were more prevalent for grandparents, with increases in men's expressiveness and women's autonomy. Compassion increased in a near linear manner for men from single adulthood to grandfatherhood. Finally, there were sex differences in tenderness, with women reporting more of this attribute in every stage except married childlessness and grandparenthood. These studies indicate that the advent of grandchildren elicited greater interest in babies in both men and women, although the effect was especially prevalent for grandmothers—an involvement that was "commensurate with her history of well developed nuturing skills" (Feldman & Nash, 1979, p. 431).

In conjunction with the conceptual shifts and a better understanding of the implications of demographic changes, these data on the increases in interest in infants as a result of achieving grandparenthood provide a convincing argument that our prior neglect of grandparents is clearly unjustified. Grandparents have the potential for significantly influencing the developing child.

The extent to which grandparents have been studied as significant influences on family functioning and subsequent child development is also a function of methodological limitations. First, our ability to measure the influence of grandparents is, at least in part, attributable to the difficulties associated with any research involving old people (Riegel, Riegel, & Meyer, 1967), such as elevated morbidity and mortality rates, as compared with younger subjects. Second, only recently have theoretical models and accompanying statistical analytic techniques emerged that permit the meaningful evaluation of units of analysis beyond the dyad. In grandparental research, although the grandparent–adult-child and grandparent–grandchild dyads are of interest, the interaction patterns among the three generations are of interest as well. In turn, these require an analysis of triads and even larger groups of interactive agents (Lewis & Feiring, 1982; Parke *et al.*, 1979). A third reason for the child developmentalist's lack of attention to this issue stems, in part, from the failure to adequately distinguish direct and indirect patterns of influence and, in turn, to develop adequate techniques for distinguishing and measuring these two types of influence on children's development (Belsky, 1981; Lewis & Weinraub, 1976; Parke, 1981a). Finally, intergenerational research must reflect the differential perceptions and motivations of both members of a two-generation dyad by collecting data from each agent (Bengston & Cutler, 1976). A complete picture of the psychological impact of grandparents can be achieved only by the inclusion of information obtained from grandparents, parents, and grandchildren, a pragmatically difficult task.

THE NATURE OF GRANDPARENTAL INFLUENCE

Social influence in families can take both direct and indirect paths (Cochran & Brassard, 1979). By *direct influence*, we mean the process by which one individual influences another by directly acting on the other. In the direct case, the grandparent can directly influence a grandchild through face-to-face interaction. In contrast, *indirect influence* refers to the process whereby one person or agency influences another through the mediation of another person or agency. In this case, family members may be influenced indirectly by altering the behavior of one member, which, in turn, may alter the behavior of another person in the family. For example, by providing a mother with relief from child care, through babysitting, the grandmother may alter the subsequent interaction patterns of the mother and child. Figure 1 both illustrates the role of the grandparents in the child's social network, and, in addition, shows that grandparents can affect children directly by the nature of their face-to-face contact with their grandchildren, as well as indirectly by the frequency and form of support that they provide other members of the child's social network. In addition, the influence process in the grandparent–adult-child and grandparent–grandchild relationships is bidirectional; that is, the influence flows in both directions (Cochran & Brassard, 1979; Lewis & Rosenblum, 1974; Troll & Bengston, 1979). This bidirectionality does not necessarily imply that influence and/or support is balanced. For example, Streib (1958) examined reciprocity patterns in support between older parents and their adult children. He found that

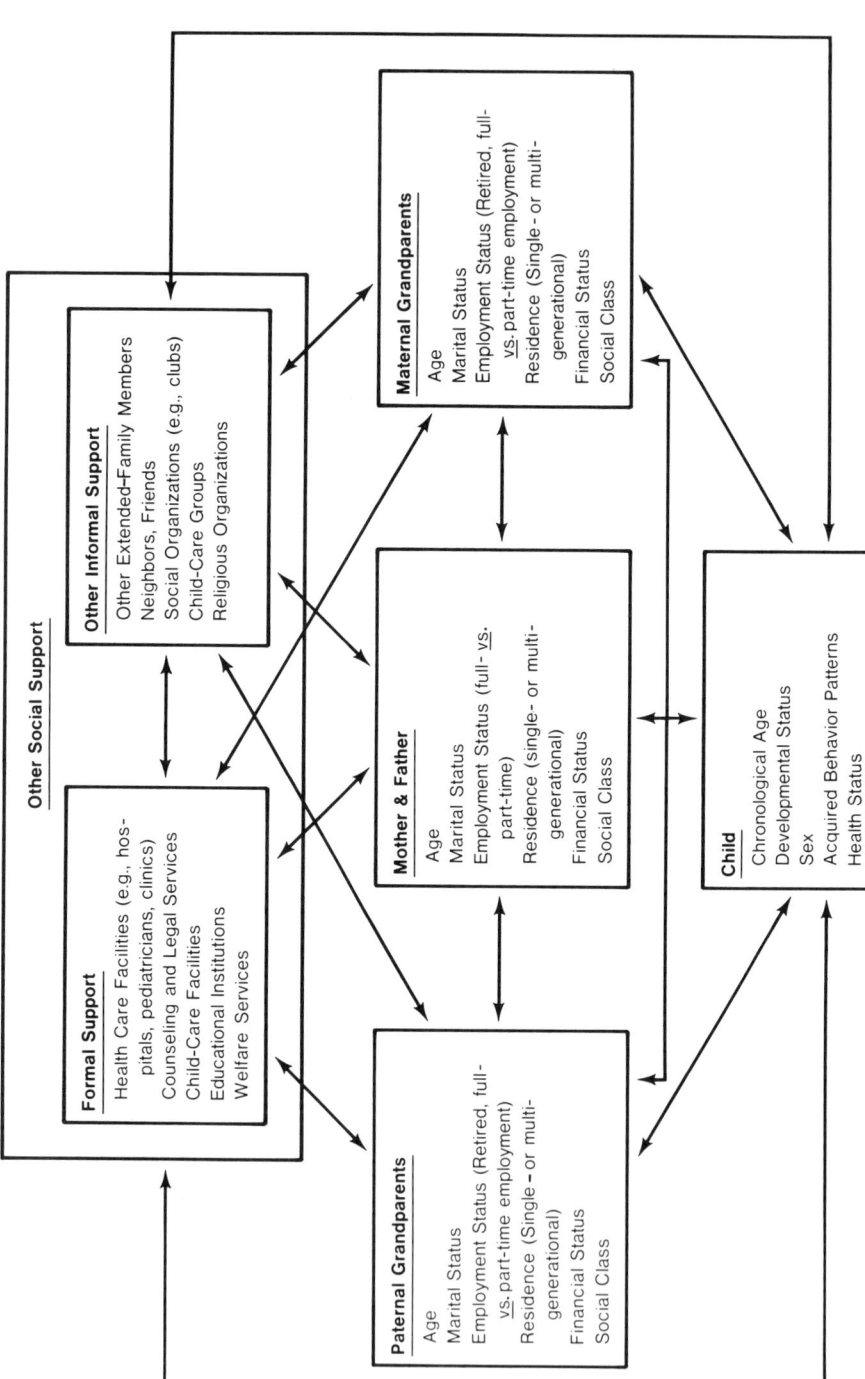

FIG. 1. The role of the grandparents in the child's social network.

aged parents help their adult children to a greater extent than the adult children support their parents. Across the life span, of course, these reciprocity patterns may shift, and older grandparents may assume greater dependency on their children. Similary, the amount and direction of influence between grandchildren and grandparents shifts as the child develops.

Moreover, grandparental influence can be either beneficial or harmful; a high degree of connectedness with extrafamilial networks, including grandparents, is not necessarily always positive either for the grandparents themselves, when they are the recipients of contact from their adult children (Berghorn, Schafer, Steeve, & Wiseman, 1977) or for parents who have contact with extended kin, such as grandparents (Berger & Fowlkes, 1980; Wahler, 1980). One of the major issues in current research on the impact of social support networks, including grandparents, on family adaptation is to identify the conditions under which contact is, in fact, perceived as supportive. A promising candidate as a major intervening variable is the degree of control that the recipient of support maintains over the frequency and timing of the contact.

Finally, grandparents are important sources of influence and support for families on a routine daily basis (Pearlin & Schooler, 1978) as well as during times of crisis and transition (Mogey, 1976–1977). Both the maintaining and the stress-buffering functions of grandparents warrant consideration.

Although it has been suggested that grandparents have, to a large extent, been replaced by formal societal institutions, such as social service agencies (e.g., day-care centers) and the mass media, or by informal social networks, with regard to functions such as caretaking, acting as role models, and giving child-care advice to parents (Hess & Waring, 1978; Kornhaber & Woodward, 1981), we argue that there remain many ways in which grandparents influence both parents and grandchildren. As discussed earlier, there still exists a great deal of intergenerational contact. Moreover, other sources of informal support, such as neighbors and friends, appear to be relied on only when kin, such as grandparents, are not available (Croog, Lipson, & Levine, 1972; Rosow, 1967). Support agents such as friends are supplements, in contrast to replacements, to the support provided by grandparents and other relatives (Lee, 1979).

In the following sections, research concerning the direct and indirect forms of support that grandparents provide their grandchildren is considered. Grandparents provide unique and different qualities of interactive experience for grandchildren, which, in turn, expand the range and variety of sitmulation that infants and children receive. At the same time that grandparents demonstrate some behavior patterns that are different from the parents' interactive behavior, it is also quite likely that there is considerable overlap between grandparent–grandchild interaction and parent–child interaction. As Lamb (1979) suggested, this "redundancy" of behavior of various socialization agents may be functional: "A socialization system that incorporates redundancy has a greater probability of success than does one that relies on a number of socializing agents each solely responsible for providing specific and necessary information or experiences" (p. 94). In sum, both the unique and the overlapping or redundant aspects of the grandparental contribution needs to be considered.

Grandparents as a Direct Social Influence

Grandparents directly influence their grandchildren by serving both as general role models for grandchildren (Hentig, 1946) and more specifically as caregivers and playmates, functions already served by the parents. The direct functions of grandparents vary, however, in important ways across the developmental stages of the child and for grandmothers and grandfathers.

For infants and young children, the grandparent can function as a caregiver or "surrogate parent." The proportion of grandparents who play this role varies markedly across social class, ethnic group, intact and nonintact families, and cohorts. Only a small percentage of grandparents within middle-class intact families play this role; for example, Neugarten and Weinstein (1964) found that only 10% of the grandmothers and none of the grandfathers functioned in this way. In certain historical periods, such as wartime conditions, a larger percentage of grandmothers—regardless of social class—may fill a vacant family role (because of the absence of a father on military duty) (Hentig, 1946). More generally, grandparents are likely to assume this caretaking role in times of stress or crisis, such as the illness or death of a family member (Albrecht, 1954; Hentig, 1946; Streib, 1958). Secular changes in childbearing patterns may alter the extent to which grandparents function in this role. Noteworthy are two trends that have very different effects on grandparental participation: the marked increase in adolescent pregnancies (Gallas, 1980) and the more recent trend toward later childbirth among career-oriented couples (Daniels & Weingarten, 1982).

Early and late first births and the consequent early and late onset of grandparenthood affect the role of grandparents in very different ways. Early timing, such as adolescent childbirth, increases the likelihood that the mother is unmarried; according to a recent estimate, about 35% of all births to teenagers occurred out of wedlock (Alan Guttmacher

Institute, 1976). For a variety of reasons, such as possible lack of financial independence, inexperience in child-care skills, lack of preparation for the parental role, the possible continuation of education, and, in some cases, the lack of a father to assist in child care, the direct role of the maternal grandmother is likely to be increased. Often, in such circumstances, the adolescent mother and the infant join or remain in the maternal grandparental household. Not insignificantly, the grandmother is relatively young (35–45) when she assumes this role and, in some cases, may serve simultaneously as mother to her own children and as surrogate mother to her grandchild. In situations where the adolescent mother is unmarried, the burden and the opportunity for grandparenting are largely on the maternal side; in such cases, the direct role of the paternal grandparents may be limited or even nonexistent, because, at least in part, of the father's continuing active career participation.

In contrast, the late timing of parenthood may yield a very different grandparental role. Both parents and grandparents are older. Late-timed grandchildren are more likely to be planned and prepared for and to be born into an intact, two-parent home and a more stable financial situation, to parents who are more sophisticated in their parenting knowledge and skills (although not necessarily more prepared for the actual tasks of caregiving). In general, their dependence—emotionally, financially, and informationally—on their own parents is likely to be much less than in the case of the adolescent parent. Both maternal *and* paternal grandparents are legitimate and available sources of direct support for the late-timed grandchild; in turn, this availability increases the size of the social network for the child. On the other hand, the age of the grandparent in this case may shift the types of direct interaction. The amount of grandparental participation in direct caretaking may be less in the case of late-timed grandchildren. In fact, the parents may be giving care to their own parents as well as to the new baby; in this case, the direction of support may flow from the parent to the grandparent rather than vice versa (Shanas, 1979).

These extremes of the timing-of-parenthood continuum illustrate the importance of considering the timing of the onset of both parenthood and grandparenthood in studying the possible impact of grandparents on their grandchildren.

However, this daily caregiver role is not a predominant one for most grandparents, partly because of the use of institutions for child care. Since 1970, the percentage of children in day care has risen dramatically and many of the nuturing and caretaking functions of grandparents are being performed by professionals outside the extended-family network (Zigler & Gordon, 1982). In part, this is due not only to the increasing proportion of others who work regularly but also to the increase in the proportion of women in the grandparental generation in the work force.

Although, the nurturing and caregiving function may be on the wane, other functions that are less dependent on schedules, large quantities of time, or geographic proximity still remain. Just as fathers serve as *playmates* for their children (Lamb, 1982; Parke, 1981a), grandparents often do as well. In their classic study of the grandparent role, Neugarten and Weinstein (1964) interviewed 70 sets of non-live-in grandparents to determine their style of enactment of the role of grandparent and found that 20% of the grandmothers and 17% of the grandfathers were characterized as "fun seekers" who played with their grandchildren for enjoyment, without concern about authority or discipline. Because this study classified grandparents in terms of their *predominant* style of interaction, the proportion of grandparents who do play with their grandchildren was probably seriously underestimated. Moreover, there are as yet no systematic data concerning *how* grandparents play with their grandchildren. Just as fathers and mothers differ in their style of play (Clarke-Stewart, 1980; Power & Parke, 1982; Yogman, 1982), it would be surprising if grandparents did not show differences in their play styles. Specifically, one might expect children to be given a more active and directing role in their play with their grandparents because of the assumed greater acceptance and tolerance of a wider range of children's behavior by grandparents. Style may differ as well in that grandparents may show more imaginative and creative play and less didactic and educational play. Work by Tinsley (1983) is addressed to these issues.

Secular changes in the timing of birth may alter play patterns as well as caregiving patterns. Vigorous face-to-face play—either physical or toy-mediated—may be less frequent with older than with younger grandparents.

Based on an interview study of 300 children, Kornhaber and Woodward (1981) suggested a number of other direct roles played by grandparents. In their role of *historians*, "grandparents are living ancestors who can liberate children from the tyrannies of the present" (p. 168). Values, ethnic heritage, and family traditions are often transmitted across generations by grandparents. Moreover, the ways in which children view their parents are often modified by grandparents who tell children of their own parent's childhood; in turn, parent–child relationships may be improved. Finally, children acquire more abstract concepts of kinship and family lineage through the often informal techniques of historical tales of earlier generations. The uniqueness of this role needs to be stressed; although other institutions, such as television, movies, and books, can provide general perspectives on the past, the responsibility of the specific and unique perspectives of the family's own history cannot be abdicated to other cultural institutions.

Grandparents can serve as *mentors* for grandchildren. Unlike mentors such as schoolteachers or older peers, who often serve temporarily in this role, grandparents provide constancy and continuity: "Today's grandchildren can be grandparents for nearly half their lives" (Kornhaber & Woodward, 1981, p. 172). Role models portrayed by grandparents may possible shape children's eventual enactment of their own grandparental role. Although there are other ways in which this role can be learned, such as their parent's attitude toward their own parents (secondhand grandparents) or stereotypical portrayals created by the mass media, these will probably yield different guidelines for the execution of the grandparental role. This function is particularly important because media portrayals of grandparents, as well as aging people in general, tend to be distorted and misleading (Galper, Seefeldt, Jantz, & Serock, 1980). Grandparents also serve as role models for aging, and close ties with a grandparent may aid in rectifying inappropriate stereotypes of the aged. In all of these cases, however, the effects can be either positive or negative, depending on how successfully the current grandparent executes his or her role and on how well he or she has adapted to the aging process.

Finally, grandparents can act as direct *negotiators* or *buffers* between parent and child, which in turn could improve the parent–child relationship. Again, this function is being increasingly assumed by outside agencies, such as therapists and other professional organizations. Although buffering can take place even when grandparents live at a distance, this function is generally more likely in families who live in close geographic proximity. In summary, grandparents serve a variety of direct support functions for nuclear family members.

The Grandparent as an Indirect Influence

In contrast to parents, for whom direct face-to-face contact is the predominant vehicle of social influence, a significant portion of grandparental influence is often indirect. As noted above, indirect influence is the process whereby the grandparent may influence a grandchild through the mediation of another person in the family social network. A variety of indirect paths and types of influence can be distinguished, including the provision of emotional, physical, financial, and informational support. A number of reviewers have summarized considerable evidence of indirect effects in general (Lewis & Feiring, 1981; Lewis &

Weinraub, 1976; Pedersen 1982), although the specific indirect effects of grandparents have not received much attention.

The most common path through which grandparents indirectly affect their grandchildren is, of course, through their relationships with their adult children, the grandchildren's parents; according to one survey (Kornhaber & Woodward, 1981), grandparents themselves may not be aware of the importance of this type of influence. Forty-five percent of the grandparents reported that they did not feel that their relationship with their children affected their relationships with their grandchildren. Either the impact can be mediated through individuals, such as the mother or the father, who, in turn, may alter her or his behavior or attitudes toward the children, or grandparents can indirectly affect the child's development by modifying the quality of the marital relationship. These indirect grandparental influences are important in light of a number of recent studies that have demonstrated the impact of the husband–wife relationship on parent–infant interaction and hence on subsequent infant development (see Belsky, 1979, 1981; Goldberg, 1982; Pedersen, 1980). Parents are not the only indirect path of grandparental influence; although less commonly, grandparents could alter grandchildren's behavior through their impact on the other members of the child's social network, such as peers and other relatives. Kornhaber and Woodward (1981) reported that 65% of the grandparents in their study "knew their grandchildren's friends." Their reactions to their grandchildren's choice of friends, for example, could, in turn, modify the nature of these grandchildren's peer-group relations. Similarly, the quality of the relationship between the maternal and the paternal grandparents may affect the grandchildren. Mutual acceptance and shared activites on the part of these two lines of grandparents probably enhance the grandchild's relationships with both sets of grandparents.

A common form of support is emotional support, which may involve allaying the fears and anxiety associated with becoming a parent or serving as a counselor during times of emotional crisis, such as marital conflict or a parent–child disagreement. Grandmothers, in contrast to grandfathers, are more likely to provide emotional support for parents. Also, grandparents, especially grandmothers, indirectly influence grandchildren through advice, information, and modeling of child-rearing skills. Grandmothers serve as models for novice mothers and, depending on their own level of skill, can either aid or hinder the new parent. For example, inappropriate child-rearing tactics, such as abusive practices, may, in some cases, be acquired through the observation of grandparental models (Parke & Collmer, 1975). Grandparents not only transmit child-rearing methods but serve to monitor current child-care practices and provide corrective feedback (Parke & Lewis, 1981; Parke & Tinsley, 1982).

Finally, grandparents, especially grandfathers, indirectly influence their grandchildren through their provision of financial support to their own adult children (Albrecht, 1954). By relieving families of financial burdens, or by permitting changes in housing or recreational opportunities for children, grandparents can significantly alter parent–child relationships. However, there are few data concerning the impact of the intergenerational flow of financial resources on grandchildren.

As noted earlier, grandparents provide these types of indirect support both on an ongoing basis and during times of crisis and transition. Cross-cultural studies have provided impressive evidence of the impact of grandparents on the routine child-rearing practices of mothers. In their six-culture study of mothers in India, Okinawa, the Phillipines, Mexico, East Africa, and New England, Minturn and Lambert (1964) found that the presence of grandparents in a household, particularly a grandmother, was associated with a mother's higher emotional stability and her increased expression of warmth to her children than when the mother is the sole custodian. Moreover, the effect is unique to grandparents; the presence of other female adults, such as sisters-in-law, especially if they have children, is associated with decreased warmth. Based on an analysis of ethnographies of 101 societies, Rohner (1975) found a similar relationship between parental behavior and household composition. Mothers who had a continuous and unrelieved responsibility for child care were more likely to reject their children than mothers who had someone else in the household (grandparent and/or father) to share the burden of child care. Nor is the indirect support restricted to other cultures. In a study of middle-class American women, Abernathy (1973) examined the relationship between social network support and "sense of maternal competence"—a measure of how well mothers perceived their skills as parents. According to Abernathy (1973), *a sense of maternal competence* reflects a recognition of the malleability of children, an appreciation of individual differences, and a knowledge of how child-rearing tactics match children's maturational levels. Social network was assessed by asking each mother to designate her four most frequently seen friends and relatives. Of particular interest is the finding that the components of social networks vary in their contribution to the overall score on "sense of maternal competence," with the frequency of contact with her own mother (i.e., the grandmother) being the strongest single predictor of a woman's sense of competence. The mechanisms by which contact enhanced a woman's sense of competence as a mother included emotional support and physical sharing of child-care responsibilities.

In summary, both the identity of the individual support figure (e.g., sibling or grandparent) and the organization of the network are important predictors of maternal sense of competence. Whether or not directly assessed maternal competence shows a similar relationship to social network variables as the self-report indexes of competence needs to be addressed.

Grandparents can also be supportive during life transitions. For example, Bittman and Zalk (1978) reported increased contact during the period of pregnancy between women and their parents and men and their parents. In light of data that suggest that social support, including that provided by the expectant grandparents, is positively related for both the father and the mother to self-perceived well-being, to the father's perception of his own competence, and to marital satisfaction (Wandersman, Wandersman, & Kahn, 1980), grandparents have significant potential for positively supporting their adult children's transition to parenthood (Belsky & Tolan, 1981). This is particularly true in the case of the unwed teenaged expectant mother (Smith, 1975). Studies by Field, Badger, and others suggest that nonemployed grandmothers are often the primary caretakers of infants of single, teenaged mothers (Badger, 1981; Field, Widmayer, Stringer, & Ignatoff, 1980; Hardy, King, Shipp, & Welcher, 1981; Mills & Cairns, 1981). Temporary variations in family structure (e.g., divorce, widowhood, or the extended absence of a nuclear family member) may elicit kin aid. For example, Gibson (1972) found that single, divorced, and widowed people were more involved with extended-kin networks than married individuals. In their study of divorce, Hetherington, Cox, and Cox (1978) found that the adequacy of the mother's social support network was positively related to her effectiveness with her children. Although the father was the most significant support figure, other members of the mother's social network, including the children's grandparents, were also related to the quality of the mother–child relationship.

Finally, Furstenberg (1982) has recently begun to explore how the role of extended kin, including grandparents, shifts after remarriage. The ways in which grandparents from earlier and later marriages organize themselves into an enlarged kin network and, in turn, the impact of this larger network on grandchildren are only poorly articulated. The study of stepgrandparenting is a fertile area for future study in light of the current and projected high rates of divorce, changing custodial arrangements, and remarriage (Cherlin, 1981).

Heretofore, little attention has been paid to what are the specific types of support that grandparents provide for families with young children or to how grandparental support relates to the other types of support available.

Factors That Influence the Social Influence Function of Grandparents

A number of variables modify the extent of grandparental direct and indirect social influence on families and children. First, parents often serve as important mediators; the grandparental role is at least partially dependent on the extent to which the parents make the grandchildren accessible to the grandparents. Parents, in effect, serve as "gatekeepers" between their own children and the grandparents. Robertson (1975) offered a useful set of dimensions along which this gatekeeping function can be organized. The parents' attitude concerning the significance of grandparenting for themselves and their child determines, in large part, a variety of other aspects of the grandparent–grandchild relationship, such as the form of contact, the location of interaction, the types of activities, and the frequency of contact, some of which have been discussed earlier in this chapter. In turn, these factors shape, in significant ways, the nature of the grandparent–grandchild relationship. A variety of other aspects of the mediator role merit mention (Robertson, 1975). How often do the parents mediate and in what situations? Who serves as the initiator of interaction—the parent, the grandchild, or the grandparent? The parents' assessment of the importance of their mediator role may, in turn, affect how effectively and how often they mediate between the generations. Another issue concerns the focus of mediation, namely, who are the parents most concerned about in mediating between the generations—the grandparent or the grandchild? Finally, the extent to which the parents maintain equity between the maternal and the paternal sets of grandparents clearly affects the child's relationships with its grandparents. Through their mediational or gatekeeper role, parents define both grandchild and grandparent roles by setting the limits and by providing the structure for the grandparent–grandchild relationship.

Second, as discussed previously, grandparental support is somewhat dependent on the residential proximity of grandparents and young nuclear famiies. In spite of trends toward increasing mobility in American families, significant intergenerational family relationships exist and flourish (Shanas, 1979; Troll, 1979). Indeed, nuclear families have demonstrated a preference for living near extended-family members, such as grandparents, especially in the working class (Lee, 1979). Moreover, in a recent survey, Kornhaber and Woodward (1981) found that grandparents also prefer to be near their children and grandchildren; in fact, 70% of the grandparents who moved away from their grandchildren indicated that they would not do so again if they had a choice, as they felt out of touch with the emotional life of their families. Even when grandparents live a significant distance from the nuclear family, evidence suggests that they are still an important source of aid and support. As proposed earlier, the actual types of grandparental support provided nuclear families are somewhat constrained by the distance between the residences (Adams, 1964; Lee, 1979; Litwak, 1960). However, the amounts of such aid may be less limited by residential proximity (Troll & Bengston, 1979).

Social class and residential propinquity often interact in qualifying support patterns. Living close to extended family members such as grandparents is more common in the lower than in the middle class (Adams, 1968; Berardo, 1967; Blumberg & Bell, 1959; Brown, Schwarzweller, & Magalam, 1963; Lee, 1979; Townsend, 1957, Troll, 1971). Social class *per se* is also a significant modifier of extended-family support such as is provided by grandparents (Gordon & Knoll, 1975; Winch, Greer, & Blumberg, 1967). Adams (1964) found that middle-class family members are more likely to exchange financial help than "blue -collar" families, who more frequently exchange services. However, some recent research (see Lee, 1980, for a review) suggests that there may be less of a relationship between social class and kin relationships than is generaly cited, and that this relationship may be confounded with urban versus rural residency.

Third, the match between the ages of grandparent and grandchild is important to consider. As Troll (981) noted, in light of the fact that "the ages of grandparents may be anywhere between 40 and 120 years, and of grandchildren anywhere up to 80 years old" (p. 476), the effect of age on grandparent–grandchild relationships cannot be ignored. As discussed elsewhere in this chapter, older grandparents may be subject to more disease and disability, given that age is positively related to declining health; thus, they may be less likely to be energetic with their grandchildren. Moreover, the same grandparent may behave very differently with individual grandchildren, depending on the age and the health status of the grandparent with respect to the grandchild's age (Hagestad & Speicher, 1981). Age differences in grandparenting styles were found by Neugarten and Weinstein (1964): older grandparents were more distant and formal than younger grandparents, with their fun-seeking orientation, and they showed greater diversity in grandparenting behavior styles. Moreover, younger grandchildren were judged more appealing to grandparents in studies by Kahana and Kahana (1970) and Clark (1969).

Nor are grandparent–grandchild relationships determined solely by the grandparents themselves; children of different ages may affect the frequency and the quality of contact between themselves and grandparents. Just as it is well recognized that bidirectionality of influence is common between parents and children, it is assumed that mutual influence characterizes the grandparent–grandchild relationship as well. Two types of studies are relevant to the issue of the grandchild's influence on the grandparent–grandchild relationship: (1) research that addresses child perceptions of old people and their roles and (2) data on grandchildren's perspectives on grandparenting.

With respect to the first issue, children's perceptions of old people and their roles, Hickey, Hickey, and Kalish (1968) analyzed essays written by a large sample of third-grade children about old people, operationally defined as "like your grandparents." Although qualified by socioeconomic class, in general these children perceived the elderly as being physically feeble, but kind and friendly. Related to these results are data from another project suggesting that older adults (i.e., of the grandparents' generation) are preferred by young children for some functions more than other members of the child's social network, such as peers. In a study assessing the social preferences of 3 to 5-year-olds, using photographs of different agents (e.g., peer, older child, or adult of grandparents' generation), Edwards and Lewis (1979) found that male adults of the grandparents' age (50–70 years) were chosen more often for dependency functions, such as helping when hurt or providing directions when lost, than same-age or older children. In terms of play partners, peers were preferred over grandparent-age adults of either sex. The preference patterns were similar for parents in comparison with peers. Whether parents and grandparents are differentially preferred, however, was not directly assessed and awaits further research. The study does represent a step toward the articulation of children's perceptions of the relationships between social functions such as helping, playing, and the choice of social agent. Assessments of the origins of these perceptions or stereotypes of the elderly and their roles is also of interest. For example, how do these views vary with children's age, exposure to the elderly, having grandparents, frequency of grandparental contact, and media portrayals of grandparents and the elderly?

Some evidence is available addressing the issue of the impact of having a grandparent on children's perceptions of grandparents. Kornhaber and Woodward (1981) found that the degree of importance attached to grandparents and childrens' differentiation of grandparental roles varied with the existence and the closeness of a grandparent–grandchild relationship. Grandchildren with a close relationship with one or more grandparents viewed grandparents as important persons in their social networks and were able to provide highly differentiated functions for grandparents, such as mediators, caregivers, mentors, and family historians. Grandchildren, especially very young children, whose grandparents were relatively unavailable to them because of such factors as death, geographical separation, and divorce viewed grandparents as less valuable, felt emotionally distant from their grandparents, and were less able to provide grandparental role definitions. Developmental differences in grandchildren's perceptions of their grandparents were also found by Kahana and Kahana (1970). In an interview study of 85 children, it was found that grandparents are appreciated for their indulgent qualities by very young children (i.e., 4–5 years of age); 8- and 9-year-old children preferred the active fun-loving grandparent; and the oldest children (11–12 years old) felt that more distance between the generations was preferable. Whether or not these perceptions were associated with different behavior patterns on the part of grandparents interacting with children of different ages remains unclear. Finally, the contribution of individual differences in children's cognitive competence to these changing perceptions of grandparents merits investigation (Jordan, 1980).

Fourth, the sex of the grandparent, the parent, and the grandchild needs to be considered. A very stable finding in extended-family research is that daughters function as "kin keepers," maintaining lines of communication and contact among kin such as grandparents, parents, and grandchildren (Cohler & Grunebaum, 1981; Havighurst, 1973; Hill, 1970; Lee, 1980). In fact, sex differences in such extended-family contact as that between grandparents and nuclear families supersede even social class differences (Havighurst, 1973). One study (Troll & Turner, 1976) suggests that women take over the kin-keeping function even before they are kin; undergraduate women were more likely to be writing to their boyfriends' parents than were the men themselves—even though they were not yet married! An interesting related finding by Jordan (1980), in an investigation of children's social cognitions with respect to kin concepts, is that girls are able to understand kin concepts earlier than boys. In all social classes, the maternal grandparents give more help than the paternal grandparents (Adams, 1964). Overall, kinship ties are stronger for females (Troll & Bengston, 1979). Another related issue is the match between the sex of the grandparent and the sex of the grandchild. Grandmothers are more likely than grandfathers to have warm relationships with their grandchildren and appear to differentially attempt to influence their grandchildren's behavior more than grandfathers on the basis of the sex of the grandchild. Grandchildren do not appear to discriminate in similar ways between grandmothers and grandfathers (Troll, 1918).

Other mediators of grandparental influence are frequency of contact and satisfaction with that contact. Contact is not a linear variable: more grandparent–nuclear-family contact is not necessarily better than less. In fact, contact with support agents such as grandparents seems to be conceptualized best as a curvilinear variable (French, Rodgers,& Cobb, 1974), with moderate levels of contact being more helpful than the more extreme high and low levels. Hess and Waring (1978) termed this a "Goldilocks" effect, "whereby both too many and too few family contacts are perceived as stressful so that there is a level of 'just right'—enough to satisfy needs without evoking resentment" (p. 259). Kahn and Antonucci, 1980) suggested that, just as support is often appropriate and desired, it is sometimes the case that support such as that provided by grandparents may be inappropriate and unwanted. They presented a concept developed by French et al. (1974) termed *goodness of fit*, representing the match between preferred support and offered support. The extreme example of high frequency of contact is intergenerational residence-sharing. Several investigators have found this to be a source of conflict (Cohler & Grunebaum, 1981; Cohler, Grunebaum, Weiss, & Moran, 1971; Komarovsky, 1967; Lee, 1979; Staples & Smith, 1954).

In summary, a variety of variables—including the sex of the grandparents, the relative age of the grandparents and the grandchildren, the nature of the grandparent–parent relationship, residence patterns, geographic separation, and contact patterns—affect the amount and the type of grandparental influence.

THE IMPACT OF GRANDPARENTS ON CHILDREN'S DEVELOPMENT

Network members, such as grandparents, can directly affect children by providing experiences and opportunities different from those provided by the parents, functioning as unique role models and cognitive and social stimulators or, indirectly, as providers of support for the parents. Three kinds of evidence are available: studies of families in which the grandparent is a co-parent or is part of a live-in extended family; studies of the impact of extrafamilial social networks, including grandparents, on mothers and children; and experimental investigations involving grandparents as stimulatory agents for infants and children or support figures for parents.

There is substantial evidence suggesting that a live-in grandparent, or a grandparent functioning as a co-parent, can have a substantial direct and indirect impact on children. In a cross-cultural study by Rohner (1975), parental acceptance of children was found to be higher in a household in which mothers received child-care aid from another adult (father or grandparent), and, in turn, the accepted children tended to be more self-reliant than rejected children and to draw on their own resources to meet their physical and emotional needs. Similar results were found by Werner and Smith (1982) in a study of the factors that promote resiliency in biologically at-risk infants and children. Following a group of children from birth to 18 years, these investigators found that the girls who successfully adapted came from households where there were other adults, including grandparents, available to help with the children. These girls, in contrast to their peers, coped better and scored higher on standard cognitive tests at 10 and 18 years of age.

Finally, a study by Kellam, Ensminger, and Turner (1977) of the effect of variations in family structure on child development provides further evidence that grandparents can play an important role in children's development. First-grade children from lower-income families headed by a mother and a grandmother scored just as well as children from families headed by a mother and a father and much better than children in families head by only a mother on measures of psychological well-being (i.e., clinical ratings on a variety of scales such as anxiety, depression, bizarre behavior, emotional flatness, and hyperkinesis) and social adaptability. Social adaptability was defined by teacher ratings evaluating the children's adaptive skills, such as peer relations, maturity, ability to respond to rules and authority, and nonadaptive behaviors, such as learning problems, with or without social adjustment problems (e.g., aggressiveness or shyness). These data suggest that one important role that grandparents can effectively play is that of a co-parent, substituting, in this case, for an absent father.

Indirectly, grandparents who are not living in the same residence have an effect on their grandchildren's development by providing support for the parents. A number of studies have suggested relationships between social support and the adequacy of family functioning (Cochran & Brassard, 1979). In studies of abusive families, it has been found that these families are often socially isolated and lack adequate social support systems (Belsky, 1980; Garbarino & Gilliam, 1980; Parke, 1981b; Parke & Collmer, 1975).

Unger (1979) examined the influence of social networks on the mother–infant interaction of white low-income mothers and their young infants (under 6 months old). Using the HOME Scale (Elardo, Bradley, & Caldwell, 1975) to assess interaction, Unger found that mothers who experienced high levels of stress were more actively involved with their infants when they had weekly contact with friends and relatives than when they were infrequently in contact. Parents also appeared more responsive to their infants if they were receiving material resources from their network members. In a related study that also employed the HOME Scale, Pascoe, Loda, Jeffries, and Earp (1981) found a positive relationship between the level of maternal social support and selected subscales for the HOME Scale in a group of families with 3-year-old children. Unfortunately, neither Unger or Pascoe *et al.* examined both intrafamilial (i.e., spouse support) and extrafamilial social support separately; therefore, an evaluation of the relative importance of differing sources of support, such as grandparents, is not possible. However, these studies are important in view of the previously established positive relationship between the HOME Scale and mental test performance (Bradley & Caldwell, 1976). These investigations suggest that extrafamilial social support, including grandparental support, may affect infant and child developmental outcomes indirectly, through modifying the nature of parent–child interaction patterns.

Other evidence linking children's social development and perhaps their later cognitive development to the availability of social support to their parents comes from Crockenberg (1981), who examined the relationship between the perceived adequacy of the social support available to the mother at 3 postpartum months and infant attachment at 1 year. She found clear relationships between the adequacy of the mother's social support and the security of the infant–mother attachment; low social support was associated with high resistance and high avoidance on the part of the infants in the Ainsworth Strange Situation and with anxious rather than secure attachment. Of particular interest was the qualifier that support had its strongest effect on irritable babies—a finding that suggests that "the availability of social support is particularly critical when the family is under particular stress" (Crockenberg, 1981, p. 862). In view of the importance of the quality of attachment in later social and cognitive functioning, this finding is of special significance: in comparison to securely attached infants, insecurely attached babies are less competent in later peer–peer interactions (Lieberman, 1977; Waters, Wippman, & Sroufe, 1979); are less able to establish friendly relationships with strange adults (Main & Weston, 1981); engage in less complex exploratory behavior (Main, 1973); and show less involvement, persistence, and enjoyment in problem-solving situations (Matas, Arend, & Sroufe, 1978). Unfortunately, Crockenberg did not examine the independent contribution of different sources of support (fathers, older children in the family, and others; extended family, neighbors, friends, professionals) on infant–mother attachment. Grandparents are clearly a potentially important source of support for mothers, but their relative importance in maternal–infant attachment remains to be assessed.

However, these studies of the naturally occurring use of extended-family networks are only suggestive and leave unanswered a central question. Because all of these studies were nonexperimental, the causal direction of influence is not clear. Perhaps, grandparents participate more in well-adjusted families, who, in turn, have well-developing children.

Fortunately, some experimental evidence in support of the potential role of grandparents in promoting children's development is available from Saltz's study of "foster" grandparents. Saltz (1973) examined the impact of aged adults in the role of foster grandparents on the progress of normal institutionalized children. The children, some of whom were infants, were cared for by male and female elderly aides in a "quasi-family" relationship characterized by emotional attachments. The foster grandparents rocked, fed, talked to, and played with the infants for four hours a day, five days a week, for up to three years. The social and cognitive development of children with "foster grandparents" was significantly higher than that of a control group of children who did not receive this personalized care. Because the effects of this caregiving cannot be directly attributed to the age of the caregivers, but must also be explained by the extra amount of affect, stimulation, and nurturance that the children received, the quality of the caregiving provided by agents of differing ages warrants systematic investigation. Other research by Saltz (1971) underscores the impact of performing a grandparental role on the grandparents. Saltz found that the program benefited not only the infants and the children; the opportunity to perform this caregiving function enhanced the self-esteem and the feelings of self-worth of the foster grandparents as well.

Grandparents and other members of informal social networks often work in concert with formal support systems in altering family's and children's development. Links between formal and informal support systems can assume a variety of forms, such as (1) strengthening the informal network through formal intervention (Powell, 1979); (2) mobilizig existing social networks in times of stress (Rueveni, 1979); and (3) using informal network members to help individuals utilize formal support services (Olds, 1981). Thus, grandparents and other members of the informal social network can function in conjunction with formal support systems in determining the quality of family functioning.

The Prenatal/Early Infancy Project developed by Olds (1981) provides an illustration of how formal and informal support can perform together. This investigation has developed a model in which the formal system identifies and trains members of the parent's informal social network to assist "at-risk" families. Specifically, a nurse/home visitor provides home-based education for improving pregnancy management and early infant development; involves "significant others" (often the grandparents), who participate in the home visits with the mothers in order to create a supportive informal environment for behavioral change on the part of the parents; and finally, links families with other health and support services in the community. Preliminary results from the pregnancy phase indicate that the mothers who had the nurse/visitor were aware of more formal community services, and a larger number (65%) were enrolled in childbirth education classes than mothers who were not visited by a nurse (50%). In the area of informal support, the mothers in the nurse/visitor group were accompanied by a support person to childbirth education more often than non-nurse-visited mothers. Similarly, nurse-visited mothers experienced support from a significant other (a friend or relative) more frequently during labor. This finding is of particular significance in light of the recent findings of Klaus and Kennell (1981) and their colleagues that the presence of a supportive other is associated with a shorter labor and fewer complications in childbirth.

Another similar use of the combination of formal support and extended kin, such as grandparents, was reported by Berger and Fowlkes (1980). They developed an intervention program for handicapped children utilizing extended-family networks. Program staff trained extended-family members in educational programming and helped them to deal effectively with community service systems. The investigators reported preliminary data suggesting that the program had positive developmental outcomes for the children involved. Although the Olds and the Berger and Fowlkes investigations illustrate the role that grandparents and other members of the extended-kin network can play, the relative importance of different members of the network was not established.

These studies illustrate the role that extended-kin and more formal social networks together can play in aiding families. However, the details of the puzzle are only partially clear, and the specific direct and indirect ways in which grandparents and other types of supporters alter different aspects of family functioning—and, in turn, children's development—are not yet well understood.

Conclusions and Future Trends

The social support of extended-family members such as grandparents can have a significant impact on the family and on children's social and cognitive development. However, a number of issues remain unresolved.

First, the structure of contemporary American families is diverse and is at least partially contingent on changing birth and divorce rates, new custodial arrangements, and the timing of transitional life events (Cherlin, 1981). Future research would profit by a recognition of the distinctions among individual time, family time, and historical time, and their impact on the functions served by grandparents (Elder, 1976; Elder & Rockwell, 1979; Hareven, 1977). *Individual time* is each family member's life course. *Family time* is defined as the timing of transitional life events for the family as a unit. *Historical time* provides the social conditions for individual and family transitions. It is suggested that individual time and family time do not always harmonize (Elder & Rockwell, 1979; Hareven, 1977). For example, the age at which a woman becomes a grandparent is determined by her child's procreative behavior, and she has relatively little control over the timing of this transition. Moreover, individual and family time are both embedded within the social conditions and values of the historical time in which they exist (Hareven, 1977). Thus, being a grandparent at the age of 40 may be more appropriate within some historical and societal conditions than within others. And contemporary shifts in divorce and remarriage rates will, by definition, yield a variety of new forms of grandparenting: the stepgrandparent and the multiple-grandparent family (Elder, 1976; Furstenberg, 1981; Troll, 1981). Researchers (Furstenberg, 1982) are only beginning to explore the roles and the relationships that families develop with these new forms of extended kin. These shifts underscore the necessity of closely monitoring these secular trends in order to assess the extent to which our generalizations are timebound or are more universal in nature. Finally, individuals function within multiple, overlapping roles simultaneously. Within a kin system, a man can be a father, a son, a brother, an uncle, and a grandfather at the same time. This multiplicity of roles demands that the man must constantly decide how to apportion his finite resources, a decision that, at times, causes conflict. Moreover, he must make these decisions within the context of cultural values (Hareven, 1977). In sum, consideration of these multiple meanings of time will yield important insights into the grandparenting process.

Second, greater attention needs to be addressed to the relative contribution of grandparents in comparison with other members of the extended-family network, in order to determine the unique as well as the similar roles played by different parts of the family's social network.

Third, the role of grandparents probably shifts in both quality and quantity across the developmental level of the grandchildren and across succeeding grandchildren. A clearly articulated commitment is needed to an analysis of the how grandparents contribute to specific aspects of children's social and cognitive status at different points in their development.

Closely related is the necessity of a life-span perspective on grandparenting. Not only is the age of the grandparent an important factor, but the lifestyle of the grandparent—in terms of work career, for example—needs to be considered as well. Although age and career position are often loosely correlated, these factors need to be considered independently. A grandfather who retires at 55 may have a very different relationship with his grandchildren from a man who stays on the job until he is 70 (Troll, 1981).

Another issue concerns the necessity of exploring the child's perspective on grandparents. Although some researchers (Hickey et al., 1968; Kahana & Kahana, 1971) have explored this issue, further research is needed to evaluate how early the grandparents are discriminated from other social agents and how children view the relative importance of grandparents in the children's own, in contrast to the parents', social network. Closely related to these issues is children's understanding of kinship structure. More work is necessary on children's understanding not only of nuclear family structure (Jordan, 1980) but of the structure of cross-generational relationships. Finally, the links between the cognitive mastery of cross-generational kinship structure, kinship roles, and grandparent–child relationships warrant examination.

Further research on the impact of serving as a social support figure on the grandparent's own development would be worthwhile, not only for achieving a better understanding of the development of grandparents as individuals but also to provide insight into the processes that determine the dynamics of how and when grandparents choose to provide support for families and children. Short-term longitudinal studies of grandparents' transition to grandparenthood in parallel with their own adult children's transition to parenthood would be a valuable contribution. In addition, longitudinal studies tracing the developmental course of grandparenthood would provide a better predictive base for understanding grandparents' contribution to the children's own development.

It is likely that no single methodological strategy will be sufficient to clarify the role of grandparents in children's development. Instead a wide range of designs, data collection, and analysis strategies are necessary (Parke, 1979). As in other recent studies of families, a mutlimethod, multimeasure approach would be optimal (e.g., Hetherington et al., 1978).

One serious omission in the grandparent literature is the lack of direct observational investigations of how grandparent–parent and grandparent–grandchild relations change as a function of settings and tasks. After two decades of "speculative excursions, didactic exercises or sweeping conclusions derived from clinical practice" (Troll, 1979, p. 1), careful observations of what grandparents actually do would be a welcome and long-overdue addition to the literature. Observation in both natural settings, such as homes, and in controlled laboratory contexts would be worthwhile. The recent work of Feldman and Nash (c.f. 1977, 1979) concerning reactions to infants at different phases of the life cycle illustrates the utility of laboratory approaches to the study of grandparents. To obtain an adequate description of the range of activities and contexts in which grandparents and grandchildren connect, observational strategies could be supplemented with the time-use diaries (Robinson, 1977). The application of path-analytic and other causal modeling techniques, to infer more adequately the direction of influence in nonexperimental naturalistic studies, should be useful. Finally, experimental approaches to grandparenting have been virtually nonexistent. Experimental analyses of the impact of changing specific aspects of grandparental behavior, such as style of play, would be informative, along with a closer experimental scrutiny of changes in grandparent–grandchild interactional cross tasks and contexts. Nor need experiments be restricted to laboratory contexts. Field experiments in which grandparental involvement is systematicallay increased in naturally occurring settings in order to show shifts in family functioning and child development would be another possible approach. Studies utilizing these strategies are in progress by the authors of this chapter.

It should be stressed that there is no "average" grandparent. Grandparental roles are characterized by a great deal of variation and diversity. The extent to which grandparents function as social figures varies, as does their effectiveness in affecting family functioning and child development. Adequate classification studies of different types of grandparent–grandchild relations are still lacking, in spite of the pioneering work of Neugarten and Weinstein (1964). The modern grandparent has neither disappeared nor been replaced. The grandparent, in combination with other members of a family's social network, plays a potentially important but still poorly understood role in determining family functioning and children's development. The details of this process remain to be discovered.

Acknowledgments

We are grateful to Elaine Fleming for research assistance and to Mary Johnson, Jo Powell, Sally Parsons, and Liz Detweiler for their assistance in the preparation of this manuscript. Finally, we deeply appreciate detailed comments provided by Steve Asher.

References

Abernathy, V. Social network and response to the maternal role. *International Journal of Sociology of the Family*, 1973, 3, 86–92.

Abrahams, B., Feldman, S. S., & Nash, S. C. Sex role self-concept and sex role attitudes: Enduring personality characterisitics on adaptations to changing life situations? *Developmental Psychology*, 1978, 14, 393–400.

Adams, B. N. Structural factors affecting parental aid to married children. *Journal of Marriage and the Family*, August 1964, pp. 327–331.

Adams, B. N. Kinship systems and adaptation to modernization. *Studies in Comparative International Development*, 1968, 4, 47–60.

Adams, B. N. Isolation, function, and beyond. American kinship in the 60's. *Journal of Marriage and the Family*, 1970, 32, 575–597.

Alan Guttmacher Institute. *11 million teenagers: What can be done about the epidemic of adolescent pregnancies in the U.S.* New York: Planned Parenthood Federation of America, 1976.

Albrecht, R. The parental responsibilities of grandparents. *Marriage and Family Living,* 1954, *16,* 201–204.

Asher, S. R., & Gottman, J. M. (Eds.). *The development of children's friendships.* Cambridge: Cambridge University Press, 1981.

Atchley, R. C. *The social faces in later life: An introduction to social gerontology.* Belmont, Calif.: Wadsworth, 1972.

Badger, E., Burns, D., & Vietze, P. Maternal risk factors as predictors of developmental outcome in early childhood. *Infant Mental Health Journal,* 1981, *2,* 34–57.

Belsky, J. Mother–father–infant interaction: A naturalistic observational study. *Developmental Psychology,* 1979, *15,* 601–607.

Belsky, J. Child maltreatment: An ecological integration. *American Psychologist,* 1980, *35,* 320–335.

Belsky, J. Early human experience: A family perspective. *Developmental Psychology,* 1981, *17,* 3–23.

Belsky, J., & Tolan, W. J. Infants as producers of their own development: An ecological analysis. In R. M. Lerner & N. A. Busch-Rossnagel (Eds.), *Individuals as producers of their own development: A life span perspective.* New York: Academic Press, 1981.

Bengston, V. L., & Cutler, N. E. Generations and intergenerational relations: Perspectives on age groups and social change. In R. Benstock & E. Shanas (Eds.), *Handbook of aging and the social sciences.* New York: Von Nostrand Reinhold, 1976.

Berger, M., & Fowlkes, M. A. Family intervention project: A family network model for serving young handicapped children. *Young Children,* 1980, *51,* 188–198.

Berghorn, F. J., Schafer, D. E., Steeve, G. H., & Wiseman, R. F. *The urban elderly: A study of life satisfaction.* Montclair, N.J.: Allenheld Osmon, 1977.

Bittman, S., & Zalk, S. R. *Expectant fathers.* New York: Hawthorn Books, 1978.

Blumberg, L., & Bell, R. R. Urban migration and kinship ties. *Social Problems,* 1959, *6,* 328–333.

Bradley, R. H., & Caldwell, B. M. The relation of infants' home environments to mental test performance at fifty-four months: A follow-up study. *Child Development,* 1976, *47,* 1172–1174.

Bronfenbrenner, U. The changing American family. In E. M. Hetherington & R. D. Parke (Eds.), *Contemporary readings in child psychology.* New York: McGraw-Hill, 1977.

Bronfenbrenner, U. *The ecology of human development.* Cambridge: Harvard University Press, 1979.

Brown, J. S., Schwarzweller, H. K., & Magalam, J. J. Kentucky mountain migration and the stem family: An American variation on a theme by Le Play. *Rural Sociology,* 1963, *28,* 48–69.

Cherlin, A. J. *Marriage, divorce, remarriage.* Cambridge: Harvard University Press, 1981.

Clark, M. Cultural values and dependency in later life. In R. Kalish (Ed.), *The dependencies of old people.* Ann Arbor: University of Michigan Institute of Gerontology, 1969.

Clarke-Stewart, K. A. The father's contribution to children's cognitive and social development in early childhood. In F. Pedersen (Ed.), *The father–infant relationship.* New York: Praeger, 1980.

Cochran, M. M. & Brassard, J. A. Child development and personal social networks. *Child Development,* 1979, *50,* 601–616.

Cohler, B. J., & Grunebaum, H. V. *Mothers, grandmothers, and daughters: Personality and childcare in three-generation families.* New York: Wiley, 1981.

Cohler, B. J., Grunebaum, H. V., Weiss, J. C., & Moran, D. L. The childcare attitudes of two generations of mothers. *Merrill Palmer Quarterly,* 1971, *17,* 3–18.

Crockenberg, S. B. Infant irritability, mother responsiveness, and social support influences on the security of infant–mother attachment. *Child Development*, 1981, 52, 857–865.

Croog, S. H., Lipson, A., & Levine, S. Help patterns in severe illness: The roles of kin network, non-family resources and institutions. *Journal of Marriage and the Family*, 1972, 34, 32–41.

Daniels, P., & Weingarten, K. *Sooner or later: The timing of parenthood in adult lives*. New York: W. W. Norton, 1982.

Dunn, J., & Kendrick, C. *Siblings*. Cambridge: Harvard University Press, 1982.

Edwards, C. P., & Lewis, M. Children's concepts of social relations: Social functions and social objects. In M. Lewis & L. A. Rosenblum (Eds.), *The child and its family*. New York: Plenum Press, 1979.

Elardo, R., Bradley, R., & Caldwell, B. The relation of infants' home environments to mental test performance from six to thirty-six months: A longitudinal analysis. *Child Development*, 1975, 46, 71–76.

Elder, G. H. Family history and the life course. *Journal of Family History* (Vol. 1–2), 1976, 77, 279–304.

Elder, G. H. Approaches to social change and the family. *American Journal of Sociology*, 1978, 84, 1–37.

Elder, G., & Rockwell, R. The life course and human development: An ecological perspective. *International Journal of Behavioral Development*, 1979, 2, 1–21.

Feldman, S. S., & Nash, S. C. Interest in babies during young adulthood. *Child Development*, 1978, 49, 617–622.

Feldman, S. S., & Nash, S. C. Sex differences in responsiveness to babies among mature adults. *Developmental Psychology*, 1979, 15, 430–436.

Feldman, S. S., Nash, S. C., & Cutrona, C. The influence of age and sex on responsiveness to babies. *Developmental Psychology*, 1977, 13, 675–676.

Feldman, S. S., Birigen, Z. C., & Nash, S. C. Fluctuations of sex-rated self-attributions as a function of stage of family life cycle. *Developmental Psychology*, 1981, 17, 24–35.

Field, T. M., Widmayer, S. M., Stringer, S., & Ignatoff, E. Teenage, lower class, black mothers and their preterm infants: An intervention and developmental follow-up. *Child Development*, 1980, 51, 426–436.

French, J. R. P., Jr., Rogers, W. L., & Cobb, S. Adjustment as person-environment fit. In G. Coelho, D. Hamberg, & J. Adams (Eds.), *Coping and adaptation*. New York: Basic Books, 1974.

Furstenberg, F. F. Remarriage and intergenerational relations. In R. Fogel, E. Hatfield, S. Kiesler, & J. March (Eds.), *Aging: Stability and change in the family*. New York: Academic Press, 1982.

Gallas, H. B. (Eds.). Teenage parenting: Social determinants and consequences. Introduction. *Journal of Social Issues*, 1980, 36, 1–6.

Galper, A., Seefeldt, C., Jantz, R. K., & Serock, K. Children's concept of age. *International Journal of Aging and Human Development*, 1980, 12, 147.

Garbarino, J., & Gilliam, G. *Understanding abusive families*. Lexington, Mass.: Heath, 1980.

Gibson, G. Kin family network: Overheralded structure in past conceptualizations of family functioning. *Journal of Marriage and the Family*, 1972, 34, 13–23.

Gilford, R., & Black, D. *The grandchild–grandparent dyad: Ritual or relationship?* Paper presented at the 25th annual meeting of the Gerontological Society, San Juan, Puerto Rico, December 1972.

Glick, P. C. The future of the American family. *Current Population Reports*, 1979, Series P-23, No. 78.

Goldberg, W. A. *Marital quality and child–mother, child–father attachments*. Paper presented at the meeting of the International Conference on Infant Studies, Austin, Texas, 1982.

Gordon, M., & Knoll, C. E. Social class and interaction with kin and friends. *Journal of Comparative Family Studies*, 1975, 6, 239–248.

Hagestad, G. O. Parent and child: Generations in the family. In T. M. Field, A. Huston, H. C. Quay, L. Troll, & G. E. Finley (Eds.), *Review of human development*. New York: Wiley, 1982.
Hagestad, G. O., & Speicher, J. L. *Grandparents and family influence: Views of three generations.* Paper presented at the biennial meeting of the Society for Research in Child Development, Boston, April 1981.
Hardy, J. B., King, T. M., Shipp, D. A., & Welcher, D. W. A comprehensive approach to adolescent pregnancy. In K. G. Scott, T. F. Field, & E. Robertson (Eds.), *Teenage parents and their offspring*. New York: Grune & Stratton, 1981.
Hare-Mustin, R. T. Myth and assertions about the family. *American Psychologist*, 1981, *36*, 312–313.
Hareven, T. K. Family time and historical time. *Daedalus*, 1977, *106*, 57–70.
Havighurst, R. J. History of developmental psychology: Socialization and personality development through the life span. In P. B. Baltes & K. W. Schaie (Eds.), *Life-span developmental psychology*. New York: Academic Press, 1973.
Hentig, H. V. The sociological function of the grandmother. *Social Forces*, 1946, *24*, 389–392.
Hess, B. B., & Waring, J. M. Parent and child in later life. Rethinking the relationship. In R. M. Lerner & G. B. Spanier (Eds.), *Child influences on marital and family interaction: A life span perspective*. New York: Academic Press, 1978.
Hetherington, E. M., Cox, M., & Cox, R. The aftermath of divorce. In J. H. Stevens, Jr., & M. Matthew (Eds.), *Mother–child, father–child relations*. Washington, D.C.: National Association for the Education of Young Children, 1978.
Hickey, T. H., Hickey, L. A., & Kalish, R. A. Children's perceptions of the elderly. *The Journal of Genetic Psychology*, 1968, *112*, 227–235.
Hill, R., Foote, N., Aldous, J., Carlson, R., & MacDonald, R. *Family development in three generations: A longitudinal study of changing family patterns of planning and achievement.* Cambridge, Mass.: Schenkman, 1970.
Jordan, V. B. Conserving kinship concepts: A developmental study in social cognition. *Child Development*, 1980, *51*, 146–155.
Kahana, B., & Kahana, E. Grandparenthood from the perspective of the developing grandchild. *Developmental Psychology*, 1970, *3*, 98–105.
Kahana, E., & Kahana, B. Theoretical and research perspectives on grandparenthood. *Aging and Human Development*, 1971, *2*, 261–268.
Kahn, R. L., & Antonucci, T. C. Convoys over the life course: Attachment, roles, and social support. In P. B. Baltes & O. G. Brim, Jr. (Eds.), *Life-span development and behavior* (Vol. 3). New York: Academic Press, 1980.
Kellam, S. G., Ensminger, M. E., & Turner, J. Family structure and the mental health of children. *Archives of General Psychiatry*, 1977, *34*, 1012–1022.
Klaus, M. H., & Kennell, J. H. *Parent–infant bonding* (2nd ed.). St. Louis: Mosby, 1981.
Komarovsky, M. *Blue-collar marriage*. New York: Vintage Books, 1967.
Kornhaber, A., & Woodward, K. L. *Grandparents/grandchild: The vital connection*. Garden City, N.Y.: Anchor Press, 1981.
Lamb, M. E. The role of the father: An overview. In M. E. Lamb (Ed.), *The role of the father in child development*. New York: Wiley, 1976.
Lamb, M. E. Paternal influences and the father's role: A personal perspective. *American Psychologist*, 1979, *34*, 938–943.
Lamb, M. E. (Ed.). *The role of the father in child development* (2nd ed.). New York: Wiley, 1981.
Laslett, P. Characteristics of the Western family considered over time. *Journal of Family History*, 1976–1977, *1–2*, 89–115.
Lee, G. R. Effects of social networks on the family. In W. R. Burr, R. Hill, F. I. Nye, & I. L. Reiss (Eds.), *Contemporary theories about the family, Vol. 1: Research-based theories*. New York: Free Press, 1979.

Lee, G. R. Kinship in the seventies: A decade review of research and theory. *Journal of Marriage and the Family*, 1980, 42, 923–934.

Lewis, M. *The social network systems model: Toward a theory of social development.* Paper presented at the meeting of the Eastern Psychological Association, Philadelphia, April 1979.

Lewis, M. The social network systems model: Toward a theory of social development. In T. M. Field, A. Huston, H. C. Quay, L. Troll, & G. E. Finely (Eds.), *Review of human development*. New York: Wiley, 1982.

Lewis, M., & Feiring, C. Direct and indirect interactions in social relationships. In L. Lipsitt (Ed.), *Advances in infancy research* (Vol. 1). New York: Ablex, 1981.

Lewis, M., & Rosenblum, L. A. (Eds.). *The effect of the infant on its caregiver*. New York: Wiley, 1974.

Lewis, M. E., & Weinraub, M. The father's role in the infant's social network. In M. E. Lamb (Ed.), *The role of the father in child development*. New York: Wiley, 1976.

Lieberman, A. F. Preschoolers competence with a peer: Influence of attachment and social experience. *Child Development*, 1977, 48, 1277–1287.

Litwak, E., & Szeleniji, I. Primary group structures and their functions: Kin, neighbors and friends. *American Sociological Review*, 1969, 34, 465–481.

Matas, L., Arend, R. A., & Sroufe, L. A. Continuity in adaptation: Quality of attachment and later competence. *Child Development*, 1978, 49, 547–556.

Main, M. *Exploration, play and level of cognitive functioning as related to child–mother attachment.* Unpublished dissertation, Johns Hopkins University, 1973.

Main, M., & Weston, D. R. The quality of the toddler's relationship to mother and father: Related to conflict behavior and the readiness to establish new relationships. *Child Development*, 1981, 52, 932–940.

Mills, L. A., & Cairns, R. B. *Life satisfaction and grandparenting in low income elderly black women.* Paper presented at the biannual meeting of the Society for Research in Child Development, Boston, April 1981.

Minturn, L. A., & Lambert, W. W. *Mothers of six cultures: Antecedents of child-rearing*. New York: Wiley, 1964.

Mogey, J. Residence, family kinship, some recent research. *Journal of Family History*, 1976–1977, 1–2, 95–105.

Neugarten, B. L., & Moore, J. W. The changing age status systems. In B. L. Neugarten (Ed.), *Middle age and aging*. Chicago: University of Chicago Press, 1968.

Neugarten, B. L., & Weinstein, K. K. The changing American grandparent. *Journal of Marriage and the Family*, May, 1964, 26, 199–204.

Nimkoff, M. F. Changing family relationships of older people in the U.S. during the last fifty years. *The Gerontologist*, 1961, 1, 92–97.

Olds, D. L. The prenatal/early infant project: An ecological approach to prevention. In J. Belsky (Ed.), *In the beginning: Readings in infancy*. New York: Columbia University Press, 1981.

Parke, R. D. Interactional designs. In R. B. Cairns (Ed.), *The analysis of social interactions: Methods, issues and illustrations*. New York: Lawrence Erlbaum, 1979.

Parke, R. D. *Fathers*. Cambridge: Harvard University Press, 1981(a).

Parke, R. D. Theoretical models of child abuse: Their implications for prediction, prevention and modification. In R. Starr (Ed.), *Prediction of abuse*. New York: Ballinger, 1981(b).

Parke, R. D., & Collmer, C. W. Child abuse: An interdisciplinary analysis. In E. M. Hetherington (Ed.), *Review of child development research* (Vol. 5). Chicago: University of Chicago Press, 1975.

Parke, R. D., & Lewis, N. G. The family in context: A multi-level interactional analysis of child abuse. In R. W. Henderson (Ed.), *Parent-child interaction: Theory, research and prospect*. New York: Academic Press, 1981.

Parke, R. D., & O'Leary, S. Family interaction in the newborn period: Some findings, some observations, and some unresolved issues. In K. Riegel & J. Meacham (Eds.), *The developing individual in a changing world, Vol. 2: Social and environmental issues*. The Hague: Mouton, 1976.

Parke, R. D., & Tinsley, B. R. The early environment of the at-risk infant: Expanding the social context. In D. Bricker (Ed.), *Intervention with at-risk and handicapped infants: From research to application*. Baltimore: University Park Press, 1982.

Parke, R. D., Power, T. G., & Gottman, J. M. Conceptualizing and quantifying influence patterns in the family triad. In M. E. Lamb, S. J. Suomi, & G. R. Stephenson (Eds.), *Social interaction analysis: Methodological issues*. Madison: University of Wisconsin Press, 1979.

Parsons, T. The kinship system of the contemporary United States. *American Anthropologist*, 1943, *45*, 22–38.

Parsons, T. The kinship system in the contemporary U. S. In *Essays in sociological theory*. Chicago: Free Press, 1954.

Pascoe, J. M., Loda, F. A., Jeffries, V., & Earp, J. A. The association between mothers' social support and provision of stimulation to their children. *Developmental and Behavioral Pediatrics*, 1981, *2*, 15–19.

Pearlin, L. I., & Schooler, C. The structure of coping. *Journal of Health and Social Behavior*, 1978, *19*, 2–21.

Pedersen, F. A. (Ed.). *The father-infant relationship: Observational studies in the family setting*. New York: Praeger Special Studies, 1980.

Pedersen, F. A. Father influences viewed in a family context. In M. E. Lamb (Ed.), *The role of the father in child development* (2nd ed.). New York: Wiley, 1982.

Pedersen, F. A., & Robson, K. S. Father participation in infancy. *American Journal of Orthopsychiarty*, 1969, *39*, 466–472.

Powell, D. R. Family-environment relations and early child rearing. The role of social networks and neighborhoods. *Journal of Research and Development in Education*, 1979, *13*, 1–11.

Power, T. G., & Parke, R. D. Play as a context for early learning: Lab and home analysis. In I. E. Sigel & L. M. Laosa (Eds.), *The family as a learning environment*. New York: Plenum Press, 1982.

Reiss, D. *The family's construction of reality*. Cambridge: Harvard University Press, 1981.

Riegel, K. F., Riegel, R. M., & Meyer, G. A. Study of the drop-out rates in longitudinal research on aging and the prediction of death. *Journal of Personality and Social Psychology*, 1967, *5*, 342–348.

Robertson, J. F. Interaction in three generation families, parents as mediators: Toward a theoretical perspective. *International Journal of Aging and Human Development*, 1975, *6*, 103–110.

Robinson, J. P. *How Americans use time*. New York: Praeger Press, 1977.

Rohner, R. *They love me, they love me not: A world-wide study of the effect of parental acceptance and rejection*. New Haven, Conn.: HRAF Press, 1975.

Rosow, I. *Social integration of the aged*. New York: Free Press, 1967.

Rueveni, V. *Networking families in crisis*. New York: Human Services Press, 1979.

Saltz, R. Aging persons as child-care workers in a foster grandparent program: Psychosocial effects and work performance. *Aging and Human Development*, 1971, *2*, 314–340.

Saltz, R. Effects of part-time "mothering" on IQ and SQ of young institutionalized children. *Child Development*, 1973, *44*, 166–170.

Shanas, E. Social myth as hypothesis: The case of the family relations of old people. *Gerontologist*, 1979, *19*, 3–9.

Smith, E. W. The role of the grandmother in adolescent pregnancy and parenting. *Journal of School Health*, 1975, *45*, 278–283.

Staples, R., & Smith, J. W. Attitudes of grandmothers and mothers toward child rearing practices. *Child Development*, 1954, *25*, 91–97.
Streib, G. F. Family patterns in retirement. *Journal of Social Issues*, 1958, *14*, 46–60.
Sussman, M. B. The isolated nuclear family, fact or fiction? *Social Problems*, 1959, *6*, 333–340.
Tinsley, B. R. *The effect of grandparents as socialization and support agents on child development*. Paper presented at the biennial meeting of the Society for Research on Child Development, Detroit, April 1983.
Townsend, P. *The family life of old people*. London: Routledge and Kegan Paul, 1957.
Troll, L. The family of later life: A decade review. *Journal of Marriage and Family*, 1971, *33*, 263–290.
Troll, L. E. *Grandparenting*. Paper presented at the meeting of the American Psychological Association, New York, 1979.
Troll, L. E. Grandparenting. In L. W. Poon (Ed.), *Aging in the 1980's: Psychological issues*. New York: American Psychological Association, 1981.
Troll, L., & Bengston, V. Generations in the family. In W. R. Burr, R. Hill, F. I. Nye, & I. Reiss (Eds.), *Contemporary theories about the family, Vol. 1: Research-based theories*. New York: Macmillan, 1979.
Troll, L. E., & Turner, B. *The effect of changing sex roles on the family of later life*. Paper presented at the Ford Foundation Conference on Changing Sex Roles in the Family, Merrill-Palmer Institute, Detroit, 1976.
Troll, L. E., Miller, S. J., & Atchley, R. C. *Families in later life*. Belmont, Calif.: Wadsowrth, 1979.
Uhlenberg, P. Death and the family. *Journal of Family History*, 1980, *5*, 313–320.
Unger, D. *An ecological approach to the family: The role of social networks, social stress and mother–child interaction*. Unpublished master's thesis, Merrill-Palmer Institute, 1979.
Uzoka, A. F. The myth of the nuclear family: Historical background and clinical implications. *American Psychologist*, 1979, *34*, 1095–1106.
Wahler, R. G. Parent insularity as a determinant of generalization success in family treatment. In S. Salzinger, J. Antrobus, & J. Glick, (Eds.), *The ecosystem of the "sick" child: Implications for classification and intervention for disturbed and mentally retarded children*. New York: Academic Press, 1980.
Wandersman, L. P., Wandersman, A., & Kahn, S. Social support in the transition to parenthood. *Journal of Community Psychology*, 1980, *8*, 332–342.
Waters, E., Wippman, J., & Sroufe, L. A. Attachment, positive effect, and competence in the peer group: Two studies in construct validation. *Child Development*, 1979, *50*, 821–829.
Werner, E. E., & Smith, R. S. *Vulnerable but invincible: A longitudinal study of resilient children and youth*. New York: McGraw-Hill, 1982.
Whiting, B. B., & Whiting, J. W. M. *Children of six cultures: A psychocultural analysis*. Cambridge: Harvard University Press, 1975.
Winch, R. F., Greer, S., & Blumberg, R. L. Ethnicity and extended familism in upper-middle-class suburb. *American Sociological Review*, 1967, *32*, 265–272.
Yogman, M. W. Development of the father-infant relationship. In H. Fitzgerald, B. Lester, & M. W. Yogman (Eds.), *Theory and research in behavioral pediatrics* (Vol. 1). New York: Plenum Press, 1982.
Yorburg, B. The nuclear and the extended family: An area of conceptual confusion. *Journal of Comparative Family Studies*, 1975, *6*, 5–14.
Zigler, E. F., & Gordon, E. W. *Daycare: Scientific and social policy issues*. Boston: Auburn Publishing House, 1982.

9

Infants, Mothers, Families, and Strangers

Ross A. Thompson and Michael E. Lamb

Developmental theorists have long viewed the infant–mother relationship as a crucial determinant of early socioemotional development (e.g., Ainsworth, 1967, 1973; Bowlby, 1969; Erikson, 1963; Freud, 1938; Maccoby & Masters, 1970; Sears, Maccoby, & Levin, 1957). Mother–infant interactions comprise a major part of the average infant's everyday experiences, and they occur in contexts that are likely to be especially important to the baby, such as feeding, the relief of distress, and play. Thus it seems likely that the infant–mother relationship would have a greater direct influence on infants than other early relationships, at least in traditional families. This does not mean, however, that the infant–mother relationship is exclusively important. Other partners also provide valuable experiences, ranging from the physically arousing stim-

---

Earlier versions of this paper were presented at the annual meetings of the Midwestern Psychological Association, Detroit, April 30–May 2, 1981, and at the annual meeting of the American Psychological Association, Washington, D.C., August 22–27, 1982. We are grateful to the American Psychological Association and the Society for Research in Child Development, Inc., for permission to reprint some previously published material.

Ross A. Thompson • Department of Psychology, University of Nebraska, Lincoln, Nebraska 68588. Michael E. Lamb • Department of Psychology, University of Utah, Salt Lake City, Utah 84112. The research described in this chapter is based on a dissertation conducted by the first author in partial fulfillment of the requirements for the degree doctor of philosophy in the Rackham School of Graduate Studies at the University of Michigan. It was supported by a grant from the Riksbankens Jubileumsfond of Sweden to author Lamb, and by a dissertation grant from Rackham to author Thompson. While the research was conducted, Thompson was a predoctoral fellow in the Bush Program in Child Development and Social Policy at the University of Michigan and a National Science Foundation predoctoral fellow.

ulation provided by fathers (Lamb, 1977, 1981b) to the give-and-take interactions provided by peers (Lewis, Young, Brooks, & Michalson, 1975; Mueller & Vandell, 1979). Furthermore, events outside the mother–infant dyad—like the father's presence (Lamb, 1979) or supportiveness (Pedersen, 1981)—affect the quality of the interactions shared by infant and mother. For these reasons, it is unwise to view the infant–mother relationship without considering the broader socioemotional context within which their interactions occur (Lewis & Rosenblum, 1978).

In this chapter, we consider some of the ways in which the infant–mother relationship influences and is influenced by events and interactions occurring outside the dyad. The conceptual and methodological rationale behind the Strange Situation procedure, a technique used for assessing the security of infant–mother attachment relationships is first described. What is then described is research concerned with the way in which stresses and changes in family circumstances can force the renegotiation of mother–infant relationships over time, resulting in a change in the security of attachment. This evidence leads us to conclude that any assessment of infant–mother attachment reflects only the *current* status of this relationship, and that the nature of this relationship changes in response to changing family conditions. Events beyond the mother–infant dyad, in other words, affect a crucial aspect of the dyadic relationship: the security of attachment and its stability over time.

Evidence is next considered of a different kind of relationship between events occurring within and outside the dyad, and in the final section, we argue that the security of infant–mother attachment affects the way infants behave in initial encounters with unfamiliar adults. Moreover, we present evidence indicating that when the security of attachment changes over time, so too does stranger sociability. We conclude that the security of attachment and the stability of attachment relationships over time are important influences on individual differences in stranger responsiveness.

In all, therefore, there are bidirectional influences between experiences within and beyond the infant–mother relationship, a fact that underscores the need to view the dyad not in isolation, but within its broader social context.

The Security Of Attachment

In recent years, individual differences in the quality of infant–mother relationships have been fruitfully studied from the perspective of ethological attachment theory (Ainsworth, Blehar, Waters,

& Wall, 1978). According to attachment theorists, most infants develop significant relationships with their parents, but the quality or security of these relationships varies widely. These individual differences appear to be attributable to differences in the quality of the social interactions between the infants and the adults concerned—such as their emotional harmony and the responsiveness of each to the social cues of the other. Although both adults and infants have an influence on the harmony of these exchanges, most attachment theorists consider adults more influential because of their broader behavioral repertoire and their greater ability to act planfully. Research conducted by Ainsworth and her colleagues (1978), for example, has suggested that the adult's sensitivity to the infant's signals and needs is of crucial importance. When mothers were sensitively responsive and emotionally accessible, their infants developed secure attachments to them; when the mothers were insensitive or inaccessible, insecure relationships were more likely.

From the infant's standpoint, it appears that the predictable helpfulness of the sensitively responsive mother fosters the development of confidence in the adult's reliability (Lamb, 1981a,c). This is what is meant by a *secure attachment relationship,* and it is manifested in positive social responsiveness to the caregiver, particularly under conditons of stress, alarm, or uncertainty, when the caregiver is most needed. Conversely, inconsistent or unhelpful parental behavior prevents the emergence of trust and leads instead to confusion, or to expectations that the adult's behavior will be predictably aversive. An insecure attachment relationship is thus reflected in more negative or ambivalent responding by the infant.

In short, the "security of attachment" concerns the quality of a baby's expectations of the parent, which derive from these consistent features of social interaction. Remarkably enough, Ainsworth and her colleagues succeeded in designing a procedure that is suitable for assessing individual differences in the security of attachment and that is thus useful for demonstrating the relationships between infant and maternal behavior that we have just described.

The Strange Situation Procedure

The Strange Situation is a 21-minute semistandardized procedure that involves two brief separations and reunions of parent and baby (see Table 1). The procedure is designed to subject the infant to gradually escalating stress, on the assumption that the infant's expectations of the adult will be most clearly revealed when the attachment behavioral system has become activated. Stress is induced, for example, by the unfamiliar setting, by interactions with an unfamiliar adult, by separations

TABLE 1. Procedure for the Strange Situation

Duration of episode	Description of episode
3 minutes	1. <u>Mother and baby</u> Mother remains seated, responds naturally and appropriately to baby, but does not initiate interaction. *Transition:* Stranger enters the room.
3 minutes	2. <u>Mother, baby, and stranger</u> First minute: stranger sits quietly. Second minute: stranger chats with mother. Third minute: stranger tries to interact and play with baby. *Transition:* Mother leaves the room.
3 minutes (or less)	3. <u>Baby and stranger—first separation</u> Stranger remains responsive to baby; tries to comfort the infant, if necessary. Episode shortened if baby remains highly distressed. *Transition:* Mother returns to room, and stranger leaves unobtrusively.
3 minutes	4. <u>Mother and baby—first reunion</u> Mother signals arrival by talking in hallway before entering, then pauses in doorway to allow greeting from baby. Mother comforts child if necessary, then seeks to reinterest in toys. Mother remains responsive as in Episode 1. *Transition:* Mother leaves the room.
3 minutes (or less)	5. <u>Baby—second separation</u> Episode shortened if baby remains highly distressed. *Transition:* Stranger enters the room.
3 minutes (or less)	6. <u>Baby and stranger</u> Same as in Episode 3. *Transition:* Mother returns to room; stranger leaves unobtrusively.
3 minutes	7. <u>Mother and baby—second reunion</u> Same as in Episode 4.

from the parent in the company of that adult, and finally by being left alone. By examining the infant's behavior throughout this procedure—especially on reunion with the parent after the brief separations— researchers have been able to evaluate the security of individual infant–parent attachment relationships.

There are two ways in which the infant's attachment behavior is evaluated by means of Ainsworth's procedure. The first consists of episode-by-episode ratings of important dimensions of the social-interactive behavior directed toward the parent or the stranger. Behaviorally based rating scales are used to quantify the degree of (1) proximity- and contact-seeking; (2) contact maintenance; (3) avoidance of interaction; (4) resistance to interaction and contact; and (5) distance interaction. These ratings thus provide an episode-by-episode appraisal of changes in interactive activity as the Strange Situation proceeds.

The second scoring procedure is the more important of the two. It involves an overall classification of the security of the infant–adult attachment relationship that is based primarily on the nature of the infant's behavior on reunion with the parent. The baby's preseparation behavior is also considered, however, particularly with respect to the infant's play activity when the parent is present.

On the basis of these criteria, three overall classification groups are distinguished, each of them evidently involving differences in the infant's expectations of the parent's behavior. Infants who are *securely attached* (Group B) show the normative pattern of behavior: they evince positive behaviors toward the parent throughout the Strange Situation, greet the parent on reunion, and gain sufficient security from the parent's presence so that they can comfortably explore when he or she is present. Subgroups within this category denote variations within this overall pattern: B1 and B2 infants rely primarily on distal interactive modes when greeting the adult (e.g., smiling, vocalizing, and offering toys), whereas B3 and B4 infants more actively seek proximity and contact with the parent on reunion. All securely attached infants, however, exhibit pleasure at the parent's return and act as if they expect the parent to respond appropriately to their manifest needs and social bids.

Insecurely attached infants behave in a markedly different fashion. *Avoidant* (Group A) infants exhibit clear-cut avoidance of the adult (e.g., looking or turning away, ignoring his or her initiatives) throughout the Strange Situation, and especially during reunions. Their play activity is also noninteractive in quality. By contract, *resistant* (Group C) infants display strong proximity- and contact-seeking behaviors on reunion, but these are accompanied by angry, resistant behaviors, such as pushing away, hitting or slapping the parent, or rejecting his or her bids to engage in social interaction. These infants also seek proximity and contact *before* separation, which inhibits their exploration. Subgroups within the A and C groups also denote variations in the infants' behavior on reunion within these overall patterns.

Several types of evidence substantiate the validity of this classification system. First, Ainsworth and her colleagues showed (although with a small sample) that there were strong similarities between the socioemotional behaviors displayed in the Strange Situation and at home (Ainsworth *et al.*, 1978; Ainsworth, Bell, & Stayton, 1971; Stayton & Ainsworth, 1973). For example, infants who exhibited frequent contact-maintaining behaviors in the Strange Situation responded positively to being held at home; babies who showed avoidance or resistance in the Strange Situation responded negatively when held at home. Other analyses showed similarities between patterns of infant–mother interaction early in the first year at home and in the Strange Situation at 1 year of age (Ainsworth *et al.*, 1978; Blehar, Lieberman, & Ainsworth, 1977).

Second, individual differences in the security of attachment are related to important aspects of maternal behavior, in both prior and contemporaneous assessment. Several research groups have reported that the mothers of securely attached infants appear to be more sensitively responsive to infant signals, more supportive in problem-solving situations, and more emotionally expressive than the mothers of insecurely attached infants (Ainsworth & Bell, 1969; Ainsworth et al., 1971, 1978; Blehar et al., 1977; Estes, Lamb, Thompson, & Dickstein, 1981; Main, Tomasini, & Tolan, 1979; Matas, Arend, & Sroufe, 1978).

Third, individual differences in the security of attachment are also related to other dimensions of social responsiveness or personality organization, such as sociability with unfamiliar adults (Main, 1973, and below); social competence with peers (Easterbrooks & Lamb, 1979; Lieberman, 1977; Pastor, 1981; Waters, Wippman, & Sroufe, 1979); and "ego resiliency" (Arend, Gove, & Sroufe, 1979).

THE TEMPORAL STABILITY OF ATTACHMENT RELATIONSHIPS

The greatest interest in the concept of the security of attachment has been evoked by claims concerning the consistency of attachment classifications in test–retest assessments during the second year. In a widely cited study, Waters (1978) observed a sample of 50 middle-class infants and mothers in the Strange Situation at 12 and 18 months of age and reported that 48 out of 50 infants received the same overall attachment classification at both ages. This finding seemed to indicate that individual differences in the security of mother–infant attachment relationships were highly stable after the end of the first year. This report was particularly significant because most other short-term longitudinal studies have reported low, and usually nonsignificant, stability of social behaviors over time (see Masters & Wellman, 1974, for a review). Waters's findings thus suggest that the factors that contribute to individual differences in the security of attachment are highly consistent during the second year, or, alternatively, that there is a "sensitive period" sometime during the first year during which the security of attachment is determined, and that, once a pattern of infant–mother interaction has been consolidated, it is unlikely to change. Waters and his colleagues have sometimes adopted the latter view of continuity in early socioemotional relationships (see, for example, Sroufe & Waters, 1977).

There is an important caveat, however. As later reported by Vaughn, Egeland, Sroufe, and Waters (1979, p. 971), Waters's middle-class sample was highly stable, with consistent paternal employment and almost no residential changes. Although Waters (1983) has indicated that he employed no special selection procedures to obtain these sample characteristics, it remains uncertain how broadly Waters's findings may be generalized. In other words, we do not know how stable attachment classifications would be in a more representative middle-class sample.[1]

One goal of the present study, then, was to recruit an unselected, representative middle-class sample and to assess the stability of attachment classifications during the second year. When the infants were $12\frac{1}{2}$ and $19\frac{1}{2}$ months old, a sample of 43 infants and mothers were observed in the Strange Situation.[2] There were 21 boys and 22 girls in our sample; 15 of the children were firstborns. The sessions were videotaped and were later scored by highly trained observers, in the manner earlier described, using consistent scoring criteria at each age. Interrater reliabilities were consistently high. (For further details, see Thompson, Lamb, & Estes, 1982, 1983.)

In this study, we sought to obtain a relatively heterogeneous middle-class sample that was unselected with respect to social status, income level, or educational attainment. The mothers and infants were recruited from a subject pool administered by the department of psychology, and through birth announcements published in a local newspaper. Roughly 78% of the families we called agreed to participate in the study. Later, the heterogeneity of the sample was confirmed by demographic data provided by the mothers. The occupation of the major breadwinner in these families ranged from college professor or engineer to assembly worker at a local automotive plant. Similarly, the educational level ranged from graduate degrees to partial high-school training. All the families except one were intact, two-parent families. All the families were white except one, which was Puerto Rican.

---

[1] We have not included in this discussion two studies concerning the stability of attachment classifications that are sometimes cited elsewhere. Connell's (1976) dissertation, which found that 81% of his middle-class sample had the same attachment classification at 12 and 18 months, did not use Ainsworth's classification system. Rather, Connell employed a set of weighted equations derived from Ainsworth's data, which he used to classify infants in his own sample. In calculating these equations, Connell eliminated from Ainsworth's sample those infants in two of the four secure-classification subgroups (i.e., subgroups B1 and B4) because he found them difficult to classify accurately. Main and Weston (1981) found 73% stability from 12 to 18 months in a very small ($N = 15$) sample. They used an additional classification category ("unclassifiable") and stringent sample selection criteria, which very likely resulted in a somewhat select group of families. For these reasons, both stability estimates are difficult to compare validly with the estimates of those who have used the standard Ainsworth classification system (e.g., Thompson et al., 1982; Vaughn et al., 1979; Waters, 1978).

[2] There were two exceptions. One infant was initially tested three weeks before his first birthday; another was seen for the second time one week following her 20-month anniversary. Inspection of the data revealed that these infants did not differ from the others in their relevant behaviors, so they were included in the sample.

Taken together, the composition of this middle-class sample is reflected in scores on Hollingshead's Four-Factor Index of Social Status (Hollingshead, 1975). Of the families in the sample, 14 were in Hollingshead's Class I (major professional and business); 12 were in Class II (minor professional and technical); 13 were in Class III (skilled craftsmen, clerical, sales workers); and four were in Class IV (machine operators and semiskilled workers).

At both $12\frac{1}{2}$ and $19\frac{1}{2}$ months of age, the distribution of infants across the overall attachment classifications (Groups A, B, and C) was very similar and was consistent with the normative trends previously reported by Ainsworth (Ainsworth et al., 1978) and others (e.g., Main, 1973; Matas et al., 1978; Waters, 1978). At each age, for example, between 65% and 70% of the sample were securely attached. However, only 23 of the 43 infants (53%) were assigned the same overall attachment classification at both ages (see Table 2). The temporal consistency of classification subgroups was 26%. Thus, we found a high degree of change in the security of attachment during the second year.

The same patterns of consistency and change were evident in the correlations of scores on the mother-directed interactive variables at $12\frac{1}{2}$ and $19\frac{1}{2}$ months of age (see Table 3). Although the mother-directed behaviors during the preseparation episodes were nonsignificantly correlated whether or not the attachment classifications remained the same, a different pattern of findings was evident for interactive behaviors during the reunion episodes. When the overall attachment classification was temporally consistent, four of the five interactive variables were positively and significantly correlated over the two assesments. However, when the attachment status changed over time, the correlations between the interactive variables at each age were all small and nonsignificant. Reunion-based interactive behaviors are more informative, we

TABLE 2. Temporal Stability of Attachment Classification

		A_1	A_2	B_1	B_2	B_3	B_4	C_1	C_2
		\multicolumn{8}{c}{$19\frac{1}{2}$-Month classification}							
	A_1	1	0	0	0	0	1	1	0
	A_2	0	0	2	1	1	0	0	0
	B_1	0	0	1	0	0	0	0	0
$12\frac{1}{2}$-Month	B_2	1	0	2	3	3	2	2	0
classification	B_3	1	3	0	1	3	2	1	0
	B_4	0	0	0	0	2	1	2	0
	C_1	0	0	2	1	0	0	1	0
	C_2	0	0	1	0	0	0	0	1

TABLE 3. Correlations of 12½-Month and 19½-Month Scores of Mother-Directed Interactive Behaviors in the Strange Situation[a]

	Temporal stability of attachment classification			
	Unstable		Stable	
	Preseparation	Reunion	Preseparation	Reunion
Proximity seeking	−.09	−.09	−.09	−.18
Contact maintaining	−.05	−.15	.26	.54[c]
Resistance	.39	−.15	−.04	.77[c]
Avoidance	.15	−.19	−.03	.42[b]
Distance interaction	.15	.04	.17	.45[b]

[a] Preseparation episodes (Episodes 1 and 2) and reunion episodes (Episodes 4 and 7) were combined to increase the reliability of episode-by-episode scores.
[b] $p < .05$.
[c] $p < .01$.

assume, because they figure prominently in the appraisal of the security of attachment.

These findings thus corroborate the patterns of stability and change in overall attachment classifications, and they indicate that these changes were not due simply to the unreliability of the classification system. Instead, it is clear that changes in the security of infant–mother attachment may occur with great frequency in an unselected middle-class sample. The mother–infant relationship is, in short, a more dynamic one than we had earlier believed.

As noted earlier, infant behavior in the Strange Situation appears to be lawfully related to prior patterns of mother–infant interaction. If this is so, then our findings concerning frequently occurring changes in the security of attachment imply that they have been preceded by changes in patterns of mother–infant interaction at home. The crucial issue, of course, is to determine whether one can predict which relationships will remain the same over time, and which are more likely to change. This issue is the focus of the next section.

FAMILY CIRCUMSTANCES AND CHANGE IN THE SECURITY OF ATTACHMENT

Ours was not the first study to find a high degree of change in attachment relationships during the second year. In an earlier investigation, Vaughn et al. (1979) explored the stability of attachment classifications from 12 to 18 months in a sample of 100 socioeconomically disadvantaged infants and mothers. Like us, they found that just over half (62%) of their sample received the same classification at both ages. Moreover, Vaughn and his colleagues reported a relationship between the direction of change in the security of attachment and the frequency of stressful events occurring during the six-month period between the two assessments. The mothers of infants who changed from securely to insecurely attached reported a greater number of stressful family events between the two assessments than did the mothers of securely attached infants who remained consistent over time. In this socioeconomically disadvantaged sample, therefore, a higher frequency of stressful family events biased infants toward developing insecure attachment relationships with the caregivers.

We wondered whether the same relationship would be evident in middle-class families. Clearly, this question has important implications for understanding the conditions that foster stability or change in the security of attachment. If the occurrence of family stress also biased middle-class infants toward insecurity, for example, it would suggest that stress consistently affects mother–infant interaction adversely, regardless of ameliorating circumstances. On the other hand, if the same pattern is not evident in middle-class samples, it would suggest that the effects of family stresses on the security of attachment may be modulated by the nature of the events, their severity, the availability of support systems, and other conditions that distinguish middle-class families from lower-income families.

Moreover, the *kinds* of family events that are associated with change in attachment status also merit consideration. Because Vaughn and his colleagues employed a composite life-events index, they could show only that the frequency of stressful events was associated with a change in attachment relationships, not whether certain kinds of events exerted a greater influence than others. At least three kinds of events seem likely to influence the consistency of attachment classifications over time: (1) *critical experiences*, such as a major separation from the mother; (2) *circumstances with enduring or recurrent effects*, such as the mother's return to work and/or the assumption of regular caregiving responsibilities by others; and (3) *specific changes influencing the family as a whole*, which may have fewer direct effects on mother and baby, such as moving to a new home. Each of these events could create stress for the baby by forcing adaptation to a changing environment, unfamiliar relationships, or different patterns of caregiving. Thus, any of these experiences could precipitate changes in the quality of mother–infant interaction and thus affect the security of attachment. We wondered whether these different kinds of family events would have similar influences on the stability of attachment relationships, and thus, we asked the mothers in our sample to complete a family-events questionnaire following each laboratory assessment.

In addition to descriptive demographic information, each questionnaire included inquiries concerning the composition of the family group, the employment status and occupation of each parent, the identities of other major caregivers (if any), and the amount of time that each caregiver spent with the baby. There were also questions concerning the incidence and duration of any separations experienced by the baby, as well as the occurrence of major changes and stresses in family circumstances, such as moving to a new home, a new birth, parental divorce, major hospitalizations, parental job change, and similar events. Using these data, we then explored the influences on the security of attachment of family events occurring during two periods: (1) between the two Strange Situation assessments (i.e., between $12\frac{1}{2}$ and $19\frac{1}{2}$ months) and (2) before the first assessment (i.e., during the baby's first year). The latter period was considered because we felt that events occurring before the first assessment could also be associated with the patterns of stability and change in attachment relationships, perhaps by precipitating short-term changes in mother–infant interaction that influenced the initial Strange Situation assessment, but not the later one. Thus, we inquired into family events occurring throughout the baby's lifetime.

Chi-square analyses were used to examine the relationships between the occurrence of specific life events and patterns of stability and change in the security of attachment. In general, we found that the occurrence of one or more important changes in family circumstances during the first year was associated with a change in the baby's attachment classification between the two assessments ($\chi^2 = 4.98$, $df = 1$, $p < .05$). The relationship between changing family circumstances and attachment stability was even stronger when family events occurring at any time before the $19\frac{1}{2}$-month assessment were considered ($\chi^2 = 6.58$, $df = 1$, $p < .025$). In both cases, changes in the security of attachment were *bidirectional;* there was no tendency for family stresses to produce insecure mother–infant attachment relationships as they had in Vaughn's sample. In other words, some infants changed from insecurely to securely attached, whereas other attachment relationships changed from secure to insecure.

Next, we used the questionnaire responses to determine which *kinds* of family changes and stresses were most strongly associated with change in the security of attachment. Two kinds of family influences seemed most important. First was the mother's return to work, which in this sample was often associated with a change in the father's employment status (see Table 4). Ten of the mothers returned to work during the baby's first year, and in eight cases, the security of attachment changed

TABLE 4. Selected Family Circumstances and Their Temporal Relationships to Stability of Attachment Classifications

	Before 12½ months	Between 12½ and 19½ months
Maternal employment		
Initially insecure—change ($N = 10$)	5 (50%)	2 (20%)
Initially secure—change ($N = 10$)	3 (30%)	2 (20%)
Initially insecure—stable ($N = 3$)	0	0
Initially secure—stable ($N = 20$)	2 (10%)	1 (5%)
Regular nonmaternal care		
Initially insecure—change ($N = 10$)	3 (30%)	2 (20%)
Initially secure—change ($N = 10$)	4 (40%)	2 (20%)
Initially insecure—stable ($N = 3$)	2 (67%)	0
Initially secure—stable ($N = 20$)	0	0
Major separations from the mother		
Initially insecure—change ($N = 10$)	4 (40%)	4 (40%)
Initially secure—change ($N = 10$)	3 (30%)	4 (40%)
Initially insecure—stable ($N = 3$)	0	1 (33%)
Initially secure—stable ($N = 20$)	4 (20%)	7 (35%)

($\chi^2 = 4.10$, $df = 1$, $p < .05$). Between the two assessments, five mothers returned to work, and in four cases, the attachment relationship changed. In sum, more than half (12 of 20) of those infants whose classification changed over time had mothers who had returned to work by the time of the 19½-month assessment, whereas this was true of less than 15% (3 of 23) of those whose classification remained the same ($\chi^2 = 10.28$, $df = 1$, $p < .005$). Again, these changes in attachment relationships were bidirectional: some infants shifted from securely to insecurely attached, and others moved in the opposite direction.

The second factor associated with change in attachment status—one that often accompanied maternal employment—was the baby's experience of regular nonmaternal care for 15 hours per week or more (see Table 4).[3] "Nonmaternal care" includes, for example, caregiving by father,

---

[3] This figure was chosen on an *a priori* basis as a good estimate of what "regular" caregiving from *any* source would be to a baby. Fifteen hours weekly is, for example, approximately the amount of time an infant or toddler would spend in a weekday morning childcare or play group.

a relative, a babysitter, or a child-care agency. Nine infants experienced regular nonmaternal care during the first year; in seven of these cases, the security of attachment changed ($\chi^2 = 2.98$, $df = 1$, $p < .10$). Between the two assessments, four babies entered into a regular nonmaternal caregiving arrangement, and in every case, the security of infant–mother attachment changed. In sum, by $19\frac{1}{2}$ months, more than half (11 of 20) of those infants whose classification changed over time were regularly in the care of persons other than their mothers, whereas this was true of fewer than 10% (2 of 23) of those infants whose classification remained the same ($\chi^2 = 11.09$, $df = 1$, $p < .001$). Once again, these changes in attachment status were bidirectional.

By contrast, other family changes and stresses—including the occurrence of a major (≤ 24 hours) separation from the mother[4] (see Table 4) or moving to a new home—were *not* associated with change in the security of attachment. The family events most strongly associated with change in attachment relationships were, in short, those that were most likely to affect the ongoing quality or quantity of mother–infant interaction. Interestingly, these events were not consistently related to attachment classifications *per se* at each age. In other words, family events were more likely to influence the temporal stability of attachment relationships than to be associated with A-, B- or C-group classification.

In sum, normative family changes and stresses seem to affect the security of attachment in middle-class homes, but the effects are somewhat different from those in lower-income families. In Vaughn's families, stresses and changes fostered the development of insecure attachment relationships. In our middle-class sample, by contrast, family events fostered change in attachment relationships, but secure relationships were as likely to become insecure as insecure relationships were to become secure. In other words, there was no directional bias.

What can account for this difference? Clearly, the kinds of stresses encountered by the families in these studies differed, and this difference may help explain their divergent effects on the stability of attachment relationships. In our middle-class sample, these family events included changes in parental employment status, moving to a new home, birth of a baby, and hospitalization. However, none of the mothers faced overwhelming financial or legal difficulties or experienced a rapidly fluctuating family unit. In contrast, the kinds of experiences encountered by Vaughn's families probably involved much more severe legal, financial, ecological, and health-related concerns. It may be, therefore, that more severe stresses not only introduce change in the security of attachment but also bias that change toward attachment insecurity.

---

[4] The influence of much longer separations could not be assessed because they were experienced by few infants in our sample.

To be sure, however, *all* of these events are likely to be stressful to infants or toddlers. Even in middle-class homes, the mother's return to work or the move to a new home may be only mildly disruptive to the family as a whole, but they are likely to be especially stressful for younger children because they affect particularly important aspects of the infant's daily experiences: the familiarity of the physical setting, for example, or the familiarity and predictability of caregiving arrangements. The same is true, of course, of stressful life events encountered by lower-income families. Thus, middle-class and lower-income homes may not differ significantly in the extent to which stressful events directly affect infants.

Rather, the differential influence of stressful events on the stability of attachment relationships in middle- and lower-class families may be related to their effects on the mother and others in the family. In other words, changing life circumstances and stresses, and the contexts within which these events occur, may either foster or impair the mother's coping capacities and thus affect the ongoing quality of mother–infant interaction. Moreover, whether the mother copes adaptively may be importantly influenced not only by her personal resources, but also by the availability of interpersonal and material resources that exist both within and outside the family. Lower-income and middle-class families are likely to differ significantly in the availability of these supports.

In socioeconomically disadvantaged families, family conditions are more likely to exacerbate than to reduce the level of stress experienced as a result of changing family circumstances. There are several reasons. First, because the stressful events that occur in lower-class families are more severe, there may be fewer ways of coping constructively with them. Thus, they are likely to persist over time. Second, because many of these circumstances are enduring, the parent's physical and emotional resources are likely to become strained or depleted, a situation that further hinders adaptive coping. Finally, these parents are likely to lack extrafamilial sources of interpersonal or material support because one of the hallmarks of "families at risk" is their social isolation (e.g., Garbarino & Sherman, 1980; Parke & Collmer, 1975). In contexts such as these, therefore, stressful changes in family circumstances are more likely to affect the quality of mother–infant interaction adversely; the result is a bias toward the development of insecure mother–infant attachment relationships.

In middle-class homes, by contrast, the stressful events themselves are likely to be less severe and enduring and are thus amenable to a wide range of constructive responses. In some cases, in fact, they may result in changes that are beneficial to the quality of infant–mother interaction. Because the family stresses are less overwhelming, the parents may be able to modulate their potentially negative impact on the child by providing support during transitional experiences. Further, they can turn for assistance to a variety of extrafamilial sources (e.g., nearby relatives, neighbors, and church and community groups), and they benefit from a greater degree of financial stability. In middle-class homes, therefore, the effect of changes in family circumstances on the ongoing quality of mother–infant interaction is likely to vary (much more than in lower-income homes), depending on the underlying circumstances, the nature of the events, the availability of external resources, the parents' attempts to minimize their stressful impact on the child, and so on. As a result, changing family circumstances may sometimes create conditions that foster an improvement in the quality of the infant–mother relationship, as well as modifying the potentially deleterious effects of these events on dyadic interaction. For this reason, family events are likely to foster change in the security of mother–infant attachment, but there is no clear directional bias to the changes that occur.

The circumstances surrounding a mother's return to work provide a good example of these processes. In this middle-class sample, several of the mothers decided to resume regular employment outside the home, and the resumption of employment was commonly associated with changes in the security of attachment. The reasons underlying a mother's return to work varied, however. Some mothers had always intended to resume employment after a brief interruption for pregnancy and childbirth. For others, the return to work was prompted by an unexpected change in the husband's job status. The mother's attitudes toward her new employment status—and thus, her modified caregiving role— may thus have been profoundly influenced by the circumstances underlying this decision, and these attitudes seem likely to influence the quality of subsequent mother–infant interactions (Lamb, Chase-Lansdale, & Owen, 1979; Lamb, Owen, & Chase-Lansdale, 1980). Some employed mothers make special efforts to regularly engage the child in play, whereas in other cases, sheer physical exhaustion may render these interactions more infrequent and perfunctory in quality. As a result, maternal employment may not necessarily result in insecure attachment relationships. The direction of change (secure to insecure, or the reverse) may be influenced by many factors, such as the circumstances underlying the decision to return to work, the mother's attitudes toward her new employment and caregiving status, and support from the marital partner and others.

For a variety of reasons, therefore, changing family circumstances predictably have a negative influence on the security of attachment in lower-class families, whereas their effects in middle-class families may be either positive or negative. This conclusion helps explain the one major difference between our results and those of Vaughn and colleagues (1979). Importantly, however, *both* studies showed that the stability of infant–mother attachment relationships was profoundly influenced by events occurring in the broader family context, particularly those events most directly relevant to the ongoing quality of mother–infant interaction. Stated differently, stability or change in caregiving arrangements determined whether the quality of the attachment relationship would change or remain stable over time. In short, both studies showed that the stability of the infant–mother relationship is inextricably linked to events occurring in the broader social network. Even though the quality of parental behavior has a major impact on the security of infant–mother attachment, our results suggest that the quality of parental behavior may change over time in response to changing family circumstances, and this change may provoke a renegotiation of the infant's expectations of the adult.

These findings have important implications of our understanding of the long-term ramifications of individual differences in the security of attachment. Contrary to the view that there is a critical period during which the security of attachment is determined (e.g., Matas et al., 1978; Sroufe, 1979), our findings show that a baby who is insecurely attached may not remain that way: a more secure attachment may develop when changes in family circumstances foster a renegotiation of the mother–infant relationship. By implication, furthermore, even if circumstances cause a secure relationship to become insecure, later changes may permit the mother and the infant to establish a more secure relationship once again.

In sum, it is clear that the security of the infant–mother attachment may change much more frequently than was formerly believed (cf. Sroufe, 1978, 1979; Waters, 1978). Of course, in the absence of important changes in family and caregiving circumstances, previously established patterns of interaction between infant and mother are likely to continue unchanged. But when life stresses occur and circumstances change, these interactional patterns must be renegotiated, and the result may be a change in the security of attachment. This means that the security of attachment should not be viewed as the consequence of infant–mother interaction during an early sensitive period, but as a reflection of the *current* status of their relationship. The current status, we have shown, is influenced by changing family circumstances that influence mother and infant both directly and indirectly. In other words, the infant–mother relationship is a dynamic relationship that is susceptible to influence by events in the broader family context.

Security of Attachment and Stranger Sociability

If it is true that the security of attachment changes frequently over time, what implications does this change have for the relationships between attachment status and other dimensions of social responsiveness? Do individual differences in attachment relationships have, in other words, broader ramifications for the baby's sociability to other partners? If so, how are changes in attachment relationships over time related to changes in other socioemotional variables? In posing these questions, we shift the direction of our inquiry concerning the mother–infant dyad. Whereas in the previous section we focused on the way in which events *outside* the dyad affected the infant–mother relationship, we now consider the ways in which events *within* the dyad affect the infant's responses to other people. More specifically, we consider how the security of attachment influences the infant's responsiveness to other social partners.

According to attachment theory, infants are likely to generalize the interactive skills and tendencies acquired in the context of mother–infant interactions to encounters with other social partners as well (Lamb, 1981c; Main, 1973). Securely attached infants, for example, may generalize the trust and confidence derived from interaction with their mothers to initial experiences with other persons. Conversely, insecurely attached infants should respond more negatively to other people because of a history of inconsistent or unsatisfying interactions with their mothers.

By and large, the research evidence tends to support these formulations, particularly where interactions with infant peers are concerned. Pastor (1981), for example, found that infants who were securely attached at 12 and 18 months of age were more sociable with peers in a play session at 20–23 months than were insecurely attached infants. Waters et al. (1979) reported that infants who were securely attached at 15 months of age were more socially competent with peers in preschool settings at age $3\frac{1}{2}$. Pursuing a slightly different tack, Easterbrooks and Lamb (1979) showed that the different subgroups within the secure (Group B) classification were also associated with different degrees of sociability with peers. At 18 months, infants in Subgroups B1 and B2 engaged in more frequent and more sophisticated interaction with same-age peers in an unstructured play assessment than did infants in Subgroups B3 and B4. (Insecurely attached infants were not tested.) The latter infants relied on proximal interactive modes when they were with their mothers, and this reliance may have inhibited successful peer encounters in the mother's presence.

Only one researcher has examined the relationship between the security of attachment and the infant's reactions to unfamiliar adults outside the Strange Situation. Main (1973) assessed the attachment status at 12 months of age and then observed the same infants at 20 months in a Bayley examination and during a play session with an unfamiliar adult. She reported that the securely attached infants were more cooperative, exhibited greater "game-like spirit" during the Bayley examination, and were also more friendly and cooperative with the adult playmate than were insecurely attached infants. Main did not examine subgroup differences, however.

In all, therefore, there appear to be consistent and predictable relationships between the security of infant–mother attachment and the infant's responsiveness to unfamiliar social partners. Unfortunately, however, much of this research is difficult to interpret because it relies on relationships between assessments of the security of attachment at one age and measures of sociability or social competence in a later assessment. The interpretation of these findings thus depends on assumptions regarding the consistency of attachment classifications over time, and the information needed to assess these assumptions is typically unavailable. Clearly, this information is crucial to understanding these research results. If predictive associations were evident only when the security of attachment remained stable, this finding would suggest that sociability is related to the *current* status of infant–mother attachment, rather than to the security of attachment during some special early period. If predictive relationships were apparent whether or not attachment status remained the same, however, a somewhat more complicated view of developmental continuity would be required. It is thus unclear how other dimensions of socioemotional responsiveness are influenced by stability and change in attachment relationships.

To address this issue, we included a brief assessment of stranger sociability at each age immediately before the Strange Situation procedure, using a modified version of a procedure developed by Stevenson and Lamb (1979). The procedure involved a series of playful initiatives conducted by a female stranger in the mother's presence. The inital social bids occurred while the infant was seated on the mother's lap: the stranger first offered an attractive toy to the baby and then attempted to initiate a give-and-take exchange. Then, the infant was placed on the floor, and the baby's initial response to floor freedom was observed. The stranger then moved to the floor, again offered a toy, and then initiated a turn-taking exchange. After a few moments of play, the stranger attempted to pick up the baby, before finally leaving the room.

The entire assessment was thus brief (five to six minutes in duration), informal, and nonstressful. Sociability was consistently assessed before the Strange Situation because the Strange Situation usually causes some distress to the baby, whereas the sociability assessment does not. Thus, no infants began either assessment in a distressed state. From videotaped records, the infant's inital response to each successive initiative was scored on behaviorally based 1- to 5-point scales, with 1 indicating withdrawal or other negative reactions, and 5 indicating an outgoing, friendly response. A summary assessment was also conducted, and the sum of these nine ratings constituted the infant's score for stranger sociability (see Lamb, 1982; Thompson & Lamb, 1982, for further details).

Mean stranger sociability scores were very similar at each age ($12\frac{1}{2}$ months: 34.53; $19\frac{1}{2}$ months: 34.95), and the $12\frac{1}{2}$-month and $19\frac{1}{2}$-month scores were significantly correlated ($r = .40$, $p < .01$). Thus the infants in general did not become more or less sociable to the stranger during the second year, and individual differences in stranger responsiveness were moderately stable over time.

There was a consistent relationship between the stranger sociability scores and attachment subgroup status at each age. At $12\frac{1}{2}$ and $19\frac{1}{2}$ months, the infants in the B1 subgroup received the highest sociability scores ($12\frac{1}{2}$-month, M = 41.00; $19\frac{1}{2}$-month, M = 41.75), whereas the lowest scores were obtained by infants in subgroups A1 ($12\frac{1}{2}$-month, M = 30.33; $19\frac{1}{2}$-month, M = 31.00) and C2 ($12\frac{1}{2}$-month, M = 23.00; $19\frac{1}{2}$-month, M = 20.00). Because the number of infants in some of the subgroups was small, we combined across subgroups in the manner suggested by the findings of Easterbrooks and Lamb (1979) and Blehar and colleagues (1977) (i.e., A vs. B1,2 vs. B3,4 vs. C) before testing the reliability of the subgroup differences (see Table 5). We found that these differences were significant at $19\frac{1}{2}$ months, $F(3,39) = 3.02$, $p < .045$, and were marginally significant at $12\frac{1}{2}$ months, $F(3,39) = 2.50$, $p < .075$. A series of planned pairwise comparisons revealed that at each age, B1,2 infants were significantly more sociable than B3,4 infants. They were also more sociable than Group C babies at $19\frac{1}{2}$ months, but the difference was not significant at $12\frac{1}{2}$ months.

The relationship between stranger sociability and the security of attachment was further underscored when we examined the correlations

TABLE 5. Stranger Sociability Scores for Attachment Groups for $12\frac{1}{2}$-Month and $19\frac{1}{2}$-Month Assessments

Attachment group	Mean $12\frac{1}{2}$-month sociability score	Mean $19\frac{1}{2}$-month sociability score
Avoidant (A)	33.29	33.00
Secure (B) Subgroups 1,2	38.36	39.21
Secure (B) Subgroups 3,4	32.12	32.73
Resistant (C)	33.50	33.12

between sociability scores and ratings on the interactive variables in the Strange Situation (see Table 6). At each age, the sociability scores correlated negatively with contact maintenance and positively with distance interaction with the mother during the reunion episodes. As one might expect, however, the relationships between sociability and stranger directed interactive behaviors in the Strange Situation were more impressive. At both 12½ and 19½ months, the sociability scores correlated positively with the preseparation- and the separation-episode scores for distance interaction, and with the preseparation proximity-seeking scores. In addition, at 19½ months, the sociability scores were negatively related to the preseparation ratings of infant resistance and avoidance of the stranger. In sum, the strongest and most consistent correlates of stranger sociability were the measures of mother and stranger-directed distance interaction throughout the Strange Situation procedure.

Taken together, these results indicate that stranger sociability is strongly and consistently associated with the security of infant–mother attachment. The greater sociability of infants in subgroups B1 and B2

TABLE 6. Correlations between Stranger Sociability Scores and Interactive Behaviors for 12½-Month and 19½-Month Assessments[a]

	Mother preseparation	Mother reunion	Stranger preseparation	Stranger separation
		12½ Months		
Proximity seeking	−.20	.20	.36[b]	.15
Contact maintaining	−.32[b]	−.38[b]	.00	.07
Resistance	.14	.05	.13	−.01
Avoidance	.01	.04	−.28	−.26
Distance interaction	.21	.50[c]	.59[f]	.52[e]
		19½ Months		
Proximity seeking	.12	−.17	.37[b]	.07
Contact maintaining	−.27	−.35[b]	.00	−.27
Resistance	−.12	−.04	−.44[d]	−.09
Avoidance	−.26	−.16	−.50[c]	−.26
Distance interaction	.28	.40[c]	.46[d]	.60[f]

[a] Preseparation episodes (Episodes 1 and 2), reunion episodes (Episodes 4 and 7), and separation episodes in the company of the stranger (Episodes 3 and 6) were combined to increase the reliability of episode-by-episode scores.
[b] $p < .05$.
[c] $p < .01$.
[d] $p < .005$.
[e] $p < .001$.
[f] $p < .0005$.

parallels Easterbrooks and Lamb's (1979) findings regarding peer sociability, and the correlations between sociability scores and interactive variables in the Strange Situation suggest the existence of an underlying dimension of distal interaction linking both constructs.

These relationships are both sensible and interpretable. During the Strange Situation, B1 (and, to a lesser extent, B2) infants rely primarily on distal modes of interaction with the mother: smiling, vocalizing, showing toys, and the like. This social style may foster amiable responsiveness to unfamiliar adults in at least two ways. First, these infants can check back to their mothers for emotional reassurance (cf. Mahler, Pine, & Bergman, 1975) without having to break off interaction with the stranger and return to the mother. Second, because proximity to a stranger increases the likelihood of wariness (Sroufe, 1977), babies who are accustomed to interacting with adults across a distance can more readily engage strangers in positive social interaction than can infants who are accustomed to more proximal interactive modes. For these reasons, B1,2 infants have a social style that is most likely to yield positive, sociable interactions with unfamiliar adults.

The opposite is true of A1 and C2 infants. The former conspicuously avoid their mothers, particularly during reunions, and the latter mingle passivity in social interaction with angry and resistant behaviors. In general, infants in both subgroups show fewer positive social behaviors toward their mothers than do infants in other classification subgroups. When these tendencies are generalized to other social partners, A1 and C2 infants have a social style that is least likely to yield positive, sociable interactions with unfamiliar adults.

Stranger sociability and the security of infant–mother attachment thus appear to be related developmental phenomena. This finding is particularly significant in view of the general failure of researchers to find meaningful and consistent relationships between stranger responsiveness and the amount of the baby's prior experience with *non*familial partners, such as babysitters, daycare, and similar experiences (Beckwith, 1972; Clarke-Stewart, Umeh, Snow, & Pederson, 1980; Harmon, Morgan & Klein, 1977; Morgan & Ricciuti, 1969; Stevenson & Leavitt, 1980). In this study, we obtained similar findings: maternal questionnaire reports of the range of the infant's peer experiences, the amount of experience with nonfamilial caregivers, the number of separations from the mother in the company of others, and similar variables were not related to stranger sociability scores at either age (see Thompson & Lamb, 1982, for further details). Individual differences in stranger sociability thus appear to be more strongly affected by interactions *within* the family (including the security of the infant–mother relationship) than by *extra*familial experiences. In short, the quality of mother–infant interaction is strongly related to the nature of the baby's social exchanges with partners outside the dyad.

In this middle-class sample, however, the security of the infant–mother attachment relationship changed in nearly half the families we studied. How did these changes affect the stability of the sociability scores over time? We found that when attachment relationships were stable, the 12½-month and 19½-month sociability scores correlated very highly ($r = .74$, $p < .0001$). When the security of attachment changed over time, however, the sociability scores were not significantly correlated ($r = -.18$). The same trend was evident when we compared the degree of change in sociability scores over the two assessments when the security of attachment was stable and when it was not. When the attachment status remained consistent, the average absolute change in the sociability scores was 4.22; when the attachment relationships changed, the average degree of change in stranger sociability was 7.45 ($t = -2.30$, $df = 41$, $p < .03$). The infants who exhibited the smallest change in sociability scores were those who were securely attached at 12½ months and remained so at 19½ months. Those who showed the greatest change in sociability were those who changed from securely to insecurely attached.

Thus, the strong association between the security of attachment and stranger sociability was reflected not only in their consistent contemporaneous relationships, but in their temporal interrelationships as well. Individual differences in mother–infant relationships and the quality of stranger responsiveness changed in an intercoordinated fashion over time. When changes in family conditions fostered a renegotiation of mother–infant interaction, this appears to have affected the quality of the baby's sociability toward unfamiliar adult partners as well. Conversely, although the sociability scores were significantly correlated over time for the sample as a whole, the magnitude of this correlation was greatest when the security of attachment was consistent over time. In other words, the sociability scores were highly stable when the security of attachment was also stable over time.

These results demonstrate that the quality of the mother–infant relationship has meaningful and significant ramifications for other dimensions of socioemotional responsiveness, a conclusion that is consistent with the findings of the other studies reviewed earlier. Importantly, these results indicate also that although the security of attachment is likely to provide important information about concurrently assessed dimensions of socioemotional responding, it is likely to have *predictive* value only when the attachment status is consistent over time, because stability and change in the security of attachment is strongly related to consistency and change in other socioemotional variables. Thus, when attachment relationships change (as they did in nearly half the cases in this sample), predictive relationships are likely to be nonsignificant, even though contemporaneous associations are strong and consistent. In short, stability and change in mother–infant interaction importantly influence the consistency of individual differences in social responsiveness to other partners—a finding that underscores the strong relationships between events and experiences occurring within and beyond the mother–infant dyad.

Conclusion

The results reported here show that the mother–infant relationship is linked to broader social networks in at least two ways. First, mother–infant interaction is affected by changes and stresses occurring within the family as a whole, particularly those with the greatest direct impact on mother and baby. Second, the security of infant–mother attachment also has important implications for the baby's responsiveness to other, nonfamilial social partners. In other words, the mother–infant dyad is itself affected by broader social experiences and, in turn, has an effect on them.

This finding both corroborates and extends our understanding of the meaning of individual differences in the security of infant–mother attachment. It corroborates earlier research findings indicating that individual differences in the security of attachment originate in consistent features of mother–infant interaction. At the same time, it broadens our appreciation that these interactive differences derive not only from the mother's personality and caregiving style, but also from the ongoing experiences of mother and baby within the family, and it places these differences in a temporal context. Individual differences in attachment relationships do not endure unchanged but are affected by changes in the family environment. In short, the security of attachment assessed at any point in time, as well as its temporal stability during the infancy years, provides meaningful information not only about what is going on between mother and baby, but about how they function within the family as a whole.

Changes in the security of attachment also have important implications for other aspects of socioemotional responsiveness in infancy. Our findings corroborate earlier indications that individual differences in the security of attachment extend beyond the mother–infant dyad to affect the baby's sociability with other partners as well. We extend earlier findings by showing that the stability of individual differences in stranger responsiveness is significantly influenced by consistency or change in the security of the mother–infant attachment. When the security of attachment is stable, so too are differences in stranger sociability; when the attachment relationships change, the sociability scores do also. This association seems to occur as a result of the generalization of interactive skills and tendencies that the infant learns in the context of mother–infant interactions.

To be sure, individual differences in stranger responsiveness are importantly influenced by other variables, such as infant temperament (e.g., Lamb, 1982; Stevenson & Leavitt, 1980; Thompson & Lamb, 1982). But whereas temperamental dimensions tend to be highly stable over time, the security of infant–mother attachment changes quite often, and these changes foster parallel changes in stranger sociability. Because previous studies have shown poor relationships between differences in stranger responsiveness and the infant's prior social experiences outside the family, it appears that the quality of family relationships—particularly with the mother—constitutes one of the most important experiential determinants of stranger sociability during infancy.

In all, it is clear that mother–infant interaction both influences and is influenced by social networks that exist outside of the dyad. Such a conclusion, although neither surprising nor unexpected, helps us to view the mother–infant relationship in the context of the broader social ecology of infancy.

Acknowledgments

We are very grateful for the assistance provided by Lisa Colvin, Kate Craig, Susan Dickstein, Melinda Feller, Leslie Fried, Michael Iskowtiz, Margaret Madden, Susan Piconke, Jane Stein, Jamie Steinberg, Susan Strzelecki, and Michele Tukel. David Estes and Catherine Malkin provided special help with testing and scoring. We are also appreciative of the technical assistance provided by Lee Davis. Finally, we wish to acknowledge a special debt to the families who participated in this study.

References

Ainsworth, M. D. S. *Infancy in Uganda: Infant care and the growth of love.* Baltimore: Johns Hopkins University Press, 1967.

Ainsworth, M. D. S. The development of infant-mother attachment. In B. Caldwell & H. Ricciuti (Eds.), *Review of child development research* (Vol. 3). Chicago: University of Chicago Press, 1973.

Ainsworth, M. D. S., & Bell, S. M. Some contemporary patterns of mother–infant interaction in the feeding situation. In A. Ambrose (Ed.), *Stimulation in early infancy*. London and New York: Academic Press, 1969.

Ainsworth, M. D. S., Bell, S. M., & Stayton, D. J. Individual differences in strange-situation behavior of one-year-olds. In H. R. Schaffer (Eds.), *The origins of human social relations*. London and New York: Academic Press, 1971.

Ainsworth, M. D. S., Blehar, M. C., Waters, E., & Wall, S. *Patterns of attachment*. Hillsdale, N.J.: Erlbaum, 1978.

Arend, R., Gove, F. L., & Sroufe, L. A. Continuity of individual adaptation from infancy to kindergarten: A predictive study of ego-resiliency and curiosity in preschoolers. *Child Development*, 1979, 50, 950–959.

Beckwith, L. Relationships between infants' social behavior and their mothers' behavior. *Child Development*, 1972, 43, 397–411.

Blehar, M. C., Lieberman, A. F., & Ainsworth, M. D. S. Early face-to-face interaction and its relation to later infant–mother attachment. *Child Development*, 1977, 48, 182–194.

Bowlby, J. *Attachment and loss, Vol. 1: Attachment*. New York: Basic Books, 1969.

Clarke-Stewart, K. A., Umeh, B. J., Snow, M.E., & Pederson, J. A. Development and prediction of children's sociability from 1 to $2\frac{1}{2}$ years. *Developmental Psychology*, 1980, 16, 290–302.

Connell, D. B. *Individual differences in attachment: An investigation into stability, implications, and relationships to structure of early language development.* Unpublished doctoral dissertation, Syracuse University, 1976.

Easterbrooks, M. A., & Lamb, M. E. The relationship between quality of infant-mother attachment and infant competence in inital encounters with peers. *Child Development*, 1979, 50, 380–387.

Erikson, E. *Childhood and society* (2nd ed.). New York: Norton, 1963.

Estes, D., Lamb, M. E., Thompson, R. A., & Dickstein, S. *Maternal affective quality and security of attachment at 12 and 19 months*. Paper presented to the biennial meeting of the Society for Research in Child Development, Boston, April 1981.

Freud, S. *An outline of psychoanalysis*. London: Hogarth, 1938.

Garbarino, J. & Sherman, D. High-risk neighborhoods and high-risk families: The human ecology of child maltreatment. *Child Development*, 1980, 51, 188–198.

Harmon, R. J., Morgan, G. A., & Klein, R. P. Determinants of normal variation in infants' negative reactions to unfamiliar adults. *Journal of the American Academy of Child Psychiatry*, 1977, 16, 670–683.

Hollingshead, A. B. *Four-factor index of social status*. Unpublished manuscript, Yale University, 1975.

Lamb, M. E. Father–infant and mother–infant interaction in the first year of life. *Child Development*, 1977, 48, 167–181.

Lamb, M. E. The effects of the social context on dyadic social interaction. In M. E. Lamb, S. J. Suomi, & G. R. Stephenson (Eds.), *Social interaction analysis*. Madison: University of Wisconsin Press, 1979.

Lamb, M. E. Developing trust and perceived effectance in infancy. In L. P. Lipsitt (Ed.), *Advances in infancy research* (Vol. 2). Norwood, N.J.: Ablex, 1981.

Lamb, M. E. The development of father–infant relationships. In M. E. Lamb (Ed.), *The role of the father in child development* (rev. ed.). New York: Wiley, 1981(b).

Lamb, M. E. The development of social expectations in the first year of life. In M. E. Lamb & L. Sherrod (Eds.), *Infant social cognition*. Hilsdale, N.J.: Erlbaum, 1981.(c)

Lamb, M. E. The origins of individual differences in infant sociability and their implications for cognitive development. In H. W. Reese & L. P. Lipsitt (Eds.), *Advances in child development and behavior* (Vol. 16). New York: Academic Press, 1982.

Lamb, M. E., Chase-Lansdale, L., & Owen, M. T. The changing American family and its implications for infant social development: The sample case of maternal employment.

In M. Lewis & L. Rosenblum, (Eds.), *The child and its family.* New York: Plenum Press, 1979.

Lamb, M. E., Owen, M. T., & Chase-Lansdale, L. The working mother in the intact family: A process model. In R. R. Abidin (Ed.), *Parent education and intervention handbook.* Springfield, Ill.: Charles C Thomas, 1980.

Lewis, M., & Rosenblum, L. A. (Eds.). *The development of affect.* New York: Plenum Press, 1978.

Lewis, M., Young, G., Brooks, J., & Michalson, L. The beginning of friendship. In M. Lewis & L. A. Rosenblum (Eds.), *Friendship and peer relations.* New York: Wiley, 1975.

Lieberman, A. F. Preschoolers' competence with a peer: Relations with attachment and peer experience. *Child Development,* 1977, *48,* 1277–1287.

Maccoby, E. E., & Masters, J. C. Attachment and dependency. In P. H. Mussen (Ed.), *Carmichael's manual of child psychology* (Vol. 2). New York: Wiley, 1970.

Mahler, M. A., Pine, F., & Bergman, A. *The psychological birth of the human infant.* New York: Basic Books, 1975.

Main, M. *Exploration, play, and cognitive functioning as related to child-mother attachment.* Unpublished doctoral dissertation, Johns Hopkins University, 1973.

Main, M., & Weston, D. R. The quality of the toddler's relationship to mother and to father: Related to conflict behavior and the readiness to establish new relationships. *Child Development,* 1981, *52,* 932–940.

Main, M., Tomasini, L., & Tolan, W. Differences among mothers of infants judged to differ in security. *Developmental Psychology,* 1979, *15,* 472–473.

Masters, J. C., & Wellman, H. W. The study of human infant attachment: A procedural critique. *Psychological Bulletin,* 1974, *81,* 218–237.

Matas, L., Arend, R., & Sroufe, L. A. Continuity of adaptation in the second year: The relationship between quality of attachment and later competence. *Child Development,* 1978, *49,* 547–556.

Morgan, G., & Ricciuti, H. Infants' responses to strangers during the first year. In B. M. Foss (Ed.), *Determinants of infant behavior* (Vol. 4). London: Methuen, 1969.

Mueller, E. C., & Vandell, D. Infant–infant interaction. In J. D. Osofsky (Ed.), *Handbook of infant development.* New York: Wiley, 1979.

Parke, R.D., & Collmer, C. W. Child abuse: An interdisciplinary analysis. In E. M. Hetherington (Ed.), *Review of child development research* (Vol. 5). Chicago: University of Chicago Press, 1975.

Pastor, D. L. The quality of mother-infant attachment and its relationship to toddlers' initial sociability with peers. *Developmental Psychology,* 1981, *17,* 326–335.

Pedersen, F. A. Father influences viewed in a family context. In M. E. Lamb (Ed.), *The role of the father in child development* (rev. ed.). New York: Wiley, 1981.

Sears, R. R., Maccoby, E. E., & Levin, H. *Patterns of child rearing.* Stanford, Calif.: Stanford University Press, 1957.

Sroufe, L. A. Wariness of strangers and the study of infant development. *Child Development,* 1977, *48,* 731–746.

Sroufe, L. A. Attachment and the roots of competence. *Human Nature,* 1978, *1*(10), 50–57.

Sroufe, L. A. The coherence of individual development: Early care, attachment, and subsequent developmental issues. *American Psychologist,* 1979, *34,* 834–841.

Sroufe, L. A., & Waters, E. Attachment as an organizational construct. *Child Development,* 1977, *48,* 1184–1199.

Stayton, D. J., & Ainsworth, M. D. S. Individual differences in infant responses to brief, everyday separations as related to other infant and maternal behaviors. *Developmental Psychology,* 1973, *9,* 226–235.

Stevenson, M. B., & Lamb, M. E. Effects of infant sociability and the caretaking environment on infant cognitive performance. *Child Development*, 1979, *50*, 340–369.

Stevenson, M. B., & Leavitt, L. A. *Associations among temperament, sociability and social experiences in one-year-olds.* Paper presented to the biennial meeting of the International Conference for Infant Studies, New Haven, Conn., April 1980.

Thompson, R. A., & Lamb, M. E. Stranger sociability and its relationships to temperament and social experience during the second year. *Infant Behavior and Development*, 1982, *5*, 277–287.

Thompson, R.A., Lamb, M. E., & Estes, D. Stability of infant-mother attachment and its relationship to changing life circumstances in an unselected middle-class sample. *Child Development*, 1982, *53*, 144–148.

Thompson, R. A., Lamb, M.E., & Estes, D. Harmonizing discordant notes: A reply to Waters. *Child Development*, 1983, *54*, 521–524.

Vaughn, B., Egeland, B., Sroufe, L. A., & Waters, E. Individual differences in infant–mother attachment at twelve and eighteen months: Stability and change in families under stress. *Child Development*, 1979, *50*, 971–975.

Waters, E. The reliability and stability of individual differences in infant–mother attachment. *Child Development*, 1978, *49*, 483–494.

Waters, E. The stability of individual differences in infant attachment: Comments on the Thompson, Lamb, and Estes contribution. *Child Development*, 1983, *54*, 516–520.

Waters, E., Wippman, J., & Sroufe, L. A. Attachment, positive affect, and competence in the peer group: Two studies in construct validation. *Child Development*, 1979, *50*, 821–829.

10

How Parents Influence Their Children's Friendships

ZICK RUBIN AND JONE SLOMAN

>Mr. and Mrs. Nelson, deciding among several houses to buy, finally settle on the one on the street where they noticed a group of 4-year-old children at play; their son Peter is 4.
>
>Mrs. Freeman, who wishes that her daughter Sally would make closer friendships, learns that Sally has played with Robin in school. She calls Robin's mother to invite Robin over to play, so the girls can get to know each other better.
>
>Mr. Ramsay listens to his son Howard's account of a fight he has had with his best friend Steven and then tells Howard how *he* would have handled the situation.

Friendships are among the central ingredients of children's lives, from as early as age 3 and, in some cases, even earlier. Children's relationships with their peers directly affect their well-being, provide an opportunity to learn and practice social skills, and may establish enduring patterns of relating to others (cf. Rubin, 1980). Parents typically recognize the importance of children's friendships and often try to help their children establish and maintain rewarding friendships. Even parents who have no specific intention of influencing their child's friendships can hardly avoid doing so, through the settings they choose to live in, their reactions to the child's social behavior, and the values they convey through their

---

ZICK RUBIN AND JONE SLOMAN ● Department of Psychology, Brandeis University, Waltham, Massachusetts 02254. This research was made possible by a grant to Zick Rubin from the Foundation for Child Development.

own relationships with others. Whether or not these parental influences are intended, and whether or not they are recognized by the parents themselves, they are among parents' most important legacies to their children.

These parental influences on children's friendships have been largely ignored by researchers. To be sure, parent–child relationships have long been a central concern of students of child development, and in recent years, children's friendships and peer relationships have also become topics of active interest. For the most part, however, the two sorts of relationships have been treated as separate spheres. Parent–child relationships and peer relationships have typically been studied in different settings and by different investigators. Relatively few studies have examined children's relationships with both their parents and their peers, and little attention has been given to the ways in which one sort of relationship may influence the other. As Hartup (1979) wrote, "Quite possibly the most serious oversight in the literature on social development is the absence of information concerning the interdependencies existing between experiences in one social world and experiences in others" (p. 944).

In this chapter we undertake a preliminary examination of the ways in which parents influence their children's friendships and peer relationships. We will describe five different modes of influence, with illustrations concerning parents' impact on the friendships of their 4-year-old sons and daughters:

1. *Setting the stage.* By making decisions about what neighborhood the family lives in, what school or day-care setting the child is in, and whom the parents themselves associate with, parents define to a large extent the pool of potential playmates for their children, thus setting the stage for their social interactions and relationships.

2. *Arranging social contacts.* Parents often arrange social contacts outside school for their children, performing such functions as scheduling visits between playmates, enrolling children in organized activities, and chauffeuring children from one social setting to another.

3. *Coaching.* Parents commonly give their chidren advice and guidance about whom they should have as friends and how they should conduct their social interactions and relationships.

4. *Providing models of social relationships.* Parents influence their children's friendships by serving as models, through their own social interactions and relationships, of how friendships should be conducted.

5. *Providing a home base.* The quality and the security of parent–chid relationships (the "home base") can influence children's readiness to explore peer relationships and affect their orientation toward such relationships.

There are individual studies in the research literature that relate to one or another of these modes of influence. With respect to stage setting, for example, Berg and Medrich (1980) have explored the impact of neighborhoods on 12-year-old children's social relationships, and many investigators have examined the impact of different school settings on children's peer relationships (e.g., Smith & Connolly, 1981). With respect to arranging social contacts, O'Donnell and Stueve (1983) have investigated social-class differences in parents' involvement in their school-age children's organized activities. With respect to coaching, Sloman (1981) has examined the impact of mothers' active involvement or noninvolvement during play sessions on the nature of their toddlers' play. With respect to modeling, the observations of Berg and Medrich (1980) point to ways in which children's social relationships may be patterned after their parents' relatively formal or informal styles of social interaction. And with respect to providing a home base, Lieberman (1977), among others, has found links between the security of preschool children's bonds with their mothers, on the one hand, and their social competence with peers, on the other hand. In addition, Rosenblum and Plimpton (1979) have illustrated several of these modes of influence in their description of adults' effects on infant–peer interaction among two species of macaques.

Despite these relevant observations in the literature, there have been no efforts to consider parental influence on children's friendships within a broader framework. We offer our five categories as a convenient typology for organizing the wide range of parental influences. These categories are illustrated with material from interviews with the parents of 4-year-olds, and much of this paper can be taken as a discussion of aspects of preschoolers' social relationships. We believe, however, that the same categories are applicable to parents' influence on the social relationships of children of all ages, from infancy through adolescence. Thus, we hope that our framework will be useful in research that focuses on the ways in which parents' influence on their children's social lives develops over time.

The illustrative data for this analysis come from semistructured interviews with 18 parents (16 mothers and 2 fathers) of 4-year-old children, as well as one pilot interview with the mother of a 5-year-old. Interviews with the children themselves were also conducted. The sample was recruited through a combination of word-of-mouth contacts, referrals from community day-care centers and nursery schools, and departmental subject pools, with the goal of obtaining a diverse sample. Half of the children were boys, half were girls; half were firstborns, half later-borns; half were from working-class homes and half from middle-class homes. Twelve of the children lived with both of their parents, four children lived in single-parent homes, and two lived with one parent and a stepparent. Six of the mothers, as well as the two fathers, worked outside the home at least three days a week. The interviews were conducted by Jone Sloman in the summers of 1980 and 1981. Each interview took about 1½ hours and was conducted in the respondent's home.

In the next section of the paper, we discuss parents' values about their children's friendships, these values often underlie the parents' attempts to influence their children's social lives. In the succeeding five sections, we examine in turn each of the five modes of influence that have been specified—setting the stage, arranging social contacts, coaching, modeling, and providing a home base. We then consider the ways in which the parents' previous experience with child rearing, as well as their own childhood experiences, may affect the sorts of influence that they exert. Finally, we consider some implications of our analysis for the understanding of children's social development.

Parents' Values about Their Children's Friendships

Parents have their own values, whether explicit or implicit, about various aspects of children's peer relationships: the importance of such relationships, their benefits and liabilities, the sorts of children who make good (or bad) friends, the ways in which friendships should be conducted, and the degree to which they should be guided and controlled by parents.

1. *How important are peer relationships for young children?* Until fairly recently, many psychologists espoused the view that young children's relationships with their peers tended to be relatively superficial and unimportant, vastly overshadowed by children's relationships with their parents. In recent years, however, psychologists have moved to the view that friendships are likely to be of great importance to preschool children (see Rubin, 1980).

A few of the parents in our sample felt that friendships remain relatively unimportant for preschool children:

> I tend to feel that little kids form close relationships only to adults. (Mr. Shaw)

> Right now, kids aren't a big part of Jake's life. When he starts school they will be. (Mrs. Perry)

But most of the parents we interviewed shared the emerging view that friendships were of great importance to their children. They saw satisfying peer relationships as contributing to their children's well-being:

> I'm happy she has Andrea. They can be a handful, but they are such good friends. (Mrs. McKay)

And they saw peer relationships as necessary to teach their chidren essential social skills:

> He is smart but he is not going to be happy unless he learns how to deal with others, and that is what I think is important about friends—even when you are 4. (Mrs. Dorr)

Parents also mentioned the value of peers in helping young children to develop a needed measure of independence from their parents and in preparing them for group situations in school.

For the most part, indeed, parents seem to take for granted that friendships are beneficial to their 4-year-olds. As a result, a lack of interest in friends may be seen as a danger sign. Mrs. Dorr, for example, had been upset by Matthew's apparent lack of interest in other children. She was greatly relieved when, in part as a result of her encouragement, he began to be more sociable.

> Sometimes I say to my husband when he comes home, "Matthew said he wanted to have somebody over," and we both smile because it is so much of a relief that he wanted to as opposed to not caring.

Whereas most of the parents whom we interviewed focused on the benefits of friendship, several of them also were wary of an overemphasis on sociability. Some parents noted that children who are overly concerned with their social life may slight other aspects of their development:

> I think with my first child I pushed socializing too much. When she got to school, she thought it was going to be just one big social occasion again. She couldn't settle down to schoolwork. (Mrs. Foley)

In addition, some parents bemoan the fact that peers are likely to expose the child to new behaviors and values that run counter to the behaviors and values that the parents desire to inculcate:

> I detected prejudices that he didn't have himself that he was picking up. Race things and boy–girl things. These two guys he hung around with were real macho-men. Needless to say, I hope my boys don't turn out to be macho-men. (Mr. Shaw)

This sort of concern about the clash between parental and peer values is usually discussed in connection with later childhood and adolescence, but our interviews demonstrated that it is a concern of the parents of preschool children as well.

2. *What sort of friends are "good"—and "bad"—for the child?*

> The other day a girlfriend and I were sitting around discussing whose kids we liked and whose we didn't. We thought we were really picky. But you do get very fussy about who they play with. (Mrs. Chaney)

Parents frequently have strong views about which other children are "good" or "bad" playmates for their own children. Other parents are less "picky" than Mrs. Chaney, taking—or at least professing—the view that any child whom their child enjoys being with is an appropriate playmate. Nevertheless, the large majority of parents do prefer particular sorts of friends for their children.

Parents generally prefer to have their children play with others of similar social background and values. To be sure, many parents would also like their children to get to know others from a wide range of social backgrounds. As Mr. Kimball said, "I feel that he has to know that there are lots of kinds of people out there, and you have to fend for yourself given that there are all these people." In practice, however, parents typically do more to encourage friendships with other children from similar than from widely different backgrounds.

Most of the parents we interviewed were open to the possiblity of their children having friends of both sexes. With today's heightened emphasis on nonsexist child-rearing, many parents—especially middle-class parents—recognize that there may be value in changing the prevailing pattern of same-sex friendships. Nevertheless, the parents in our sample remained most likely to expect and to promote same-sex friendships. This preference was not so often stated explicitly as it was taken for granted that same-sex friendships are more desirable or appropriate. "I wish somebody would move around here that was 'a nice little boy,'" Mrs. Chaney said, thus betraying the view that "a nice little girl" would not be as suitable for her son. And the deep-seated feeling that same-sex friends are more "natural" may rise to the surface if a child shows a preference for friends of the other sex:

> The other kids he gravitates to are mostly girls. I think at this age it doesn't really matter. But it does bother me a little on a gut level. I would like him to be a little more boyish. A little more like Barry. If it weren't for Barry [his best friend], I guess I would really worry. (Mrs. Dorr)

Many parents also expressed the view that it is valuable for children to have social contact with older and younger children, as well as with children of the same age. Our parents noted that children can learn various skills from older children and can learn to be more nurturant and compassionate by playing with younger children. Nevertheless, our parents generally believed that friendships between children of the same age were most likely to be successful. Mrs. Tierney expressed what seemed to be the modal view:

> I think anyone Michael's own age is a favorite. Kids younger are boring, and kids older are exciting. He can be good with younger kids, but it doesn't last long. He is slightly awed and afraid with older kids, but it is thrilling. He is most comfortable with his own age group. As long as he is around younger

and older children at least some of the time, I think it's good to have most
of his play with children of about the same age. It just seems to work best.

In addition to their views about general categories of potential playmates, parents are also likely to have opinions about specific children as friends for their child. One child is approved of because of his good looks, another disapproved of because of her awkward manner. In many cases, parents act on such preferences in attempting to influence their child's social relationships.

3. *How should children's friendships be structured?* In addition to their views about whom their child should have as friends, parents have ideas about how these friendships should be structured. Most centrally, a child's network of friendships may be relatively "exclusive" or "inclusive." In the "exclusive" case, the child may have only one or two friends, but these friends may be quite close. In the "inclusive" case, the child may have many friends or playmates, but these friendships may tend to be superficial. To get an indication of parents' values on this dimension, we presented them with the following vignette:

Nick and Tim play together almost all the time at nursery school and they rarely play with any other children. What should their teacher do?

(A) She should encourage them to separate some of the time and play with other children.

(B) She should not encourage them to separate.

A slight majority of the parents we interviewed (11 out of 18) said that the friends should not be separated. These parents emphasized the special value that a close friendship may have for children:

Unless it was really extreme, this friendship is obviously serving a need for the children, so I would just let it be. (Mrs. Hoffman)

Those parents who believed that the close friends should be separated emphasized the need to learn to get along with a larger group of others:

When you get to know just one other child, you get to know just one other style of playing. And I don't think that is good. You have to know how to get along with a lot of people in this world, and learning that begins early. (Mrs. Dorr)

The parents also expressed attitudes about other aspects of children's friendships. Some parents believed, for example, that it is valuable for children to have friendships that endure for long periods of time, whereas other parents did not see any special value in long-term friendships for young children. Some parents felt that children should be able to get together with their friends spontaneously, whereas other were just as happy to have their contact scheduled ahead of time. And some parents viewed the child's worlds of family and of friends as essentially separate spheres, whereas other parents placed a premium on integrating these spheres. These preferences often seemed to correspond to the patterns of friendship that the parents had adopted in their own lives.

4. *To what extent should parents direct or control their young children's social lives?* Parents have widely differing views about the extent to which they should guide or direct their preschool children's social interactions and relationships. Some of the parents we interviewed had few reservations about actively regulating their children's friendships:

> There have been times when there were kids I tried to keep her away from temporarily. You might as well exercise some control over that while you can. Later on, you don't get the opportunity. (Mrs. Foley)

Other parents took more of a *laissez-faire* position:

> You can't pick who your kids' friends are, anyway. It just wouldn't work. They have some rights. They have feelings, too, and you've got to trust these sometimes. (Mrs. Mendoza)

These views are clearly related to broader values about children's emerging independence over the course of childhood. Most parents agree that children must eventually learn to make decisions for themselves, including decisions about friends. Parents differ, however, about the proper balance of control and independence for children at particular ages.

All of the values that we have discussed—parents' views of the importance of friendships, their appraisals of "good" and "bad" friends, their ideas about how friendships should be structured, and their feelings about the degree to which parents should control their young children's friendship—are closely related to the attempts at influence that parents actually make. It should be noted, however, that parents do not typically embark on the journey of child rearing with an existing set of clearly held values, which they then proceed to act on. To the contrary, parents often hold their values only in a fuzzy or implicit form. In many cases, the parents' values and expectations emerge from their experience with their children's friendships, rather than being a preexisting set of views.

Setting the Stage

The neighborhood in which a child lives, the day-care program or nursery school that he or she attends, and the people with whom the child's parents socialize are the major determinants of whom the preschool child has available as potential playmates. Whereas an older child can begin to reach beyond these contexts, a 4-year-old remains almost entirely dependent on these contexts for his or her pool of friends. Parents' decisions about homes, schools, and family friends thus set the stage for the child's friendships.

The Neighborhood

For a preschool child, without much mobility and with limited school contacts, the neighborhood one lives in has a large impact on whom the child plays with and how his or her play is structured. Some neighborhoods have very few children, and some have many:

> Its a long street, but it's like retirement city. Two kids. That's it. (Mrs. Dorr)

> Its like Sesame Street. Children everywhere. There is always someone around if you want them. (Mrs. O'Neill)

Some neighborhoods are socioeconomically and ethnically homogeneous, and others are heterogeneous. Some are dangerous and inconvenient for play, and others are safe and conducive to play:

> Before we moved here, we lived in a second-floor apartment on a busy street. Rebecca was so little that every time she wanted to go outside and play, I'd have to stop what I was doing and take her out, which of course cut down on her playtime. But here, there are no steps, no traffic, and this house seems to be the geographic center of the action. The minute she comes home, she's out on the street, and I have to pull her off the street to put her to bed. (Mrs. Hoffman)

Parents choose neighborhoods to live in for many reasons: finances, proximity to workplaces, and so on. Although parents are often concerned about their community's public school system, children's informal social opportunities are only rarely an overriding consideration. Indeed, some parents in our sample acknowledged that their neighborhood was good for them, but bad for their children. Nevertheless, many parents did take into account the social opportunities for their children when they chose a home:

> Just look at this street—kids, swingsets, swimming pools everywhere. It a kid's paradise. That's one of the main reasons we moved. Before, we were in another section of town. We had a big beautiful house, but there weren't many kids to play with. (Mrs. O'Neill)

Other parents would like to move to a better neighborhood for the children's sake, if only they could:

> I hate this place for Daniel's sake. There aren't children his age, the street is too busy. The yard is very small, and there is no place to play. I would love to move, but we just can't right now. (Mrs. Chaney)

Whether a neighborhood is "good" or "bad" for a child is a subjective matter, however. For example, some parents consider it important to have same-sex and same-age playmates in the neighborhood, whereas other parents are glad to have a greater degree of diversity. Parents also affect the way the child uses the neighborhood and the opportunities that the neighborhood provides. Some parents define the "neighborhood" that the child is permitted to venture in by herself or himself more narrowly, others more broadly:

> There are some children up at the corner, but—this sounds terrible—I don't let him go up there. Some kids at 4 are able to go everywhere—you know, walk up and down the street on their own. But I like to keep him here. I don't think he is old enough for that yet. (Mrs. Donley)

> Up until a year ago, I was somewhat cautious about Amy. But now I let her go. Open the garage door and off she zooms. I worry about that sometimes. But I believe that she needs that—to be off on her own and to find her own way. (Mrs. O'Neill)

Similarly, parents set rules and convey values about whom in the neighborbood the child should play with. Many parents prohibited play with neighborhood children whom they considered undesirable:

> There is one family that lives by us—I guess there is a disciplinary problem in the family. Because of that, I don't want my kids playing with them. I told the parents, and I told Katie and Annie that they weren't to be around them. (Mrs. McKay)

But another parent, who lived in a low-income housing project, thought it was valuable for her child to be exposed even to "tough" children:

> You've got to be streetwise up here, and she's learned how to be a real toughy. She's learned how to take care of herself. She's never one to come crying home. She'll get mad, stick up for herself, or just get away from the kids. Guess you wouldn't get that in no Newton or someplace. (Mrs. Mendoza)

Parents also affect the child's use of the neighborhood by virtue of their own friendliness—or lack of it—toward other children:

> Stanley [my husband] does enjoy playing with kids. I'm not sure he really realizes the role he plays in bringing kids to our house, but I think some of the fathers on the block make the kids uncomfortable. They won't play there if the father is home. But I think the fact that Stanley has been open and warm to the kids, whether consciously or unconsciously, has made things easier for Rebecca in the neighborhood. (Mrs. Hoffman)

And, as we discuss later in this section, parents' own social contacts with adults in the neighborhood have an impact on their chidren's social possiblities. Thus, the parents' perceptions and values interact with the objective features of the neighborhood in determining the children's opportunities for friendships.

Nursery Schools and Day-Care Settings

Many parents, including all the parents in our sample, decide to place their children in nursery school or day-care programs by the time they are 3 or 4 years old. There are many reasons for doing this, including preparation for school, intellectual enrichment, and the parents' need for assistance with child care. But one of the most prevalent reasons for placing a child in a particular nursery school or day-care setting is to provide social opportunities for the child. Through experience with others in a structured group situation, the child may learn social skills that are not demanded in other settings:

> Learning how to work your way into a group when it's hard, instead of just showing up and being the center of attention, is important. (Mrs. Hoffman)

And the school setting may be seen as helping the child to attain greater independence from her or his parents:

> I had to let him know he could be away from me. It was hard, but I just felt as if I had to do it. By the end of the first year, he wanted no part of me and wanted to be with the other kids. (Mrs. Davis)

In choosing a specific day-care or nursery-school setting, parents must take many factors into account, including the school's location, its physical facilities, the tuition, the teachers, the philosophy, and the number, ages, and backgrounds of the children. Many of these factors are directly relevant to the sorts of social relationships that the child will form. One frequent consideration is the degree to which the school is structured (with scheduled and focused activities) or unstructured (permitting a great deal of free playtime). Which type of setting parents chose often depended on whether they placed a greater priority on the learning of intellectual or of social skills. Parents who were more concerned about intellectual enrichment were likely to choose a structured setting, whereas parents who were more concerned about social skills were more likely to choose an unstructured setting:

> The school is more structured than most of the programs in the area. More learning of letters and that kind of thing. And that is why I am sending her. She doesn't need all that social stuff. (Mrs. O'Neill)

> The school he was going to is fairly unstructured. They have a lot of options for them, which is good because my feeling was that his socialization was more important than his learning things. (Mrs. Dorr)

These parental intuitions about the effects of different settings are supported by research. In a large-scale study of the impact of the nursery-school environment on children, Smith and Connolly (1981) found children's attention span to be increased in a structured setting, whereas skills in resolving conflicts were enhanced in a nonstructured setting.

Another consideration is the number of children within the setting. A small group may provide a more comfortable environment and more group play, whereas a larger group may provide more choices of companions and more dyadic play (see Smith & Connolly, 1981). We presented our parents with the following situation:

> Mrs. S. is choosing between two nursery schools for her child. School A has 6 children and one teacher. School B has 24 children and four teachers. Which do you think she should choose?

Although many of the parents felt that a smaller group might be more appropriate before age 4, most of them (12 out of 18) believed that a 4-year-old child would gain most from a larger group. Mrs. Lewis explained why she was transferring Jennifer from a smaller to a larger group, now that she was 4:

> A group of 5 is different from a group of 20. I want her to be able to form a relationship with someone when there are a lot of choices. At school now and at home, it is always such a small group that it is always group play. I think next year, in the larger school, she will form some special friendships. She will be able to handle a crowd and still decide what she wants to do.

Day-care and nursery-school settings also vary with respect to the age range of the children. Of 18 parents in our sample, 11 said that they would prefer a mixed-age setting (with children ranging from ages 2–6) for their 4-year-olds, to a same-age setting (with all 4-year-olds). The parents who preferred the mixed-age setting suggested that children can acquire a wider range of social experiences and skills by interacting with both younger and older children. The parents who chose the same-age setting typically reasoned that it would be better preparation for school. As Mrs. Davis stated, "That's who they will be with throughout their schooling, so they might as well get used to that from the beginning."

Finally, each preschool setting has its own distinctive philosophies and policies. For example, some settings make a point of admitting children of different socioeconomic and ethnic backgrounds, whereas others are more homogeneous, reinforcing the homogeneity that is likely to be found in the neighborhood. Some schools deliberately discourage sex-segregated play; some advocate a particular style of handling conflicts; some take special note of children's need for privacy. Parents often take such policies into account when they choose a setting for their child, and in many cases, parents have the opportunity to help shape the school's policies. These policies, in turn, can have a large impact on the friendships that children form and the patterns of interaction that they learn.

The Parents' Social Network

Parents also set the stage for their children's friendships through their own social networks. Many of our parents reported that their children's best friends were the children of their own friends. In some of these cases, the children also lived in the same neighborhood or attended the same school, but in many cases, they did not. Parents bring their children into contact with their friends' children by taking them with them on visits or participating in joint activities that create the potential for friendship:

> Matthew and his friend David have been together since they were about 2. I am very friendly with his mother, and we get together a lot, which is why they have been around each other so much. (Mrs. Dorr)

Once children have become friendly, moreover, if the parents are friends it is easier to make arrangements for the children to get together. Along these lines, Schiavo and his co-workers have found that preschool classmates were most likely to have contact during the summer vacation and to remain friends over a one-year period if their parents were friends than if they were not (Schiavo & Solomon, 1981; Schiavo, Solomon, Evers, & Cohen, 1982).

Of course, friendship among parents does not guarantee that their children will become friends:

> There is one little boy, who is 4, too, whose mother is a good friend of mine, but Silvia doesn't like to be with him too much. I don't know why. What can you do? She knows who she likes. (Mrs. Mendoza)

And parents' attempts to push a friendship between their children may meet with resistance:

> We were friends, so we forced the kids on each other and it backfired. After we stopped doing that and got them together only when they asked, they did fine. (Mrs. Donley)

Because of the links between parents' and children's social networks, the size of the parents' social network may expand or limit their chidren's social possiblities. Within the neighborhood, for example, outgoing parents may make it easier for the child to meet others, and reserved parents may make it harder:

> I'm not the type of person who can just go knock on someone's door and say, "Hi, how are ya." So I don't know a lot of the neighborhood kids or that many kids at all around here. There are kids around, but I just haven't been out on the street. I tend to stick to myself, and I know that affects him. (Mrs. Tierney)

The length of time that parents have spent in a community can also enlarge or restrict children's choice of friends. One child in our study was in the joint custody of her divorced parents and spent alternating three-week periods with the two parents. The father had remained in the same community, and the mother had moved to a new area. Mr. Miller discussed the impact of this arrangement on Alison's friendships:

> The parents of all these kids have known me and Alison since the separation or before. So I feel it will be easy to maintain the friendships. But she doesn't talk about her friends in Rhode Island that much. Her mother hasn't lived there that long, so I doubt that they have developed the same kinds of friendships with the adults as we have. (Mr. Miller)

Whereas positive relationships between parents may help create or cement friendships, negative relationships between parents may discourage children's friendships:

> I objected to the way the husband treated the wife. It just became too painful to watch, so I withdrew from contact with them. Once in a while, the little girl will call and we'll say we'll get together, but I just can't bring myself to do it. It's sad because the kids, the woman, and I were all quite close for a while. (Mrs. O'Neill)

In addition, the parents' animosity may be perceived and spill over to the children:

> I think my resentment of the mother comes through to Matthew. That may be why he doesn't really like James. (Mrs. Dorr)

Just as parents' friendships can affect their children's social relationships, children's friendships can affect the social relationships of their parents. Parents often meet and become friends through their children (see O'Donnell, 1983). Indeed, some parents are highly dependent on their children for their own social contacts:

> We are not really involved with people very much except as the kids are concerned. I guess that is why the kids' friends' parents have become my friends. Killing two birds with one stone. (Mrs. Vernon)

In addition, preexisting friendships between parents are likely to become closer if the children are also friends. On the other hand, the lack of affinity between their children can weaken the friendship between parents.

The extent to which parents' and children's friendships are linked varies a great deal from case to case. In some cases, parents and children draw their friends from entirely different domains, and in other cases, there is much more integration of their social worlds. These variations often reflect the parents' broader values about family and community. Some parents put a high priority on the family as a whole being part of a community of friends and having two or more generations socialize together. For example, a central part of the Hoffmans' social life was getting together with other families with children close to their daughter's age. Similarly, the Shaws spent most of their leisure time with their "family-oriented" church group. These parents believe that when the children's friends are "family friends" or part of a community, the friendships are likely to be richer and more enduring.

Such community-based friendships may substitute for the loss of an extended family in a mobile society in which adult siblings and their children often have little contact. In some cases, however, extended families do have considerable contact, and cousins are likely to be special friends. This was most likely to be the case among working-class families who had lived in the same area for a long time.

Arranging Social Contacts

In addition to setting the stages on which children conduct their social relationships, parents can have a large impact on their young children's friendships by directly arranging social contacts for them. Parents serve as booking agents, getting on the phone with other parents to set up visits between children; as hostesses, with an ample supply of milk and cookies for visiting guests; as social planners, arranging birthday parties and picnics; as activities coordinators, enrolling their children in gymnastics classes and art programs; and as chauffeurs who transport their children from one activity to the next.

For many parents, arranging social contacts for their children is an involving and time-consuming endeavor. "People complain about doing a lot of running around for their kids' social lives when they are adolescents," Mrs. Vernon reflected, "but it has already started for me. Every day we are walking to someone's house or there is a group of them here. And then the birthday parties . . ." Indeed, preschool children are typically more dependent on their parents for making such arrangements than are older children, who can begin to make their own arrangements for local activities.

Some parents are more involved in arranging social contacts than others, however. For example, when the family lives in a safe neighborhood where the child can get together with friends on his or her own relatively few arrangements may be necessary. "Just set out a plate of peanut butter sandwiches, and you're set for the whole day," Mrs. Hoffman said. When the neighborhood does not allow the child to get together with friends on his or her own, however, the parent may feel the need to make arrangements for the child:

> It would be so nice to have someone to play with close by. I've had to send him to the Montrose program and the library so there would be kids to play with. Otherwise, I wouldn't have pushed those things. (Mrs. Chaney)

In many regions, there are even seasonal variations in social arrangements, with children being able to see their friends outdoors on their own during the spring and summer months, but dependent on their parents to arrange visits during the winter.

Some of our parents regretted the continuing need to arrange social contacts of their child, both because of the time and effort it entailed and because they wished their children's social lives could be more spontaneous:

> It is really terrible. There are no occasions when you can just walk out and find kids to play with. You have to make a date, make arrangements, or schedule play. I hate that at this age. We have tried to compensate for that by all the programs he is in, but it is not the same. Those are more structured and focused. It's not like just killing time with someone, making your own play. (Mrs. Dorr)

But whereas some parents share Mrs. Dorr's feeling that children's play should not be overly planned or structured, others place a positive value on prearranged and structured social interaction. Indeed, such variations are apparent in the parents' own social lives: whereas some may drop in on friends on the spur of the moment, others prefer dinner parties that are planned weeks ahead of time.

There is some evidence that upper-middle-class people are more attuned to prearranged social interactions and that working-class people are more attuned to spontaneous interaction (Berg & Medrich, 1980). Along similar lines, O'Donnell and Stueve (1983) reported that middle-class mothers were more likely than working-class mothers to arrange for their school-age children to take part in structured classes and other activities outside school. This tendency for middle-class mothers to enroll their child in more extracurricular activities was also apparent in our sample.

For all the labor it may involve, parents' roles in arranging social activities for their children enables the parents to exert significant control over their childern's social lives. Parents may arrange more or fewer activities as a means of regulating the amount of social contact a child has. For example, Mrs. Kimball, who wanted her child to be more outgoing, found out from Johnny's teachers whom he had been playing with in school and then called the mothers and invited those chidren to visit. In addition, parents frequently try to arrange contact between their child and children they approve of and to limit contact with children they disapprove of. Mrs. Dorr, who worried about her son's preference for older children, tried to arrange get-togethers with children of the same age. Mrs. Foley, who was concerned about her daughter's dislike of boys, made a point of having boys over. Mrs. Lewis, who disliked aggressive play, stopped trading off in her baby-sitting pool with the mother of a boy who was always playing with guns. As we have already noted, moreover, these parents often arranged contacts between their children and the children of their own friends, both because they approved of these children and because the contacts were easy to arrange. If the parents did not know each other, on the other hand, visits were harder to arrange and less likely to take place (see Schiavo & Solomon, 1982).

Although the opportunity to arrange social contacts gives parents considerable influence, it does not give parents total control over this segment of their children's social lives. Several parents told us of their valiant attempts to help their child make a new friend, only to have their plans foiled when the children did not hit it off. In addition, much of the arranging that parents do for their 4-year-olds is done at the child's behest, rather than on the parents' initiative:

> Sunday, Matthew spent all morning calling all of his friends to find someone to come over and play with him. I had to go through the whole list of nursery-school kids to find someone. (Mrs. Dorr)

Thus, although the role of social arranger often gives the parent power over the child's social life, it also frequently casts the parent as a willing or unwilling executor of the child's own wishes.

COACHING

Parents make their most explicit attempts to shape their children's friendships by serving as coaches. *Coaching* refers to parents' suggestions, injunctions, expressions of approval or disapproval, and other statements or actions that are intended to influence their children's social interactions and relationships. For example, parents may encourage their children to be more sociable:

> Every day, when I drop him off at camp, I ask him if he will try to talk to one new child. Every day, he says he forgot, but he'll try the next day. (Mrs. Kimball)

. . . to interact more smoothly:

> Getting them into something inadvertently—like having them help me stir if I am cooking—sometimes gets them over that hump and back in rhythm. (Mrs. Hoffman)

. . . to be tolerant of others:

> Mark thinks Andrew is a baby and he asks about that sometimes. I just try to explain to him that everybody is different and some people don't learn so fast. (Mrs. Bennett)

. . . or to resist peer pressure:

> I'll tell Matthew not to do something, and then Jessica will say to go ahead and do it. Then I'll say, "Why did you do it?" and he'll say, "Jessica told me to do it." I say, "If someone tells you to jump off the roof of a building, are you going to jump off the roof?" And he would. Jessica told him to. We have to work on that. (Mrs. Dorr)

Parents coach their children both on the spot, when the child is in a social situation with peers, and behind the scenes, when the child is not in a social situation. Many of our parents expressed ambivalence about on-the-spot coaching, however. On the one hand, they would like to help grease the social wheels for their children, to help them get to know others, to help them out of impasses or arguments. On the other hand, most parents believe that their children must learn to engage others in interaction, to make friends, and to handle conflicts by themselves. If the parent tries to "help" too much, she or he may encourage dependency and impede the development of independent social skills.

Many of the parents' comments about on-the-spot coaching centered on the delicate balance between helping and interfering. For example, we presented parents with the following vignette:

> Matthew [Sally] is with his [her] parents at his [her] father's office picnic. Most of the children are playing together, but Matthew [Sally] is staying with his [her] mother and father.
>
> (A) His [her] parents should go over and help him [her] to get to know the other children.
>
> (B) His [her] parents should let him [her] go over to the other children on his [her] own whenever he [she] is ready to.

Regardless of which choice the parents made, their explanations typically reflected their awareness of the two sides of the issue. About half the parents interviewed (10 out of 18) said that in such a situation the parents should introduce the child to other children:

> Actually, that situation just happened at Bob's company picnic. There was Amy looking around for something to do. I saw a man with a child about her age and asked Bob to introduce them. It might have taken half the day for them to gravitate to each other, but as it turned out, they were inseparable for the whole day. (Mrs. O'Neill)

But almost all of the parents who made this choice emphasized that the introduction as simply to give their child "the option." Once the introduction was made, it would be up to the child to decide whether to play with the other children. On the other side of the coin, many parents said that they had learned that their child was more likely to interact successfully with other children in such situations if the parents left him or her alone:

> What we've done is gone up and introduced him, and the more we do that, the more resistant he is to playing. If we do nothing, sooner or later he gets bored with adults and will go to the kids on his own. (Mrs. Kimball)

Parents face a similar dilemma when they observe their children getting into fights or arguments with others. On the one hand, they often feel it necessary to help end the conflict. Such intervention can take a variety of forms, from suggesting solutions to separating the children physically. But most of our parents made clear that they would prefer not to get involved in such situations. In response to a vignette about two friends who had begun to argue a lot, 11 of 18 parents indicated that the parent should let the children work out their conflict for themselves. A central principle underlying this recommendation is the feeling that the children must learn for themselves how to resolve conflicts. As Mrs. Lewis said:

> I think the experience of figuring out the issues and the other person's feelings is good. So I think I would talk to my child before and after a session with that other child, but I would not do anything during their time together.

Our parents' recommendations in this situation were strikingly related to the sex of their child (which was always the same as the sex of the children in the vignette presented to them). Only one of the nine parents of girls (11%) indicated that the parent should intervene, whereas six of the nine parents of boys (67%) indicated that the parent should step in. The parents of girls seemed to have more confidence in their children's ability to resolve conflicts peacefully than the parents of boys did. This difference in parents' perceptions may reflect patterns of socialization in which girls (and women) learn to be more effective "interpersonal relations specialists" than boys (and men) do (see Parsons & Bales, 1955).

Whereas the majority of our parents were reluctant to intervene during the course of an argument, many of them said that they would discuss the situation with the child afterward. In some cases, the parents used such opportunities to make direct statements of values:

> Once, in nursery school, somebody hit him. He asked me later, "Why did he hit me?" I just flat-out told him that the youngster wasn't too bright. If he feels he has to go around hitting people to get his way, than he is pretty dumb. (Mrs. Kimball)

In other cases, parents try to engage the child in a dialogue about ways of resolving the conflict constructively:

> I stop myself from asking all the questions I would like, and we'll talk it through, and I'll throw out some possible solutions. Invariably, she'll say none of these things will work, but I'll ask her the next day, and she will have used one of the strategies I suggested. (Mrs. Hoffman)

Individual parents have their own distinctive styles of coaching. Whereas some parents do not hesitate to coach their children actively, whether on the spot or behind the scenes, other parents believe that active coaching attempts are too intrusive. Some parents go about their coaching efforts, at least some of the time, as part of a considered, intentional program to shape their child's social relationships ("We are working on his being social"), whereas others go about these efforts in a more spontaneous, unplanned way.[1] The parents' temperaments, values, experience with child rearing, and available time all contribute to these variations. In addition, different children seem to need or want different sorts of coaching by their parents.

There is little systematic information about the effectiveness of different approaches to coaching. It seems likely, however, that different sorts of coaching are most effective for different goals. It may be, for example, that unilateral injunctions can be most effective in producing immediate changes in behavior (see Zahn-Waxler, Radke-Yarrow, & King, 1979) but that dialogues in which the child plays an active role can be more effective in shaping lasting values about social relationships.

Providing Models of Social Relationships

A less explicit, but nevertheless crucial, way in which parents influence their children's social relationships is by serving as models. Children observe their mothers and fathers interacting with one another, with their friends, and with the children themselves and then adopt aspects of these parental styles—and the values that underlie them—in their own social relationships.

The most obvious sort of modeling involves the child's adoption of specific parental behaviors or strategies. For example, Jennifer Lewis made use of strategies she had learned from her mother to get her younger brother to share his toys with others. Mrs. Lewis reported:

> Adam might have difficulty sharing sometimes, and Jennifer will take care of that. I've heard her mimic me quite effectively: "Adam, do you want Greg to go home so you can have this toy all by yourself? No? Well, then, I think it would be better to share it if you want Greg to be here with us."

Similarly, Mrs. Hoffman, who works as a counselor, felt that Rebecca had picked up her "psychological" style of relating to people and their

---

[1] Whether planned or not, there are often similarities between parents' coaching efforts and the procedures employed in more systematic programs of social-skills training developed by psychologists (e.g., Coombs & Slaby, 1977; Stocking, Arezzo, & Leavitt, 1980).

problems. When two other children were fighting, Rebecca acted as a "playgound therapist" to help them resolve their difficulties.

In addition, children often seem to "take after" their parents with respect to more general social styles or temperaments. For example:

> She really isn't at all shy. She really has the attitude that everybody wants to be her friend. That is the way I feel about people—a stranger is a friend you haven't met. She seems to have picked up some of that. (Mrs. O'Neill)

> Neither my husband nor I am terribly aggressive or outgoing. So I'm not surprised that Michael is somewhat shy and cautious, too. (Mrs. Tierney)

Many explanations for such similarities of social style or temperament are available, including possible genetic influences on temperament, as well as environmental and cultural factors that influence both parents' and children's social styles. Nevertheless, we suspect that modeling often contributes to such resemblances, as children observe their parents' modes of interacting with others and adopt some of these modes in their own interactions.

It seems likely that girls tend to model their mothers' patterns of friendship, whereas boys tend to adopt their fathers' patterns. Thus, girls may observe their mothers have intimate relationships with a small number of friends and may become primed to adopt such an intimate pattern of friendship themselves. Boys may observe their father's congenial relations with a large network of friends but may also note their relative lack of intimate exchange and disclosure (e.g., Booth, 1972). Such observations may help to perpetuate sex differences in the friendships of the younger generation. Nevertheless, many of the parents we interviewed believed that their 4-year-olds had adopted aspects of the social styles of both of their parents.

Even though the impact of modeling on children's friendships remains to be systematically demonstrated, many parents are convinced that they shape their children's social lives by their example. Mr. Shaw, for one, said that he would rather influence his children by example than by preaching values to them. In particular, he felt that since the time he gained sole custody of his children, he had provided a new model of friendship for them:

> I've realized how much I need enduring friendships, and I think I've projected that to them. We've plugged back into friends and relatives who went out of our lives or were rejected during my marriage. They have a model of my openness to relationships and to other people.

Of course, parents are not the only people after whom children may pattern their social relationships. Other adults, older siblings and friends, and television characters may all serve as models of social relationships.

But parents may well remain the most influential models for their children and, by their example, can have a large impact on the sorts of relationships that their children will establish.

Providing a Home Base for Peer Relationships

Parents also influence their children's friendships by providing a "home base" that affects the children's readiness and ability to establish social relationships outside the home. Several recent studies suggest that secure parent–child attachments can help children to interact successfully with peers. Lieberman (1977) found that 3-year-old children whose attachments to their mothers were rated as secure were more responsive to other children and engaged in longer social interactions than did children who were not securely attached to their mothers. Similarly, Waters, Wippman, and Sroufe (1979) found that the rated security of 15-month-old children's attachments to their mothers predicted the degree to which these children were socially active and popular in nursery school when they were $3\frac{1}{2}$. A secure attachment to the mother (or, presumably, to the father) may embolden the young child to explore his or her social environment and, as a result, to discover and become engaged in the world of peers. A child without a secure parental attachment, in contrast, may be too timid and anxious to be able to interact successfully with peers.

Within our sample, the importance of parents as a home base was suggested most notably in cases in which the children seemed to be insecure about their relationships with their parents. For example, Mr. Shaw speculated that Ben's initial lack of interest in other children at his day-care center resulted from his reaction to his parents' divorce. "It was a full year," Mr. Shaw recalled, "before he developed specific friends and the kids became the focus rather than the adults." Mr. Miller had a similar impression of Alison's day-care experience after her parents' divorce:

> I don't think she formed any really enduring relationships with the other kids. I think Alison was very much in need of adult affection at that point, and other kids were often competitors for attention rather than possible friends.

Such reactions are in line with research that has demonstrated disruptions in preschool children's peer relationships in the aftermath of parental divorce (Hetherington, Cox, & Cox, 1979).

Even within two-parent families, children who are "clingy" or insecurely attached to their parents may also tend to be less socially adventurous. Mrs. Tierney reported that Michael had been too dependent on her to be able to approach other children in unfamiliar settings:

> He has always been very clingy. Like last year, he was signed up for a gymnastics program at the "Y." He would not leave my side. It was awful. We'd go week after week, and he would sit right by me. All the other kids were out there, having a great time, and there he would sit.

As Mrs. Tierney went on to note, young children typically grow out of such feelings of dependency or insecurity. In the case of Michael, she now feels "the worst is over." But the security of children's relationships with their parents may continue to influence their relationships with peers throughout the course of childhood.

The quality of the parent–child relationship may also be related to the effectiveness of the other modes of parental influence that we have described. For example, parents may be able to coach their children about social relationships most effectively in the context of a warm and secure parent–child relationship. Moreover, there is some evidence that parents who have been able to establish a secure relationship with their child are more likely to provide social experiences with peers (Lieberman, 1977) and to respond appropriately while their child is with a playmate (Pastor, 1981).

In the broadest terms, we suspect that secure family relationships provide not only secure bases, but also positive orientations toward peer relationships. Mrs. McKay speculated on how Katie's open relationship with her helped to teach Katie about relationships more generally:

> We talk a lot about things going on within our family. Like last night Katie said to me, "You're being crabby, Mom." So we talked about how you handle that, and how you still love them. I think if you can understand that from your mother, you can understand that from other people. One relationship carries over into others.

Mrs. Lewis extended the point:

> I think you have to have good relationships to form relationships. I feel very good about Jennifer's relationships with me, her father, and her brother and sister. That must help make her able to deal with others.

Effects of Experience on Parents' Values and Strategies

What accounts for the variations among parents in their values about their children's friendships and in the strategies they use to influence them? These values and strategies undoubtedly have many sources, including the parents' temperament, social and economic situation, and cultural background. Within the confines of our small sample, it was difficult to identify any systematic effects of such factors. Two sorts of factors did stand out, however, as affecting parents' stances toward their children's friendships. Both of these factors concerned the impact of experience on the parents' values and strategies: the parents' previous experiences with child rearing and the parents' own childhood social experiences.

The impact of previous experience with child rearing was especially striking. Half of the 4-year-olds in our sample were firstborns and half had one or more older siblings. Those parents who had one or more older children tended to be more relaxed about their 4-year-olds' social relationships than were the parents who had no older children. The "experienced" parents less frequently arranged social contacts for their 4-year-olds and were less likely to monitor and structure their play. "It's just so different," Mrs. Bennett said. "I'm more relaxed with the second, more confident about what he can do on his own. I'm less watchful."

Our sample of parents attributed their more relaxed attitude about their later-borns' friendships to a gradual realization that children generally turn out "all right" socially, even without any heroic parental efforts on their behalf. Thus, Mrs. Dorr worried about her firstborn son's apparent lack of social interest and took pains to arrange visits with friends and to enroll him in a variety of organized group activities. Mrs. Chaney took a less active stance toward her third-born son's social life:

> My friend was into doing a lot of children's things, and I just wasn't into that with my third. She put Brad into all kinds of preschool groups—gymnastics, library, swimming, everything that came along. I just wasn't into that. It could be because that was her first child. You lose interest in some of these things. It's terrible to say, but you are so enthusiastic with the first child. So enthusiastic and so worried. And by my third, I felt like he was going to be a perfectly nice, healthy, talented, smart little boy without getting into all that.

In addition, the parent often has less time to devote to a later-born child's social life—or to indulge in worrying about it—than she had for her firstborn's. "When you have as many kids as I do," declared Mrs. Davis, the mother of five, "you don't worry about getting them to play or watching them play. If you have some time, you take it."

Older siblings may also assume functions that enable the parent to play a less active role in the later-born child's social life. Whereas a firstborn child must find children outside the home to play with, a later-born can sometimes be integrated into the social world of an older sibling:

> With Jennifer, I did a lot more to have her around other kids. You know, with a first child, you tend to go, go, go—try to get them into everything. But I don't feel the same with Adam [her 2-year-old son]. He has had Jennifer and Jennifer's friends. (Mrs. Lewis)

Indeed, older siblings may often influence their younger brothers' and sisters' friendships in most of the ways that we have discussed—as stage setters (through their own social networks), arrangers of social contacts, coaches, models, and contributors to the quality of relationships within the family.

Some parents may regret the lesser attention that they give to the social relationships of their later-born children. But it is doubtful that later-borns suffer any social disadvantage as a result. In fact, studies have suggested that later-borns turn out to be more socially skilled, on the average, than firstborns (e.g., Miller & Maruyama, 1976). Perhaps the parents' greater degree of equanimity about their later-born children, as well as the social advantages that may be gained by having an older sibling, serves these children in good stead.

Parents' values and strategies about their children's friendships also derive from their own childhood social experiences. Some of our parents felt that their own early expriences gave them a better understanding of their child's current experience. In some cases, the parents tried to help their child avoid the social difficulties that the parents themselves had encountered. For example, Mrs. Kimball told us of her efforts to lessen Johnny's shyness by inviting other children to their home, where he could interact with others in a nonthreatening setting. She went on to report:

> I was painfully shy throughout most of my schooling. I think that is why I am so aware of it with Johnny. I realize all the pain the shyness caused me. Once it gets ingrained, no one can help you through it. So if I can start early with Johnny, I may spare him some of that.

Similarly, Mrs. Lewis's decision to transfer Jennifer to a larger preschool program was based in part on her desire to head off difficulties that Mrs. Lewis herself had experienced:

> I grew up in a very small town in Indiana. You knew everybody and went straight through school with the same 15 or 20 kids. I never needed to adapt to a lot of different people or learn how to make friends. When I got to college, I went into shock. I just didn't know how to select friends, to make friends, to deal with people I hadn't known over a lifetime. I didn't want the same thing to happen to Jennifer.

In other cases, the parents' own childhood experiences led them to feel confident about their own children's social adjustment. Mrs. Hoffman told us, for example, that she rarely felt a need to coach Rebecca about making friends. Because Mrs. Hoffman herself had never had social difficulties as a child, she assumed that Rebecca would follow the same pattern.

Parents' reliance on their own childhood experiences as a guide to their child-rearing practices may not always be justified. For example, parents may assume that their child's social resources and inclinations closely resemble those of the parents, even though the child may actually be quite different. Nevertheless, we suspect that the ability of parents to recall their own childhood social experiences can be of great help in their understanding and enchancing of their children's social lives.

Conclusion

We have described and illustrated five ways in which parents influence their children's friendships: choosing neighborhoods, school settings, and social networks that set the stage for children's friendships; arranging social contacts; coaching; providing models of social relationships; and providing a home base of relationships within the family. There are additional ways in which parents influence their children's friendships that do not fit neatly into any of our five categories. For example, parents may influence their children's appearance, manners, material resources, or personalities in ways that affect their sociability or their attractiveness. Parents may also intercede on their children's behalf with third parties, such as siblings or teachers, who, in turn, help to shape the children's social relationships. Despite such additional possibilities, we believe that our five categories provide a useful typology for organizing and examining the wide range of parents' influences on their children's friendships.

To further our understanding of these parental influences—and, ultimately, to provide information that may be of practical value to parents—researchers need to learn more about the actual impact of various parental policies and actions on their children's social relationships. As we have seen, parents often shape their children's friendships in intended ways, but parental interventions may also fail or backfire. The actual effect of particular parental actions or exhortations undoubtedly depend on a complex set of factors, including the child's temperament, his or her level of cognitive and emotional development, the nature of the parent–child relationship, and the social and cultural context in which the parents and the children live. We hope that our preliminary examination of modes of influence will be a helpful starting point for future research that begins to specify these effects of parental influences on children's social lives and well-being.

In particular, there is a need to study such influences developmentally. We believe that the general modes of social influence that we have described continue to operate throughout the course of childhood and adolescence. As young children grow older, however, both parents and children must continually adjust and negotiate the balance between attachment and separation and between control and independence. Moreover, the specific forms and effects of parents' influence on their children's social lives change as the children grow up. Thus, a 14-year-old may not be as dependent as a 4-year-old child on neighborhood friendships, nor may she require or permit her parents to schedule extracurricular activities for her the way they did when she was 4. Yet the 14-year-old's parents may continue to stage and regulate their child's friendships through their decisions and advice, whether in concert or in combat with their child, about schools, summer trips, clothes, crowds, and curfews. The models of social relationships that parents provide and the quality of relationships within the family may be as influential to the 14-year-old as to the 4-year-old—or more so—albeit in somewhat different ways.

As children grow from infancy through adolescence, and even into adulthood, the social worlds of the children and of their parents continue to affect each other. These influences should receive the attention of students of children's social development, for they are central to the ways in which people come to relate to their fellow human beings.

Acknowledgments

We are grateful to Jay Belsky, Michael Lewis, and Ann Stueve for their comments on an earlier draft of this chapter.

References

Berg, M., & Medrich, E.A. Children in four neighborhoods: The physical environment and its effect on play and play patterns. *Environment and Behavior*, 1980, 12, 320–348.

Booth, A. Sex and social participation. *American Sociological Review*, 1972, 37, 183–192.

Coombs, L. L., & Slaby, D. A. Social-skills training with children. In B. B. Lahey & A. E. Kazdin (Eds.), *Advances in clinical child psychology* (Vol. 1). New York: Plenum Press, 1977.

Hartup, W. W. The social worlds of childhood. *American Psychologist*, 1979, 34, 944–950.

Hetherington, E. M., Cox, M., & Cox, R. Play and social interaction in children following divorce. *Journal of Social Issues*, 1979, 35(4), 26–46.

Lieberman, A. F. Preschoolers' competence with a peer: Relations with attachment and peer experience. *Child Development*, 1977, 48, 1277–1287.

Miller, N., & Maruyama, G. Ordinal position and peer popularity. *Journal of Personality and Social Psychology*, 1976, 34, 615–624.

O'Donnell, L. The social worlds of parents. In L. Lein & M. B. Sussman (Eds.), *The ties that bind: Men's and women's social networks* (an issue of *Marriage and Family Review*). New York: Haworth Press, 1983.

O'Donnell, L., & Stueve, C. A. Mothers as social agents: Structuring the community activities of school-age children. In H. Lopata & J. H. Pleck (Eds.), *Research in the interweave of social roles* (Vol. 3): *Families and jobs*. Greenwich, Conn.: JAI Press, 1983.

Parsons, T., & Bales, R. F. *Family, socialization and interaction process.* Glencoe, Ill.: Free Press, 1955.

Pastor, D. L., The quality of mother–infant attachment and its relationship to toddlers' initial sociability with peers. *Developmental Psychology,* 1981, *17*(3), 326–336.

Rosenblum, L. A., & Plimpton, E. H. The effect of adults on peer interaction. In M. Lewis & L. A. Rosenblum (Eds.), *The child in its family.* New York: Wiley, 1979.

Rubin, Z. *Children's friendships.* Cambridge, Mass.: Harvard University Press, 1980.

Schiavo, R. S., & Solomon, S. K. *The effect of summer contact on preschoolers' friendships with classmates.* Paper presented at the annual meeting of the Eastern Psychological Association, New York, April 1981.

Schiavo, R. S., Solomon, S. K., Evers, C., & Cohen, W. *Maintenance of friendships among preschoolers.* Paper presented at the annual meeting of the Eastern Psychological Association, Baltimore, April 1982.

Sloman, J. *Parents and peers: Maternal impact on toddler peer interaction.* Paper presented at the biennial meeting of the Society for Research in Child Development, Boston, March 1981.

Smith, P., & Connolly, K. *The ecology of preschool behavior.* Cambridge, England: Cambridge University Press, 1981.

Stocking, S. H., Arezzo, D., & Leavitt, S. *Helping kids make friends.* Allen, Tex.: Argus Communications, 1980.

Waters, E., Wippman, J., & Sroufe, L. A. Attachment, positive affect and competence in the peer group: Two studies in construct validation. *Child Development,* 1979, *50,* 821–829.

Zahn-Waxler, C., Radke-Yarrow, M., & King, R. A. Child rearing and children's prosocial initiations toward victims of distress. *Child Development,* 1979, *50,* 319–330.

11

The Determinants of Parental Competence
Toward a Contextual Theory

JAY BELSKY, ELLIOT ROBINS, AND WENDY GAMBLE

In most analyses of parental competence, assumptions regarding what constitutes good and poor parenting are usually left implicit. As a consequence, consideration of this much discussed and investigated topic frequently stimulates heated debate. In this chapter, we hope to avoid such emotionally charged argument by approaching this focal area in much the same way that evolutionary biologists address those specific areas of inquiry that pique their interest, that is, by focusing on the concept of adaptation. More precisely, for the purposes of this chapter, competent parenting is defined as that style of child rearing that enables the developing person to acquire the capacities required for dealing effectively with the ecological niches that she or he will inhabit during childhood, adolescence, and adulthood.

Recognizing the relativistic assumptions built into definitions of parental competence, in this chapter we address the issue in terms of one particular context: the dominant subculture in the United States today. Thus, in what follows, what is currently known about competent parenting in middle-class American society is considered, by first focusing

---

JAY BELSKY, ELLIOT ROBINS, AND WENDY GAMBLE • College of Human Development, Pennsylvania State University, University Park, Pennsylvania 16802. Work on this chapter was supported by grants from the National Science Foundation (No. SES-8108886) and the March of Dimes Birth Defects Foundation (Social and Behavioral Sciences Branch, No. 12-64), (Jay Belsky, Principal Investigator).

on the characteristics and the consequences of parenting. The major emphasis of this chapter is, however, the determinants of parental competence. In our discussion of the processes that influence parenting, we stress the fact that individual differences in parenting are *multiply determined* by a variety of forces both within and beyond individual parents and their families in which they function. Indeed, in foreshadowing our eventual conclusions, we propose that three primary sets of determinants exist: the personal resources of parents, the child's characteristics, and the social sources of stress and support. And we propose that these sets of determinants interact in systematic ways to determine the likelihood that parents, and thus their children, will function competently. More specifically, we argue eventually, working from a systems perspective, that stress in any one subsystem of influence can be buffered by support in some others, but, importantly, that these three domains of influence are not equally influential. Although weakness in many always places the parent–child system in greater jeopardy than weakness in one, and although weakness in one always creates more risk than no weakness at all, certain subsystems of influence are considered more powerful forces of determination than others.

Before initiating our analysis of these multiple determinants of parental competence, and following a discussion of parental influences on child development, two theoretical perspectives in ascendance in contemporary developmental psychology are reviewed. These inform our analysis of the social forces that promote or undermine competent parenting. After outlining the primary contributions of the life-span perspective and of the ecology of human development, we offer a general model of competent parenting. Next, we examine the role that parents' own developmental histories, the child's characteristics (e.g., temperament), and the social forces within and beyond the family (e.g., marriage, the social network, and work) play in shaping parenting practices.

Parental Competence: What We Know and Do Not Know

The Consequences of Parenting

The study of parenting has a long history within psychology. Although Freudian theory and retrospective interviews with parents generated a great many data in the 1940s and 1950s, thoughtful analyses of this work led to the realization that the dimensions of parenting under study (e.g., strict vs. lenient toilet training), as well as the methods of inquiry employed in these investigations, were limited in their ability to generate consistent findings regarding the salient characteristics and the developmental consequences of parenting (Caldwell, 1964; Orlansky, 1949). In fact, the eventual willingness of investigators to recognize these limits made possible the progress in understanding that has taken place during the past two decades.

In the infancy period, detailed observational studies reveal that cognitive-motivational competence is promoted by attentive, warm, stimulating, responsive, and nonrestrictive caregiving (e.g., Belsky, Goode, & Most, 1980; Carew, 1980; Clarke-Stewart, 1973; Lewis & Goldberg, 1969; Yarrow, Rubenstein, & Pedersen, 1975). The positive influence of such parental (primarily maternal) styles has also been noted with respect to healthy socioemotional development (Clarke-Stewart, 1973). Infants whose mothers are nurturant, responsive to their needs, and accepting of their limits as immature organisms tend to develop secure, as opposed to resistant or avoidant, attachment relationships with their caregivers, and to be cooperative and communicatively skillful (Ainsworth, 1973; Ainsworth, Bell, & Stayton, 1971; Stayton, Hogan, & Ainsworth, 1971). Moreover, it is just such security of attachment that tends to forecast heightened cognitive and social competence during the preschool years (Main, 1973; Matas, Arend, & Sroufe, 1978; Pastor, 1981; Thompson & Lamb, 1981; Waters, Wippman, & Sroufe, 1979).

The work of Baumrind (1967, 1968, 1971, 1972, 1975) demonstrates that during the preschool years it is high levels of nurturance and control on the part of parents that foster competence in early childhood. *Competence* is defined as the ability to engage peers and adults in a friendly and cooperative manner, as well as the capacity to be instrumentally resourceful and achievement striving. In Baumrind's studies, the authoritative or competence-inducing parent is contrasted with the authoritarian and the permissive parent. Whereas the authoritarian parent tries to shape, control, and evaluate the behavior and the attitudes of the child with a set of absolute standards, stresses obedience to parental authority, and favors forceful punitive measures to curb self-will whenever parent and child are in conflict, permissive parents generally fail to present themselves to the child as active agents with a responsibility for shaping or modifying the child's present and future behavior. Instead, permissive parents accept children's behavior in a nonpunishing, accepting, and affirming manner, permitting their children to govern their own lives.

The parent most likely to induce competence (i.e., the authoritative parent) recognizes the child's need for control and the child's individuality. These nurturant and controlling caregivers thus tend to encourage give-and-take between parent and child, but at the same time, they recognize—and accept—their ultimate responsibility to exert influence on and control over their offspring. In sum, the authoritative parent rears the child in a manner that views the rights and duties of parents and children as complementary.

As children grow older, during the school-age years, patterns of parenting similar to those identified by Baumrind continue to promote social and intellectual competence. The use of induction or reasoning, consistent discipline, and the expression of warmth have been found repeatedly to be positively related to self-esteem, internalized controls, prosocial orientation, and intellectual achievement during the elementary-school years (e.g., Coopersmith, 1967; Hoffman, 1961; McCall, 1974).

This summary and our reading of the literature that it reflects indicate that consistent and systematic associations have been found between patterns of parenting and child functioning, thereby suggesting that parenting influences child development. Such a conclusion in no way need imply that the parents are the sole determinants of individual differences in children or that the process of socialization is such that the parents are the sole agents of influence in the family. Not only is it most certainly the case that peers, teachers and relatives also influence children, but it is also true that children themselves shape the behavior that others direct toward them and that influences their own development. Such child effects do not mean, though, that the flow of influence between parent and child is equal. Even though the necessary studies have not been done, at least in the childhood years the degree of control and the long-term influence that parents exert on children exceeds that which children exert on parents. This greater power of parents would seem to be the case because children come to the parent–child relationship in such a relatively undeveloped form, whereas parents bring to the experience of parenting a nature and an identity that have already been greatly influenced by their own developmental experiences.

In considering the results of the previously reviewed efforts to identify competence-inducing parenting practices, it would appear that, across childhood, it is parenting that is *sensitively* attuned to children's capabilities and to the developmental tasks they face that promotes the kinds of developmental outcomes thought important: emotional security, behavioral independence, social competence, and intellectual achievement. In infancy, this sensitivity translates into being able to read babies' often-subtle cues and to respond appropriately to their needs within reasonably brief periods of time (Lamb & Easterbrooks, 1980; Lewis & Goldberg, 1969). In childhood, sensitivity means continuing the warmth and affection provided in the early years, but increasing the demands for age-appropriate behavior. Parents must be willing and able to direct their children's behavior and activities without squelching their developing independence and industry.

The developmental change in competent parenting represents a response to the changing nature of the developing human organism. For example, as the child's cognitive skills develop, reasoning and demands for delayed gratification become effective disciplinary strategies. Similar efforts would be useless, of course, with an 8-month-old. Similarly, as the child becomes increasingly able to regulate his or her own behavior, parental control provides a useful scaffold to support such emerging competence. Ultimately, the sensitive, and thus competent, parent must be willing to wean the child from this overt control to permit the testing of personal limits through the exercise of internalized rules and regulations. Indeed, by the time the child reaches adolescence, the competent parent has set the stage so that the child has the psychological building blocks to encounter successfully the transition from childhood to adolescence.

The Determinants of Parenting

Surprisingly little attention has been devoted to ascertaining the determinants of competent parenting. In fact, with the exception of studies of social class differences in parenting practices (e.g., Hess & Shipman, 1965; Lewis & Wilson, 1972), and of the effect of the child on its caregiver (e.g., Lewis & Rosenblum, 1974), relatively few investigations have attempted to account for why parents parent the way they do. It is as if little interest exists within the discipline of developmental psychology in understanding the distal or "formal" causes of child competence, and thus, empirical energies are primarily concentrated on documenting the proximal or "efficient" causes (e.g., parenting practices). Although this special concern about efficient causality is understandable, the general failure to consider the causes of individual differences in patterns of parenting shows little appreciation of the fact that efficient causal explanations lie embedded within formal causal systems and remain dependent on distal causes. Thus, even a complete account of efficient causes (i.e., parental influences on child development) can represent only a necessary, but not a sufficient, explanation of individual differences in children's competence.

Research that has sought to account for parenting practices is considered limited in several respects. Although interest in social class differences in patterns of parenting does demonstrate a concern with the determinants of parenting, the work generated by this interest provides little insight into the actual *processes* of influence. In general, investigators have been more interested in documenting that parents and children from distinct social classes differ from one another than they have been in relating actual experiences within such social categories to patterns of parental and child behavior. Researchers have then gone on to propose a wide variety of explanations for the relationships discovered between social class and parenting (Gecas, 1979). Instead of examining the effects of the mediating variables directly, these researchers have usually relied on social class as a summary construct. Thus, although a good deal of prediction is achieved, the various alternative explanations that detail how social class exerts its developmental impact are impossible to examine within this approach. Ultimately, the relationship between social class and parenting "cannot be understood until the structural variable is conceptualized as a set of psychological processes or mechanisms that cause the outcome to be explained" (Elder, 1981, p. 81).

In addition to reliance on the global social-class construct, a second approach that investigators have adopted to explain variation in parenting behavior has proved to be of extreme importance in the relatively recent reconceptualization of the socialization process itself. By focusing on the effects of the child on its caregiver, Bell (1968, 1971) and others (e.g., Lewis & Rosenblum, 1974) have gone a long way toward modifying the traditional, unidirectional, "social mold" orientation that long dominated the discipline and focused attention exclusively on parental influence on child functioning (Hartup, 1978). Because children are now recognized as influencing their parents' behavior in a manner that may feed back and be of developmental consequence to them, it seems justifiable to conclude that children may serve as "producers of their own development" (Lerner & Busch-Rossnagel, 1981). But despite this conceptual advance, little empirical evidence actually exists to document the processes whereby children do, in fact, exert an impact on their parents, which then feeds back to influence their own development. All that is really available is speculation with respect to such effects (e.g., Belsky & Tolan, 1980); simulation experiments highlighting the potential for such effects (Gewirtz & Boyd, 1976; Parke & Sawin, 1975); and descriptive observational studies documenting the manner in which a wide array of dimensions of individual difference in children (e.g., age, gender, alertness, and temperament) influences parental behavior (e.g., Cherry & Lewis, 1976; Moss, 1967; Osofsky, 1976).

Contributing Perspectives

The contextual analysis of parent–child relations that forms the basis of this chapter is informed by two theoretical perspectives that have recently achieved prominence: life-span development (Baltes, Reese, & Lipsitt, 1980) and the ecology of human development (Bronfenbrenner,

1979). It is important to note, before outlining the contributions of each perspective that influence our thinking, that these are perspectives on development rather than formal theories of development. Thus, they serve only to highlight issues for consideration; they do not provide detailed models of growth and change.

Life-Span Development

The basic tenet of life-span developmental psychology is, as its name implies, that development is a lifelong phenomenon. Rather than being restricted to the childhood years, or being most dramatic or significant during certain ontogenetic epochs, growth and change are considered possible throughout the life course. From this perspective, adulthood is viewed not simply as a product or end result of childhood, but also as a time for further development.

Within the life-span framework, life events and developmental tasks serve critical functions in stimulating and determining the nature of development during the postchildhood years (Hultsch & Plemons, 1979). And it is from such a vantage point that parenting takes on importance beyond its influence on child development. Indeed, within the life-span perspective, the onset of parenting—or the transition to parenthood, as it is often called within the discipline of family sociology—can be viewed as a significant life event with the potential for giving shape and meaning to adulthood. Especially likely to be of importance is the sense of success or failure that parents experience in rearing their children. Although, as Guttman (1975) has pointed out, we know virtually nothing about the developmental consequences of parenthood for adults, it seems likely that certain individuals reap from this experience a heightened sense of fulfillment, whereas others may come away from it with a sense of frustration or defeat (Neugarten, 1968). In any event, as Erikson's (1950) epigenetic framework suggests, one's final appraisal of the meaning and worth of one's own life is likely to be influenced by one's own parenthood experience. We suspect that the competence that one's offspring evince, as both children and adults (and perhaps even as parents), is likely to figure prominently in many persons' final appraisals of their own lives (Bernard, 1974; Campbell, 1975).

Not only do we know little about the consequences of parenthood for adults, we know precious little about its ontogenetic roots. If we assume that what transpires during earlier developmental periods has the potential for influencing the manner in which adults confront critical life events—an assumption much debated under the rubric of continuities and discontinuities in development (Brim & Kagan, 1980)—it stands to reason that to understand parenting behavior we need to concern ourselves with parents' own developmental histories. In sum, parent–child relations must be examined from within the life course of the parents as well as of their offspring. It is not sufficient to focus solely on patterns of parenting and child development. Attention must also be paid to the place of parenting in an individual's life and to its ontogenetic roots and developmental consequences. Such are the lessons that we glean from the perspective of life-span developmental psychology.

The Ecology of Human Development

The basic tenet of the ecological perspective is that development takes place "in context." But as important as is this most general principle, which happens also to be shared with the contextualist school of thought (Lerner, 1978; Riegel, 1975), the major contribution of the ecological perspective is the attention that it directs to the multiple levels of context in which the developing organism is embedded. These levels of context include immediate settings (e.g., home and school) and interrelations between such microsystems (e.g., home and day care), both of which have a direct impact on the developing person; social structures and immediate contexts that have only an indirect impact on the developing individual (e.g., the father's workplace); and broad cultural practices and belief systems that shape the values and assumptions that give meaning and direction to everyday life. In essence, Bronfenbrenner's (1979) multilevel analysis of the contextual forces at work in individual development weds related, yet distinct, social science disciplines, including psychology, with its emphasis on the individual and its concern with the immediate situation; sociology, with its emphasis on large groups and its concern with social structure; and anthropology, with its emphasis on societies and its concern with cultural variation.

A second theme emerging from the ecological perspective that informs our analysis of parent–child relations, and that permeates contextualism more generally (Lerner, 1978), is the notion of reciprocal influences between organism and environment. In the analysis to follow, when we speak of such reciprocal pathways of influence, we are referring primarily to enduring social and developmental impacts (as when parenting affects child competence which in turn feeds back and influences the parent's own sense of selfworth) rather than to more immediate and ephemeral stimulus–response–response effects (as when a parent's statement elicits a verbal response from the child, which in turn evokes additional parental comment).

Our interest in reciprocal influence is not restricted to a consideration of the immediate parent–child system, however. In fact, it extends beyond the family to the interplay between the developing individual and the multiple contexts in which she or he can be seen to be developing. Thus, in addition to a concern with how the father's work and the parents' social networks might affect parenting and, thereby, child functioning, our interest extends to the reciprocal impact of the child on these domains of its parents' lives. In light of space constraints, however, we purposefully forgo an in-depth analysis of such bidirectional processes beyond the family system.

Toward a Contextual Theory of Parental Competence

In a previous section dealing with the characteristics of parental competence, we argued that sensitivity to the developing competencies of the child is the common thread that characterizes parental competence across the childhood years. The ability to be sensitve, in and of itself, however, neither defines nor assures parental competence, because sensitivity must be practiced if caregiving is to be considered competent. In other words, the quantity of time spent in caregiving, as well as its quality, is deemed important. Thus, in addition to the capacity to behave sensitively, a necessary component of parental competence is psychological and behavioral involvement with the child. We suspect that the degree of behavioral involvement that characterizes competent parenting decreases in an absolute sense—and changes in quality—as children develop, because one characteristic of sensitivity is the provision of an increasing opportunity to be autonomous as the child grows up. Such a change in behavioral involvement in no way implies—and probably does not lead to—concomitant reductions in psychological involvement.

The foregoing analysis leads us to propose a two-component model of parental competence in which a parent, to be competent, must be both sensitive to and involved with his or her children. We further propose that the three primary determinants of parental sensitivity and involvement are patience, endurance, and commitment. Our basic model assumes that patience enables individuals to hold their feelings and impulses in check and, thereby, to decenter from their own point of view and, as a consequence, understand the world of the child and recognize his or her developmental needs. Endurance provides the energy required to cope with the often physically depleting demands of parenting. And finally, commitment is what leads parents to invest their patience and energy in the parental role. Our model further assumes that patience, endurance, and commitment reciprocally influence one another. This point is illustrated as we detail processes whereby parental resources, child characteristics, and the social sources of stress and support influence parental sensitivity and involvement.

Now that we have outlined a general working model of parental competence based on our prior analysis of the characteristics of parents whose children function successfully in American society, let us consider the influences on a parent's patience, endurance, and commitment to parenting. Characteristics of parents and children themselves are considered first, and then institutional sources of stress and support, including marriage, social networks, employment, and formal community resources (e.g., day care, church, and welfare). After examining in some detail the processes by which these agents and agencies of influence each serve to promote or to undermine parental competence, the manner in which they function collectively is considered. In conclusion, a predictive model of the kinds of developmental contexts that foster varying degrees of developmental competence in the child is offered. This working model draws heavily on a systems approach, with an emphasis on interrelated components and processes whereby certain stresses on parenting are buffered by other supports.

Individual Contributions to the Parent–Child System

The Parent: Ontogenetic Origins and Personal Resources

Parenting is only one dimension of human functioning, and like many others, it is influenced by enduring characteristics of the individual that are the product of his or her developmental history. Without doubt, parenting is also likely to be influenced by the contemporary psychological life space that the parent "inhabits.". The adult undergoing a midlife crisis, for example, is probably less able to be as sensitive to his or her children as the individual who has resolved such developmental crises of adulthood or has not yet confronted them. Although such an analysis highlights the importance of life-course issues, such concerns remain beyond the scope of the present inquiry.

Evidence to support the assumption that childhood experiences can influence parental behavior can be found in the literature on child abuse, which is replete with suggestions that incompetent parenting is transmitted across generations (see Belsky, 1980). Although much of this work is based on small clinical samples and is fraught with methodological limitations, other evidence, some emanating from national surveys, exists linking exposure to violence in one's family of origin with its subsequent use (wife abuse and physical punishment of children) and approval (Carroll, 1977; Erlanger, 1979; Owens & Straus, 1975).

In addition to studies of the origins of violent and aggressive behavior, evidence exists suggesting a link between early developmental experience and subsequent parenting. For example, Hall, Pawlby, and Wolkind (1979) recently observed that mothers separated from their own parents before age 16 were less positively involved with their 5-month-olds than were mothers who had not experienced such separations (for similar data see Frommer & O'Shea, 1973a,b) and Sroufe and Ward (1980) recently reported that mothers observed to engage in seductive behavior with their toddlers had experienced incestuous relations with their own parents as children. Even though studies such as these suggest an antecedent–consequence relationship between developmental experience and subsequent parenting, they shed little light on why such associations exist.

With respect to this important issue of process of transmission, we suggest, as have others, that an individual's early experiences influence the personal resources that she or he has available as a parent. We deliberately eschew the term *personality* in this discussion to avoid becoming enmeshed in the continuing situationist–trait debate that is central to the study of personality today. Because personal resources are viewed as enduring characteristics of an individual that derive from experience, but that can be modified by subsequent experience and are thus susceptible to situational pressures with regard to their expression, such dichotomous reasoning is regarded as counterproductive.

Rohner and Rohner (1980) have offered the most detailed, process-oriented theory to link early experience with personal resources, and personal resources with parenting. On the basis of their cross-cultural research on the universal effects of parental acceptance and rejection, they concluded that

> rejected children everywhere tend more than accepted children to be: hostile, aggressive, passive aggressive, or to have problems in the management of hostility and aggression; to be dependent or "defensively independent," depending on the degree of rejection; to have an impaired sense of self-esteem and self-adequacy; to be emotionally unstable, emotionally unresponsive and to have a negative world view. We expect each of these personality dispositions to result from rejection for the following reasons. First, all of us tend to view ourselves as we imagine "significant others" view us, and if our parents as the most significant of "others" rejected us as children, we are likely to define ourselves as unworthy of love, and therefore as unworthy and inadequate human beings. In this way we develop a sense of overall negative self-evaluation, including feelings of negative self-esteem and negative self-adequacy.
>
> Adults who are rejected as children tend to have strong need for affec-

tion, but they are unable to return it because they have become more or less emotionally insulated or unresponsive to potentially close-interpersonal relations. Any of these adults who become parents are therefore much more likely to reject their own children than parents who are accepted as children. (p. 194)

In light of this analysis, it is surprising how few data are available linking personality characteristics or personal resources with parenting, though some suggestive evidence has appeared recently. In a study of parents of 4- to 6-year-olds, for example, Mondell and Tyler (1981) reported correlations between a summary personal-resource measure composed of three subscales (locus of control, interpersonal trust, and coping style) and the behavior displayed in a structured laboratory teaching task. This summary scale correlated positively and significantly with observed warmth, acceptance, and helpfulness, and negatively with verbal disapproval. And in an unpublished report, Schaefer, Bauman, Siegel, Hosking, and Sanders (1980) found significant positive correlations between the locus of control assessed during pregnancy and a summary factor score labeled "stimulation/interaction" based on observations of mother–infant interaction at 4 and 12 months of age. When considered along with the earlier cited findings of Hall *et al.* (1979) and Sroufe and Ward (1980), as well as etiological studies of child abuse, the data just presented provide support for the contention that developmental experiences influence personal resources, which themselves affect patterns of parenting. What remains unclear at present—and, in fact, unaddressed—is the age, stage, or time at which such early experiences might be most influential with respect to parenting.

The parental resources we consider most influential with respect to sensitive parental involvement are empathy/nurturance, physical health, and sex-role orientation. Empathy and nurturance provide the basis for sensitive care, probably by facilitating patience and commitment. Physical health is likely to have a direct bearing on one's endurance and therefore influences both patience and commitment. Finally, sex-role orientation may predispose an individual to view parenting as a primary or a secondary role, thereby affecting commitment. Because commitment undoubtedly affects endurance and, as a result, patience, it is important to recognize that sex-role orientation as well as empathy/nurturance and physical health are likely to influence all three primary determinants of sensitive involvement: patience, endurance, and commitment.

In concluding, it is worth reiterating a theme highlighted earlier that derives from the life-span perspective. Because development is possible in adulthood, and because parenthood is, for many, a salient developmental task, success or failure in this role can serve to influence personal resources. For example, if one's tendencies to be empathic and nurturant are rewarded through the production of competent offspring, it is likely that these tendencies will develop further. Similarly, one's involvement in parenting may lead to an unconscious reappraisal of one's sex-role orientation, as when a positive paternal experience serves to elevate the role of parenting in one's hierarchy of personal roles or when a negative maternal experience serves to elevate the role of out-of-home employment in a similar hierarchy of roles.

The Child: Influential Characteristics of Individuality

Evidence that children do influence the manner in which their parents treat them comes from a variety of sources, including studies addressing differences in the parenting of male and female infants (Cherry & Lewis, 1976; Lewis & Goldberg, 1969; Moss, 1967); studies relating neonatal characteristics to parental behavior (Osofsky, 1976); studies of maternal response to simulated infant functioning (Gewirtz & Boyd, 1976); and intervention studies demonstrating that better nourished infants are provided more autonomy than are their malnourished counterparts (Chavez, Martinez, & Yaschine, 1974). Indeed, an appreciation of children's contributions to the caregiving that they receive is now so widely shared that it has affected our way of thinking about specific social-policy–related concerns, such as child abuse (Belsky, 1980; Friederich & Boriskin, 1976; Parke & Collmer, 1975) and malnutrition (Pollitt, 1973; Rossetti-Ferreira, 1978; Zeskind & Ramey, 1978), as well as more general explanations of the developmental process itself (Sameroff, 1975).

Although demonstrations of the effects of the child abound, little effort has been made to determine exactly which characteristics of individuality exert the greatest influence on parenting (Belsky, 1981). Here, we propose four primary child characteristics: temperament, physical health, age, and gender. Temperament defines behavioral style, and we regard child temperament as neither immutable nor completely plastic. Children have behavioral proclivities that can be modified to a certain extent. In infancy, predictability and readability are important components of temperament because they make the child easier or more difficult to care for (Goldberg, 1977; Lamb & Easterbrooks, 1980). As the child grows older and becomes increasingly mobile, activity probably becomes a more central component of temperament (Buss, 1981). Because there undoubtedly exist many parameters by which behavioral variability can be classified, the most salient dimensions of behavioral style used in defining easy or difficult temperaments varies from context to context, depending on the behaviors judged important to a particular rearing environment. With this conceptualization of temperament in mind, we propose that parents who judge their children to be easy-to-rear have more patience and physical strength to deploy in their child rearing. As a result, they may be more committed to parenting and more likely to provide highly involved, sensitive care to their offspring.

The physical health of the child is another important characteristic of individuality likely to influence parental competence. Specifically, we propose that parents of healthy children not only find it easier to be patient with their children, because of the greater energy they have, but are also likely, as a consequence, to be more committed to the parental role. All these forces should promote sensitive involvement on the parents' part and, therefore, developmental competence in the child.

The child's age and gender are also important characteristics of individuality likely to shape parenting. Traditional values may lead the parents of older children, and especially boys, to be less patient. Younger children, it is important to note, are likely to require a greater investment of physical energy, and therefore, their developmental status is likely to have a strong impact on a parent's endurance. Finally, traditional values may also lead some parents, especially men, to be less committed to parenting infants and daughters.

In considering the influence of child characteristics on parental competence, it is important to keep in mind that because two of the influential child characteristics just discussed can themselves be influenced by parenting (i.e., temperament and physical health), it is likely that the competent parent will produce a child who is easier to care for, whereas the incompetent parent will produce a child whom it is more difficult to behave sensitively toward and to be involved with. Because it is also true that certain children elicit certain styles of parenting that serve to further encourage specific child behavioral styles, we can conclude with respect to reciprocities within the parent–child system that, whereas children serve as producers of their own development (Lerner & Busch-Rossnagel, 1981), parents serve as producers of their own parenting.

Parental Competence in Context: The Role of Stress and Supports

The discussion of the determinants of parental competence through this point has focused exclusively on the contribution of parent and child. The ecological perspective requires a consideration of the context of parent–child relations. In this section are considered four sources of stress and support that serve to promote or undermine parental competence: the parents' marital relationship, informal social networks, employment, and formal social resources, including day care/school, church, and health and social services. The discussion focuses first on why these contextual factors have been selected for analysis, and then the process by which these sources of stress or support function, both separately and collectively.

It is important to point out that not every one of these sources is functional in every parent's or even every family's life. Some parents, of course, do not work, and some families do not rely on day care or formal social services. We do not mean to imply, then, that every part of the discussion to follow is relevant to every parent or family. A second point that needs to be made with respect to the discussion to follow is that, for the most part, our analysis is couched in terms of support, rather than stress. At times, the implication is that the absence of such support is itself a stress, though this is not always the case.

Sources of Stress and Support

The influence that marital relations, social networks, employment, and formal resources might have on patterns parenting is suggested by diverse findings from several fields of inquiry, which together serve to justify consideration of these sources. Concern with the marital relationship, for example, grows out of the study of the fathering role. The addition of the father to the typically studied mother–child dyad does more than create a second parent–child relationship; it creates a family system comprising husband–wife as well as mother–child and father–child relationships (Belsky, 1981). The effect that marital relations can exert on parenting has now been suggested in studies of quite distinct developmental periods. In the infancy period, for example, Pedersen (1982) has found tension and conflict between husband and wife (as reported by the husband) to be associated with low levels of observed maternal feeding competence, and Belsky (1979) has reported positive associations between observed levels of spousal harmony and father involvement. In the preschool years, work by Hetherington, Cox, and Cox (1977) reveals that it is primarily when ex-spouses continue to bicker after separation that parenting and, ultimately, child functioning are undermined. In the childhood years, high interspousal hostility has been found to covary with the frequent use of punishment and the infrequent use of induction or reasoning as a disciplinary strategy (Dielman, Barton, & Cattell, 1977). Finally, although Kemper and Reichler (1976) have found that undergraduates whose parents had a satisfying spousal relationship reported receiving more rewards and less intense and frequent punishment, Rutter (1971) has observed that conflicted marriages are associated with antisocial child behavior in the latency and early adolescent years (see also Gibson, 1969; Johnson & Lobitz, 1974; Kimmel & vander Veen,

1974; Nye, 1957). In sum, these data strongly suggest that to understand parenting and its influence on child development, attention must be paid to the marital relationship.

The justification for considering the role that both formal services and informal social networks play in promoting competent parenting comes from several sources. First, and most important, there exists abundant evidence that supports of all kinds, be they formal or informal, have a positive effect on both mental and physical health—either enabling the individual to cope with stress and, thereby, lessening the risk of ill health, or enabling the individual to make a more optimal recovery from illness (Caplan, 1974; Cassell, 1974; Cobb, 1976; Mitchell & Trickett, 1980; Powell, 1979). Thus, social networks, which we define as relationships with friends, neighbors, and relatives, and assistance received from formal community services (e.g., church, welfare department, and day-care center) can serve preventive as well as remediative functions. Because competent parenting is regarded as a form of mental health, there is reason to believe that the support received (or not received) from formal social services and informal social networks affects the quality of one's parenting.

Evidence to support this contention can be found in a number of recent studies (Feiring & Taylor, n.d.,; Hetherington et al., 1977; Kessen & Fein, 1975; Toms-Olson, 1981). Abernathy (1973) found, for example, that the presence of a tightly knit social network among mothers of preschoolers was positively associated with parents' sense of competence in the caregiving role. Similarly, Zur-Szpiro and Longfellow (1981) reported that the assistance that low-income mothers received in child care and other household chores from their husbands/boyfriends was negatively and significantly associated with feeling stressed in the parenting role. At a more behavioral level, Colletta (1979) observed, in three distinct groups of mothers (single parents from low- and moderate-income households, and mothers with husbands), that total support provided by friends, relatives, and spouse was negatively associated with maternal restrictiveness and punitiveness. In fact, she was led to conclude, on the basis of her data, that "mothers receiving the least amount of total support tended to have more household rules and to use more authoritarian punishment techniques" (p. 843). In a recent study of parents of infants, material resources received from kin were found to correlate positively with maternal responsiveness (Unger & Powell, 1980). What is so intriguing about these last two sets of data is that it is just the kind of parenting that Colletta, and Unger and Powell, have related to social support that other investigators have linked to child competence (e.g., responsiveness) and incompetence (e.g., authoritarian rearing). On the basis of these findings, then, there seems to be good reason to assume that social networks, as well as formal services, play an influential, but still largely unexplored, role in the development and maintenance of competent parenting (Cochran & Brassard, 1979; Unger & Powell, 1980).

Evidence that employment, our third extrafamilial influence, can influence parenting also derives from several sources. The literature on child abuse, for example, implicates the absence of employment (i.e., unemployment) as an important contributor to incompetent parenting (Belsky, 1980). The work of Kohn and Schooler (1969) more directly implicates employment as a source of influence on parenting, as their efforts demonstrate that the characteristics of fathers' jobs causally affect the values that men hold with respect to their children's development. Similarly, the work of Kemper and Reichler (1976) shows that fathers who are more satisfied with their jobs are more likely than less satisfied fathers to rely on reasoning rather than physical punishment strategies. Of course, the study of maternal employment also provides evidence of the need for students of parent–child relations to consider the role that employment plays in influencing parenting patterns, as this literature indicates not only that working mothers place different demands on their children than do nonworking mothers, but that such parenting differences translate into developmental differences in the offspring (Hoffman, 1979; Lerner, Spanier, & Belsky, 1982). Given these diverse strands of evidence, it should be clear why we consider marital relations, informal social networks, formal social services, and, finally, employment potentially important determinants of competent parenting.

General Processes of Influence

The above-noted sources of stress or support are hypothesized to exert their impact on parenting in three general ways: (1) by providing emotional support to the parents; (2) by providing instrumental assistance; and (3) by providing social expectations (cf. Caplan, 1974; Cassell, 1974; Cobb, 1976; Cochran & Brassard, 1979; Mitchell & Trickett, 1980; Powell, 1979; Unger & Powell, 1980). In our general model of the processes by which sources of stress and support influence parenting, these three distinct types of influence map directly onto the previously described immediate determinants of sensitivity and involvement: patience, endurance, and commitment. Specifically, we hypothesize that emotional support, which communicates to the individual that she or he is loved, esteemed, and valued, influences, in a positive manner, the patience that parents have to deploy in the parental role and thus their endurance and commitment. Instrumental assistance, which involves the provision of goods and services and thereby frees up physical energy, is regarded as directly influencing endurance and, therefore, patience and commitment. And finally, social expectations, which provide information regarding the nature and the extent of parental involvement that is expected by others, are regarded as primarily influencing the individual's commitment to the parental role, which very likely has an impact on endurance and, therefore, the patience that a parent displays.

It should be clear that each form of support can derive from any or all of the sources of stress and support under consideration. Although it is self-evident that spouses and informal networks provide emotional support, it may be less apparent that employment, through the status and the social contact that a job can offer, and formal services (e.g., the church), through the positive regard received for community participation, frequently function as important sources of emotional support, too. Formal services, however, are probably more easily recognized as sources of instrumental assistance (e.g., child care), as is employment because of the financial and fringe benefits that jobs provide (e.g., salary and health insurance). Social expectations, of course, exist in all spheres of life, though they are not always easy to recognize. When, for example, a job requires mandatory overtime or overnight traveling, implicit assumptions are made regarding the family and, more specifically, the extent to which an individual can be involved in parenting.

In actuality, the degree of emotional support and instrumental assistance that an individual receives is determined, at least in part, by the degree to which she or he behaves in accord with the social expectations of any given source of influence (see also Lamb & Easterbrooks, 1980). More precisely, we hypothesize that when an individual parent behaves in a manner that is not consistent with others' expectations, the likelihood of her or him receiving emotional support and instrumental assistance is decreased. We can assume, on the basis of this hypothesis, that the mother who returns to work during her child's infancy and violates the expectations of her friends will receive less support from them. If such behavior is in accord with her spouse's desires, however, the support she receives from her mate may be a more than adequate compensation.

It is important to note with respect to each specific type of influence described that parenting can be directly or indirectly affected. In this presentation, direct effects are conceptualized as ones targeted at parental behavior, whereas indirect effects are presumed to have an impact on parenting through a mediated process. When a parent is praised, for example, by a neighbor or a teacher for her child's good behavior or for her skill in handling children, emotional support can be considered direct with respect to parenting; when a spouse lets his mate know that she is loved and cherished in general, however, we assume that such positive sentiments, though not directly targeted at parenting, nevertheless affect caregiving and are therefore regarded as indirect forms of emotional support. Similarly, the provision of babysitting services or information on child care is considered a direct form of instrumental assistance to the parenting role, whereas help in running non-child-related household errands is considered a form of indirect instrumental assistance. Finally, social expectations that speak directly to issues of parental involvement (e.g., mothers should stay at home with their babies and fathers should be the disciplinarians) are regarded as distinct from parenting, though not necessarily more influential with respect to parenting than those that only indirectly address issues of parental involvement (e.g., a father's being expected by his friends to go hunting on weekends or a businesswoman's being expected by her employer to be free to travel away from home overnight).

Interface of Sources

Because it is likely that parents have contact with more than one source of stress and support, it is important that we recognize that parenting is affected by the manner in which these sources interface. We propose, in this regard, that parental competence is undermined when social expectations across sources are inconsistent, especially when such inconsistency exists within sources deemed especially important by the individual parent. Consider the case, for example, of the husband whose wife expects him to be highly involved in fathering, but whose employer expects him to be continually available at the office. There is no way that such an individual can fulfill both sets of expectations, and failure to meet one set, or an attempt to meet each halfway, is bound to generate stress that we presume will undermine parental competence.

As implied above, it is likely that sources of influence are not all equal in terms of the degree to which their support is valued or needed by a particular parent. We propose, then, that although competent parenting is enhanced when emotional support, instrumental assistance, and social expectations consistent with one's personal orientation are available from at least one source, they are most supportive of parental competence when they derive from the source(s) most valued by the individual parent. Unclear to us at present is whether it is most beneficial to receive these three distinct types of support from one primary source or to have them distributed across sources. We do suspect, however, that the more sources one can draw support from, the more protected is one's parental competence, because the effect of loss of support from one source can be buffered by that received from another.

Parenting: A Buffered System

The model we have been building assumes that parental functioning is multiply determined, with its three major determinants being the personal resources of the parent, the characteristics of the child, and the available sources of stress and support. Having separately examined the processes by which each of these subsystems of influence affect parental competence, we turn our attention in this final section of the chapter to the manner in which these three subsystems of influence function collectively as a system of interrelated components. Our purpose here is to propose a singularly important property of the parenting system that emerges from this analysis and to offer a model capable of predicting the likelihood that a child growing up under certain conditions will be competent.

The singularly important property just alluded to, which derives directly from the analysis of parenting through this point, should be intuitively obvious and makes good evolutionary sense. *Because parental competence is multiply determined, the parenting system is buffered against threats to its integrity that derive from weakness in any single component.* This property can be concretely illustrated by the following claims: The undermining effect of a difficult child on parental competence is lessened when the parent has an abundance of personal resources. Conversely, an easy-to-rear child can compensate for limited personal resources on the part of the parent in maintaining parental competence. Analogously, an abundance of personal resources can compensate for the absence of support, whereas an abundance of support can reduce the threat to competent parenting that derives from parents' limited personal resources. Finally, the easy-to-rear child can compensate for limited support because the parents may not need as much support with such children, whereas an abundance of support compensates for the difficulty usually encountered when rearing a problem child because such children, with such support, are not regarded as being so difficult as they would be without such support.

The preceding hypotheses all describe relationships between pairs of the major determinants of parenting when one contributor is at risk (i.e., does not function to promote parental competence). When two of the three determinants of parenting are at risk, we propose that parental competence is most protected when only the personal resource subsystem functions to promote sensitive involvement and is least protected when only the subsystem of child characteristics functions to promote sensitive involvement. What this proposition implies, of course, is that if something must go wrong in the parenting system, we believe that the system would function best (defined in terms of producing competent offspring) when the personal resources of parents are the only determinants that remain intact.

Evidence to support the claim that risk characteristics in the child are the easiest to overcome can be found in the literature on high-risk infants. Premature birth, much research teaches us, does not compromise subsequent development when rearing takes place in middle-class homes in which both personal resources and support systems are likely to function effectively (Sameroff & Chandler, 1975). Thus, unless the subsystems of support or personal resources are at risk, as they are more likely to be in impoverished homes, we do not find incompetent child-functioning in the face of difficult child characteristics. Because students of parent–child relations have not examined, in any single research effort, all three of the major determinants of parental competence that we have identified, we can find no data to test our claim that personal resources have the greatest potential for buffering the parenting system.

At this juncture, we can offer, on the basis of the preceding analysis, a predictive model of the parenting system. As noted already, the system itself is composed of three subsystems (the parents' personal resources, the child's characteristics, and subsystems of support), and each functions, in our model, in one of two modes, which we label *support-stress*. Table 1 displays the relative probability of a child's developing competently when reared in the eight possible conditions that describe the total range of variability of the system's functioning.

Not surprisingly, the child is hypothesized as developing most competently when each subsystem is turned on and functioning in the supportive mode (+), and as least likely to develop competently when each subsystem is turned off and functioning in the stressful mode (−). When only two subsystems are in the supportive mode, we consider the child's chances of developing competently greatest when the subsystems of personal resources and support are turned on, and least when personal resources are the only system turned off. We have already stated our predictions regarding the conditions under which only one subsystem is in the support mode, and these are appropriately ranked in Table 1.

In the real world, as in the development of our model, these subsystems are themselves recognized as being complex and composed of a variety of components; thus, the binary label of *stress-support* is inappropriate. To be more accurate, we should speak in terms of the degree

TABLE 1. Theoretical Model of the Relative Probability of a Child's Developing Competently in All Possible Conditions of the Parenting System

Relative probability of child developing competently	Conditions of the parental subsystems[a]		
	Parent's personal resources	Subsystems of support	Child's characteristics
High ∧ ∣ ∣ ∣ ∣ ∣ ∣ ∨ Low	+	+	+
	+	+	−
	+	−	+
	−	+	+
	+	−	−
	−	+	−
	−	−	+
	−	−	−

[a] (+) stands for supportive mode (−) stands for stressful mode.

of stress-support provided by each subsystem rather than in terms of the presence or absence of stress-support. We have chosen not to do so in order to simplify our model. Because we are not at all certain about the trade-offs or the dynamic interactions that take place between subsystems, it is difficult to predict, for example, how much personal-resource support it takes to balance out child-determined stress. We do feel secure in maintaining, however, that equal quantities of each are not required to achieve a balance. Clearly, far more theoretical and empirical work is required before we can be certain about the nature of the complex relationships that exist between these subsystems.

CONCLUSION

It is imperative to recognize that all the proposals put forth in this chapter are based on a traditional theoretical assumption that is well known to be ecologically invalid, that is, that each hypothesis is true, "all other things being equal"—*ceteris paribus*. In the ecology of human development, all other things are rarely equal. Consider, for example, the fact that individuals with an abundance of personal resources are likely (1) to have children who are relatively easy to care for and (2) to successfully attract support from their spouse, their friends, their neighbors and relatives, and their jobs. Thus, not only are such individuals likely to be sensitively involved with their offspring because they possess the personal resources that we have argued facilitate parental competence, but they are likely to function in a parenting system in which determinants of parental competence other than personal resources also foster involvement. Put in more everyday language, we see that good things often go together. The converse we know to be true also, and the literature on child maltreatment demonstrates this repeatedly. Parental incompetence, numerous studies indicate, is likely to occur (1) when the parents' own developmental histories have produced individuals with a dearth of personal resources; (2) when children are at risk for developmental problems or otherwise difficult to rear; and (3) when social supports are minimal if present at all (Belsky, 1980; Garbarino, 1979; Parke & Collmer, 1975).

The fact that persons with an abundance of personal resources have a great deal going for them, whereas those without such resources have a great deal going against them, is not terribly surprising, but nevertheless, it illustrates a theme that has run through this entire essay: Individuals—in this case, parents—structure (willfully or accidentally) the contexts in which they function and, as a consequence, serve to produce their own development by the feedback processes they set in motion or help to maintain.

The same is true, of course, of children. Consider the fact that (1) children who are easy to rear enhance the personal resources of their parents by producing in them a sense of effectance in the parental role (Goldberg, 1977), (2) whereas children who are difficult to rear often undermine the support that would otherwise come from the parent's spouse, as well as from informal social networks. Because we have previously argued that effects such as these, in and of themselves, serve to promote or undermine parental competence and, therefore, developmental competence in the child, we see that children also structure the contexts in which they function and, in so doing, produce their own subsequent development.

Consideration of the plight of a family with a child prone to illness or with a child experiencing some kind of behavioral disturbance like hyperactivity provides clinical support for these claims. Consider first how difficult it must be to develop a sense of effectance as a parent when caring for such a child. Consider next how lack of success in caring for such a child can generate spousal disagreement regarding child rearing and, thereby, undermine the spousal support that might otherwise be available to a parent. And finally, consider the difficulty that such children might create with respect to the maintenance of social contacts outside the family, either because a suitable babysitter is difficult to secure, or because others do not like being around problem children.

Despite these probabilities, we recognize a need to qualify the notions just presented. In the absence of intervention, the processes described seem to have an aura of inevitability, although in specific circumstances exceptions should be expected (e.g., a difficult-to-rear child who develops well can generate a greater sense of parental effectance than an easy-to-rear child). But in a caring society, intervention is possible at many points in the causal chain. Thus, a handicapped child may actually draw social support to the family, rather than reducing it, when formal services are offered and when a humanitarian ethos prevails in a community. In conclusion, it is worth noting that our model draws attention to several key aspects of intervention with families at risk: First, many points of intervention are possible, because the determinants of competent parenting are multiple; second, in general, the intervention that strengthens as many of the parenting subsystems as possible will be most efficacious; third, individualization is needed in order to locate the optimal points of intervention in each instance; and finally, positive interventions in one subsystem, through feedback effects, are likely to enhance functioning in other subsystems. On the basis of this brief analysis, it should be evident that a basic understanding of the determinants of competent parenting carries with it important implications for science as well as social policy.

REFERENCES

Abernathy, V. Social network and response to the maternal role. *International Journal of Sociology of the Family*, 1973, *3*, 86–92.

Ainsworth, M. D. S. The development of infant–mother attachment. In B. Caldwell & H. Ricciuti (Eds.), *Review of child development research* (Vol. 3). Chicago: University of Chicago Press, 1973.

Ainsworth, M., Bell, S., & Stayton, D. Individual differences in strange-situation behavior of one-year-olds. In H. L. Schaffer (Ed.), *The origins of human social behavior*. New York: Academic Press, 1971.

Baltes, P., Reese, H., & Lipsitt, L. Life-span developmental psychology. *Annual Review of Psychology*, 1980, *31*, 65–110.

Baumrind, D. Child care practices anteceding three patterns of pre-school behavior. *Genetic Psychological Monographs*, 1967, *75*, 48–88.

Baumrind, D. *Naturalistic observation in the study of parent-child interaction*. Paper presented at the meeting of the 76th American Psychological Association Convention, San Francisco, 1968.

Baumrind, D. Current patterns of parental authority. *Developmental Psychology Monographs*, 4(1, Part 2), 1971.

Baumrind, D. An exploratory study of socialization effects on black children: Some black–white comparisons. *Child Development*, 1972, *43*, 261–267.

Baumrind, D. The contributions of the family to the development of competence in children. *Schizophrenia Bulletin*, 1975, *14*, 12–37.

Bell, R. Q. A reinterpretation of the direction of effects in studies of socialization. *Psychological Review*, 1968, *75*, 81–95.

Bell, R. Stimulus control of parent or caretaker behavior by offspring. *Development Psychology*, 1971, *4*, 63–72.

Belsky, J. Child maltreatment: An ecological integration. *American Psychologist*, 1980, *35*, 320–335.

Belsky, J. Early human experience: A family perspective. *Developmental Psychology*, 1981, *17*, 3–23.

Belsky, J., & Tolan, W. Infants as producers of their own development: An ecological perspective. In R. Lerner & N. Busch (Eds.), *The child as producer of its own development: A life-span perspective*. New York: Academic Press, 1981.

Belsky, J., Goode, M., & Most, R. Maternal stimulation and infant exploratory competence: Cross-sectional, correlational, and experimental analyses. *Child Development*, 1980, *51*, 1163–1178.

Bernard, J. *The future of motherhood*. New York: Penguin Books, 1974.

Brim, O. G., & Kagan, J. *Constancy and change in human development*. Cambridge: Harvard University Press, 1980.

Bronfenbrenner, U. *The ecology of human development*. Cambridge: Harvard University Press, 1979.

Buss, D. M. Predicting parent–child interactions from children's activity level. *Developmental Psychology*, 1981, *17*, 59–65.

Caldwell, B. The effects of infant care. In M. Hoffman & L. Hoffman (Eds.), *Review of child development research* (Vol. 1). New York: Russell Sage Foundation, 1964.

Campbell, A. The American way of mating: Marriage si; children, only maybe. *Psychology Today*, May 1975, 39–42.

Caplan, G. *Support systems and community mental health*. New York: Behavioral Publications, 1974.

Carew, J. V. Experience and the development of intelligence in young children at home and in day care. *Monographs of the Society for Research in Child Development*, 1980, *45*(6–7, Serial No. 187).

Carroll, J. The intergenerational transmission of family violence. *Aggressive Behavior*, 1977, *3*, 289–299.

Cassell, J. Psychosocial processes and "stress": Theoretical formulation. *International Journal of Health Services*, 1974, *4*, 471–482.

Chavez, A., Martinez, C., & Yaschine, T. The importance of nutrition and stimuli on child mental and social development. In J. Cravioto, L. Hambraeus, & B. Vahlquist (Eds.), *Early malnutrition and mental development*. Uppsala, Sweden: Almquist and Wiksell, 1974.

Cherry, L., & Lewis, M. Mothers and two-year-olds: A study of sex-differentiated aspects of verbal interaction. *Developmental Psychology*, 1976, *2*, 278–282.

Clarke-Stewart, K. A. Interactions between mothers and their young children: Characteristics and consequences. *Monographs of the Society for Research in Child Development*, 1973, *38*(6–7, Serial No. 153).

Cobb, S. Social support as a moderator of life stress. *Psychosomatic Medicine*, 1976, *38*(5), 300–314.

Cochran, M., & Brassard, J. Child development and personal social networks. *Child Development*, 1979, *50*, 601–616.

Colletta, N. Support systems after divorce: Incidence and impact. *Journal of Marriage and the Family*, 1979, *41*, 837–846.

Coopersmith, S. *The antecedents of self-esteem*. San Francisco: Freeman, 1967.

Dielman, T., Barton, K., & Cattell, R. Relationships among family attitudes and childrearing practices. *Journal of Genetic Psychology*, 1977, *130*, 105–112.

Elder, G. History and the life course. In D. Berfaux (Ed.), *Biography and society: The life history approach in the social sciences*. Beverly Hills: Sage, 1981.

Erikson, E. H. *Childhood and society*. New York: Norton, 1950.

Erlanger, H. Childhood punishment experience and adult violence. *Children and Youth Services Review*, 1979, *1*, 75–86.

Feiring, C., & Taylor, J. *The influence of the infant and secondary parent on maternal behavior: Toward a social systems view of infant attachment*. Unpublished manuscript, University of Pittsburgh, n.d.

Feldman, H. Changes in marriage and parenthood: A methodological design. In A. Michel (Ed.), *Family issues of employed women in Europe and America*. Leiden, The Netherlands: E. F. Brull, 1971.

Friedrich, W., & Boriskin, J. The role of the child in abuse: A review of literature. *American Journal of Orthopsychiatry*, 1976, *40*, 580–590.

Frommer, E., & O'Shea, G. Antenatal identification of women liable to have problems in managing their infants. *British Journal of Psychiatry*, 1973, *123*, 149–156. (a)

Frommer, E., & O'Shea, G. The importance of childhood experiences in relation to problems of marriage and family building. *British Journal of Psychiatry*, 1973, *123*, 157–160. (b)

Gecas, V. The influence of social class on socialization. In W. R. Burr, R. Hill, F. I. Nye, & I. L. Reiss (Eds.), *Contemporary theories about the family* (Vol. 1). New York: Free Press, 1979.

Gewirtz, J., & Boyd, E. Experiments on mother–infant interaction underlying mutual attachment acquisition: The infant conditions the mother. In T. Alloway, L. Kranes, & P. Pliner (Eds.), *Attachment behavior: Advances in the study of communication and affect* (Vol. 3). New York: Plenum Press, 1976.

Gibson, H. Early delinquency in relation to broken homes. *Journal of Abnormal Psychology*, 1969, *74*, 33–41.

Goldberg, S. Social competence in infancy: A model of parent–infant interaction. *Merrill-Palmer Quarterly*, 1977, *23*, 63–177.

Guttman, D. Parenthood: A key to the comparative study of the life cycle. In N. Datan & L. Ginsberg (Eds.), *Life-span developmental psychology: Normative life crises*. New York: Academic Press, 1975.

Hall, F., Pawlby, S., & Wolkind, S. Early life experience and later mothering behavior: A study of mothers and their 20-week-old babies. In D. Schaffer & J. Dunn (Eds.), *The first year of life*. New York: Wiley, 1979.

Hartup, W. W. Perspectives on child and family interaction: Past, present and future. In R. Lerner & G. Spanier (Eds.), *Child influences on marital and family interaction: A life-span perspective*. New York: Academic Press, 1978.

Hess, R., & Shipman, V. Early experience and the socialization of cognitive modes in children. *Child Development*, 1965, *36*, 869–886.

Hetherington, E., Cox, M., & Cox, R. The aftermath of divorce. In J. Stevens & M. Mathews (Eds.), *Mother–child, father–child relations*. Washington, D.C.: National Association for the Education of Young Children, 1977.

Hoffman, L. Maternal employment: 1979. *American Psychologist*, 1979, *34*, 859–865.

Hoffman, L. W. Mothers' enjoyment of work and effects on the child. *Child Development*, 1961, *32*, 187–197.

Hultsch, D. K., & Plemons, J. K. Life events and life span development. In P. B. Baltes & O. G. Brim, Jr. (Eds.), *Life-span development and behavior* (Vol. 2). New York: Academic Press, 1979.

Johnson, S., & Lobitz, G. The personal and marital adjustment of parents as related to observed child deviance and parenting behaviors. *Journal of Abnormal Child Psychology*, 1974, *2*, 193–207.

Kemper, T., & Reichler, M. Fathers work integration and frequencies of rewards and punishments administered by fathers and mothers to adolescent sons and daughters. *Journal of Genetic Psychology*, 1976, *129*, 207–219. (a)

Kessen, W., & Fein, G. *Variation in home based infant education: Language, play and social development.* Final report to the Office of Child Development, DHEW, August 1975.

Kimmel, D., & vander Veen, F. Factors of marital adjustment in Locke's Marital Adjustment Test. *Journal of Marriage and the Family*, 1974, *36*, 57–63.

Kohn, M. Social class and parent–child relationships: An interpretation. *American Journal of Sociology*, 1963, *68*, 471–480.

Kohn, M., & Schooler, C. Class, occupation, and orientation. *American Sociological Review*, 1969, *34*, 659–678.

Lamb, M., & Easterbrooks, M. Individual differences in parental sensitivity: Some thoughts about origins, components, and consequences. In M. Lamb & L. Sherrod (Eds.), *Infant social cognition: Empirical and theoretical considerations.* Hillsdale, N.J.: Erlbaum, 1980.

Lerner, R. Nature, nurture, and dynamic interactionism. *Human Development*, 1978, *21*, 1–20.

Lerner, R., & Busch-Rossnagel, N. (Eds.). *Individuals as producers of their own development.* New York: Academic Press, 1981.

Lerner, R., Spanier, G., & Belsky, J. The child in the family. In C. Kopp & J. Krakow (Eds.), *The child: Development in a social context.* Reading, Mass.: Addison-Wesley, 1982.

Lewis, M., & Goldberg, S. Perceptual-cognitive development in infancy: A generalized expectancy model as a function of mother–infant interaction. *Merrill Palmer Quarterly*, 1969, *15*, 81–100.

Lewis, M., & Rosenblum, L. (Eds.). *The effect of the infant on its caregiver.* New York: Wiley, 1974.

Lewis, M., & Wilson, C. D. Infant development in lower-class American families. *Human Development*, 1972, *15*, 112–127.

Main, M. *Exploration, play, and cognitive functioning as related to child–mother attachment.* Unpublished doctoral dissertation, Johns Hopkins University, 1973.

Matas, L., Arend, R., & Sroufe, L. Continuity in adaptation in the second year: The relationship between quality of attachment and later competence. *Child Development*, 1978, *49*, 547–556.

McCall, R. Exploratory manipulation and play in the human infant. *Monographs of the Society for Research in Child Development*, 1974, *39*(2, Serial No. 155).

Mitchell, R., & Trickett, E. Task force report: Social networks as mediators of social support. *Community Mental Health Journal*, 1980, *16*, 27–44.

Mondell, S., & Tyler, F. Parental competence and styles of problem solving/play behavior with children. *Developmental Psychology*, 1981, *17*, 73–78.

Moss, H. Sex, age and state as determinants of mother–infant interaction. *Merrill Palmer Quarterly*, 1967, *13*, 19–36.

Neugarten, B. Adult personality: Toward a psychology of the life cycle. In B. Neugarten (Ed.), *Middle age and aging.* Chicago: University of Chicago Press, 1968.

Nye, I. Child adjustment in broken and in unhappy unbroken homes. *Marriage and Family Living*, 1957, *19*, 356–361.

Orlansky, H. Infant care and personality. *Psychological Bulletin*, 1949, *46*, 1–48.

Osofsky, J. Neonatal characteristics and mother–infant interaction in two observational situations. *Child Development*, 1976, 47, 1138–1147.

Owens, D. M., & Straus, M. A. The social structure of violence in childhood and approval of violence as an adult. *Aggressive Behavior*, 1975, 1, 193–214.

Parke, R., & Collmer, C. Child abuse: An interdisciplinary review. In E. M. Hetherington (Ed.), *Review of child development research* (Vol. 5). Chicago: University of Chicago Press, 1975.

Parke, R., & Sawin, D. *Infant characteristics and behavior as elicitors of maternal and paternal responsibility in the newborn period.* Paper presented at the biennial meeting of the Society for Research in Child Development, Denver, April 1975.

Pastor, D. L. The quality of mother–infant attachment and its relationship to toddler's initial sociability with peers. *Developmental Psychology*, 1981, 3, 326–335.

Pedersen, F. Mother, father and infant as an interactive system. In J. Belsky (Ed.), *In the beginning: Readings on infancy.* New York: Columbia University Press, 1982.

Pollitt, E. Behavior of infant in causation of nutritional marasmus. *American Journal of Clinical Nutrition*, 1973, 26, 264–270.

Powell, D. Family-environment relations and early childrearing: The role of social networks and neighborhoods. *Journal of Research and Development in Education*, 1979, 13, 1–11.

Riegel, K. Toward a dialectical theory of development. *Human Development*, 1975, 18, 50–64.

Robins, E., Gamble, W., & Belsky, J. *The wider ecology of early infancy.* Paper presented at the biennial meeting of the Society for Research in Child Development, Boston, April 1981.

Rohner, R., & Rohner, E. Antecedents and consequences of parental rejection: A theory of emotional abuse. *Child Abuse and Neglect*, 1980, 4, 189–198.

Rossetti-Ferreira, M. Malnutrition and mother–infant synchrony: Slow mental development. *International Journal of Behavioral Development*, 1978, 1, 207–219.

Rutter, M. Parent-child separation: Psychological effects in the children. *Journal of Child Psychology and Psychiatry*, 1971, 12, 233–260.

Sameroff, A. Transactional models of early social relations. *Human Development*, 1975, 18, 65–79.

Sameroff, A., & Chandler, M. J. Reproductive risk and the continuum of caretaking casualty. In F. D. Horowitz (Ed.), *Review of child development research* (Vol. 4). Chicago: University of Chicago Press, 1975.

Schaefer, E., Bauman, K., Siegel, E., Hosking, J., & Sanders, M. *Mother-infant interaction: Factor analyses, stability and demographic and psychological correlates.* Unpublished manuscript, University of North Carolina, Chapel Hill, 1980.

Spanier, G., & Cole, C. Toward clarification and investigation of marital adjustment. *International Journal of Sociology of the Family*, 1976, 6, 121–146.

Sroufe, L., & Ward, M. Seductive behavior of mothers and toddlers: Occurrence, correlates and family origins. *Child Development*, 1980, 51, 1222–1229.

Stayton, D., Hogan, R., & Ainsworth, M. Infant obedience and maternal behavior: The origins of socialization reconsidered. *Child Development*, 1971, 42, 1057–1069.

Thompson, R. A., & Lamb, M. E. Individual differences in dimensions of socioemotional development in infancy. In R. Plutchik & H. Kellerman (Eds.), *Emotion: Theory, research and experience, Vol. 2: Emotions in early development.* New York: Academic Press, 1981.

Toms-Olson, J. The impact of housework on childcare in the home. *Family Relations*, 1981, 30, 75–81.

Tulkin, S. An analysis of the concept of cultural deprivation. *Developmental Psychology*, 1972, 6, 326–339.

Unger, D., & Powell, D. *Supporting families under stress: The role of social networks.* Unpublished manuscript, Merrill-Palmer Institute, Detroit, 1980.

Wainwright, W. Fatherhood as a precipitant of mental illness. *American Journal of Psychiatry*, 1966, *123*, 40–44.
Waters, E., Wippman, J., & Sroufe, L. Attachment, positive effect, and competence in the peer group: Two studies in construct validation. *Child Development*, 1979, *50*, 821–829.
Yarrow, L., Rubenstein, J., & Pedersen, F. *Infant and environment*. New York: Wiley, 1975.
Zeskind, P., & Ramey, C. Fetal malnutrition: An experimental study of its consequences on infant development in two caregiving environments. *Child Development*, 1978, *49*, 1155–1162.
Zur-Szpiro, S., & Longfellow, C. *Support from fathers: Implications for the well-being of mothers and their children*. Paper presented at the biennial meeting of the Society for Research in Child Development, Boston, April 1981.

12

A Transactional View of Stress in Families of Handicapped Children

PAULA J. BECKMAN

It has long been recognized that the family is among the most critical mechanisms for the transmission of culture within our society. As a result, the processes by which socialization takes place and the circumstances associated with difficulties for the family have become critical areas of investigation for social scientists. Many different approaches have been adopted by investigators in a variety of areas, and at least some of these lines of research are now beginning to converge.

One prime example is that investigators previously interested in exploring dyadic relationships are beginning to acknowledge the influence that individuals external to the dyad may have on the functioning within it. Moreover, the dyad is recognized as an integral part of a larger system that functions, at least in part, to move the child from infancy through adulthood. Attempts to extend the investigation of social relationships beyond the sometimes microscopic study of dyads are frequently frustrated by the sheer complexity encountered as soon as additional factors are considered. The introduction of additional family members, the variety of environmental circumstances in which families function, and the changes that take place as time elapses quickly become overwhelming and tempt a retreat back to a focus on less complex units. Without a framework around which to organize the myriad of elements

---

PAULA J. BECKMAN • Department of Special Education, University of Maryland, College Park, Maryland 20742.

presenting themselves, it is easy to lose sight of the way in which particular findings relate to a broader perspective.

The complexity of the family unit and the range of circumstances that influence its functioning have fascinated researchers for decades. Specifically, increasing attention has been focused on families of handicapped children. This burgeoning interest may be attributed to a number of factors, not the least of which are the dramatic changes that have taken place in social policy concerning handicapped children. Such policy changes have been designed to encourage the rearing of handicapped children in the home and to assure parental involvement and participation. Moreover, it has become increasingly clear that parents may be critical determinants of the long-term effects of intervention efforts (Bronfenbrenner, 1974). Thus, the ability of families to cope with the birth of a handicapped child has become a critical issue for researchers, educators, and clinicians alike.

The increased recognition of the critical role played by the parents and families of handicapped children, is reflected, at least in part, by the growing literature concerning stress in these families (Beckman-Bell, 1980, 1981; Bristol, 1979; Farber, 1959; Gallagher, Beckman-Bell, & Cross, 1983; Holroyd, 1974; Holroyd & McArthur, 1976). Stress is a multifaceted construct that has been defined in a variety of ways. For the purposes of this chapter, the definitions proposed by Rabkin and Struening (1976) are adopted. These authors view social stressors as life changes that alter the individual's social setting. Stress is viewed as "the organism's response to stressful conditions or stressors, consisting of a pattern of physiological and psychological reactions, both immediate and delayed" (p. 1014). Thus, family stress has been inferred from indicators such as disrupted patterns of interaction between family members, evidence of emotional problems for one or more family members, breakdown in family structure and/or organization, financial difficulties, and other similar manifestations consistent with Rabkin and Struening's definitions (Beckman-Bell, 1980; Bristol, 1979; Cummings, 1976; Cummings, Bayley, & Rie, 1966; Farber, 1959; Holroyd, 1974; Richards & McIntosh, 1973; Tavormina & Krajl, 1975). It has become increasingly clear that an understanding of the stress experienced by the families of handicapped children requires much more than a focus on the way in which individual members react to the occurrence of a single event. Rather, it involves an understanding of the process by which this complex system responds to an ever-changing set of circumstances.

One model that may provide a useful mechanism by which to view family stress is the transactional model described by Sameroff and Chandler (1975). Although these authors originally described this model as a way of viewing the factors placing the developmental outcome of infants at risk, it may also be helpful as a way of understanding family stress and the conditions that may ameliorate it. Arguing that most models attempting to predict the developmental outcome of infants are not adequate to handle the range of biological and environmental influences involved, Sameroff and Chandler contrasted "main-effect" and "interactional" models with the "transactional" model they proposed.

These authors argued that the basic assumption of main-effect models is that the child's environment and constitution exert independent influences on the child's subsequent development. In most investigations utilizing this approach, the child's outcome is evaluated within the context of one of these dimensions. Sameroff and Chandler suggest that, although this model is attractive because of its parsimony, it frequently fails to predict the developmental outcome for children. These authors argued that interactional models improve on main-effect models because they acknowledge that effective, long-term prediction must account for both environmental and constitutional influences on the child. Investigations based on this model therefore consider the interaction of those two dimensions when attempting to predict developmental outcome. However, the authors noted that this model may also be insufficient to improve our understanding of later disorders, because neither the environment nor the child's status necessarily remains the same over time, and both change as a result of their influence on one another.

In contrast, Sameroff and Chandler argued for the use of a transactional model. From this model, both biological and environmental influences are seen as plastic and as changing with time. The constants in development are the processes involved in interactions occurring between the child and the environment. The forces contributing to later malfunction operate over the course of development and not simply at one critical point.

Although this model has most frequently been applied to investigations of the child's developmental status, it may also be a useful way of viewing family stress. One primary reason is that inherent within the transactional model is a temporal component. Like the events that influence the child's developmental status, circumstances for families do not necessarily remain static over time. As a result, the amount of stress experienced is likely to vary as the family's circumstances for the family and the contexts within which the family functions change. One advantage of applying the transactional model to the families of children who either have or are at risk for problems of development is that it has the capacity to account for changes that may occur in the child's developmental status. Further, this approach makes it possible to look beyond the occurence of one particular event (such as the birth of a child with developmental problems) to the way in which the event interacts with others to influence the family.

In the remaining sections of this chapter, an effort has been made to demonstrate the utility of the transactional model as a perspective from which to view stress in families. Like the study of the infant's developmental status, the study of family stress involves a range of constitutional and environmental factors. After considering the relative influence of representative elements within each of these domains, studies that illustrate the use of a transactional approach are described. For the most part, the discussion is limited to studies concerning families of handicapped children, although, where pertinent, studies of families of nonhandicapped children are included as well.

Constitutional Factors

At the level of the family, the constitutional factors include the characteristics of the individual family members, the organizational and structural characteristics of the family, and the ongoing relationships between the family members.

Characteristics of Individual Family Members

The available evidence suggests that specific characteristics of individual members are often contributors to the amount of stress that families report. In families of handicapped children, specific characteristics of the child have been shown to be particularly important. For example, variations in the amount of stress reported by the parents have been reported as a function of the child's diagnostic category. Holroyd and McArthur (1976) found that the parents of autistic children reported more stress than the parents of children who had Down's syndrome or who were outpatients in a psychiatric clinic. Not only was the amount of stress reported as different for the families in the three groups, but the *pattern* of responses to questions measuring different types of stress was different. Similarly, Cummings and his colleagues (Cummings, 1976; Cummings *et al.*, 1966) reported that the parents of chronically ill and retarded children scored less favorably on measures of affect, self-esteem, and interpersonal satisfaction than the parents of children who were neurotic or who were part of a healthy control group. Other studies of the families of handicapped children suggest that males tend to be more stressful than females and that older children are more stressful than younger children (Bristol, 1979; Farber, 1959; Farber & Rykman, 1965).

Some additional characteristics also appear to contribute to the stress reported by parents. Beckman-Bell (1980) interviewed the mothers of 31 handicapped infants using the Holroyd Questionnaire on Resources and

Stress. Following the interview, the infants were observed when they were brought for regular visits to an intervention program. Measures of specific infant behavioral characteristics were entered into a regression equation in an attempt to predict the number of parent and family problems reported by the mothers. Of the six predictors identified, three were found to account for 78% of the variance in stress. These include the number of additional or unusual caregiving demands made by the child, the child's social responsiveness, and the number of self-stimulatory behaviors the child exhibited. Caregiving demands alone accounted for more than 66% of the variance in the stress reported. Bristol (1979) reported similar findings in a study of stress in families of autistic children. She found that among the most potent predictors of family stress were the child's degree of dependence and physical incapacitation.

Indirect support for the argument that the characteristics of children who are at risk for problems of development are associated with family stress is available in the child abuse literature. A number of authors have concluded that there is a disproportionate amount of child abuse and failure to thrive among low-birth-weight infants and infants with developmental problems (Elmer & Gregg, 1967; Parke & Collmer, 1975). Although causal interpretations have not been possible, given the nature of the data, these findings are interesting because abuse may be the most dramatic example of breakdown in family functioning.

In addition to characteristics of the handicapped child, characteristics of other family members may also be associated with the amount of stress that families experience. A number of reports suggest that limited intelligence, educational background, and verbal skills are frequently associated with both individual and family stress (Rabkin & Struening, 1976; Rosenberg, 1977). The individual histories and personality characteristics of the parents have been associated with stress. For example, Wise and Grossman (1980) reported that the ego strength of adolescent mothers measured during pregnancy was associated with the mother's subsequent adaptation to her infant. Pearlin and Schooler (1978) found that psychological resources such as self-reliance and self-esteem were helpful in coping with stress. Similarly, in their review of the stress literature, Rabkin and Struening (1976) found evidence that a number of personal characteristics of individuals mediated their perception of stress and their response to it. For example, individuals with more skills, more versatile defenses, and broader experience appeared to cope better in the face of stressful events.

Together, these studies suggest that the individual characteristics of family members are related to the amount of stress they experience. However, other studies in which the same characteristics have been studied have not always reported similar findings. For example, Bradshaw and Lawton (1978) investigated stress in families of thalidomide children using the Malaise Inventory (Rutter, Tizard, & Whitmore, 1970). This scale measures the presence of emotional problems, such as depression, that the respondent may be experiencing. These authors discovered that the mothers in their sample obtained significantly higher scores on this scale than have been reported previously in samples of mothers from the general population. The scores were about the same as those reported in other samples of mothers of handicapped children. However, higher scores were not related to the child's degree of mobility, type of handicap, communication skills, personal independence, or age. These conflicting findings suggest that the relationship between the characteristics of individual family members and stress is not necessarily a direct one. It may be that these characteristics take on their importance only when considered in light of other dimensions of the social system.

Family Organization

The specific organization and structure of a family may also be associated with family stress. For example, Holroyd (1974) compared single and married mothers of handicapped children using the Holroyd Questionnaire on Resources and Stress and discovered that single mothers reported more stress than married mothers. Beckman-Bell (1980) reported similar findings in her study of handicapped infants. In light of the previously reported evidence suggesting that the presence of additional caregiving demands may be an important contributor to family stress, these findings are particularly relevant. It may be that the stress experienced by families is especially severe when no one is available to assist the mother with her additional caregiving responsibilities. Thus, specific characteristics of individual members may take on added relevance when considered in light of other constitutional dimensions of the family.

The work of Bronfenbrenner and his colleagues (Bronfenbrenner, Avgar, & Henderson, 1977) also suggests that the number of parents living in the home may interact with other family and environmental variables to produce increased stress. Reporting the results of a pilot study of the families of 66 nonhandicapped preschool children, these authors discovered that one of the groups reporting the most stress was single-parent families who were below the professional level and were raising their first child. Again, the number of parents in the home was particularly important in light of other family characteristics and environmental variables.

This finding also highlights another feature of family organization, specifically, the number of children in the family. Freese and Thoman

(1978) studied the characteristics of mothers of normal infants as a function of several factors, including parity. They found more depression in primaparous mothers several weeks after delivery, whereas more irritability and tension were reported by multiparous mothers prior to delivery.

Unfortunately, the literature concerning the relationship between stress and the family's organizational structure have not always been consistent. For example, in the previously reported study by Bradshaw and Lawton (1978), no relationship was found between mothers' score on the Malaise Inventory and the number of parents in the home. In addition, although the mothers of only children received lower scores than the mothers of more than one child, no relationship between the overall size of the family and the scores was obtained. Similarly, Beckman-Bell (1980) found no relationship between the number of children in the family and the stress reported by the mother of handicapped children on the Holroyd Questionnaire on Resources and Stress. Finally, Jeffcoate, Humphrey, and Lloyd (1979) studied the responses of parents of preterm and full-term infants following delivery and found that although the parents of preterms generally experienced more emotional turmoil, delayed attachment, and management problems, these experiences were not related to parity, social class, or ethnic group.

The apparent inconsistency in the findings just described may be explained in several ways. One obvious explanation is that the operational definitions of stress that were utilized varied from study to study. In addition, however, the relationship between stress and the organizational characteristics of the family may be mediated by other factors as well. For example, a mother who receives a great deal of external support from relatives and friends may not find that having a large family is as stressful as a mother who lacks a strong support network. Thus, these conflicting findings highlight the possibility that although the family's organization and structure may contribute to the amount of stress its members experience, the nature and the extent of this contribution must be considered in light of the myriad of factors impinging on the family.

Relationships between Members

In addition to characteristics of individual family members and the overall organization of the family, a third element that appears to be related to family stress involves the relationships existing between various subgroups within the family. These relationships are reflected both by the patterns of interaction that take place between the members of the family and by the roles assumed by individual family members.

An extensive body of literature concerned with early socialization has focused on the interactions taking place between parents (most often the mother) and their young children (Feiring & Lewis, 1978; Osofsky & Connor, 1979; Parke, 1979). One relatively consistent conclusion that has emerged from this literature has been that interaction between infants and their caregivers is a reciprocal process to which both partners make important contributions (Bell, 1968; Lewis & Rosenblum, 1974). The available evidence suggests that the absence of such reciprocity may seriously interfere with the relationship between the child and the caregiver. For example, Fraiberg (1974, 1975) has noted that attachment between infants who are blind from birth and their mothers is often delayed. Exploring this issue, Fraiberg discovered that blind infants often do not utilize many of the same nonverbal cues that sighted infants use, such as eye contact and smiling. Attachment was facilitated by teaching the mothers to recognize and respond to alternative signals. Disruptions such as these in dyadic relationships may be an important source of stress for the parents of handicapped children. Robson and Moss (1970) reported that the inability to console her infant resulted in strong feelings of anger and frustration on the part of a mother whose infant was later diagnosed as brain-damaged.

Clearly, more evidence is needed to determine how stressful such disruptions in dyadic relationships are, not only for the members of the dyad, but for other family members as well. Lewis and Feiring (1980) have argued persuasively that family members may be influenced not only by the direct interaction they have with other members (such as interaction between the mother and child) but indirectly as well. Indirect effects may occur either when actions affect a family member even though they have taken place in her or his absence or when one family member witnesses and is affected by an interaction between other members of the system. For example, it can be speculated that although marital conflict may not involve the children directly, when it is present it may stress the entire family system. Although more research is needed to establish the relative importance of such indirect effects to the overall development of the child and the family, it is certainly a provocative area of investigation.

In addition to stress resulting from certain patterns of interaction occurring within the family, evidence suggests that the roles that family members must play with respect to others in the family are often associated with stress. One of the earliest studies in this regard was conducted by Farber (1959). He studied stress in families of severely handicapped children and found that older sisters experienced more difficulties in adjusting to the child than other siblings. Farber speculated that this may be because the older sisters were more often asked to assume some of the additional caregiving responsibilities associated with caring for the handicapped child.

Other studies have also suggested that the caregiving role is associated with more perceived stress. Jeffcoate *et al.* (1979) reported that although, in general, the parents of preterm infants reported feeling more negative emotions than the parents of full-term infants, the mothers reported significantly more problems than the fathers. The fathers of preterms reported having to engage in more caregiving than they had expected. The authors attributed much of the stress experienced by the mothers to their different role expectations and the loss of confidence and self-esteem they experienced. Bradshaw and Lawton (1978) found that the mother's satisfaction with her role was related to the amount of stress she experienced.

Clearly, the roles assumed by family members are likely to be differentially influenced by a variety of factors impinging on the family. For example, although the birth of a handicapped child may have an effect on all members of the immediate family, the individual assuming the role of primary caregiver (traditionally, the mother) may experience the most stress. In addition, however, other family members may be influenced both as a result of the demands that the handicapped child places on their roles as well as indirectly through the relationship between their roles and that of the primary caregiver. Farber (1975) has suggested that, as a family tries to cope with the presence of a severely retarded child, a series of role negotiations must take place in which changes in the responsibilities assumed by one member result in changes in the roles assumed by other members. These changes occur over time through a process that Farber described as minimal adaptation. Minimal adaptation implies that a family will make the smallest changes possible in order to adapt to a crisis.

The studies described in this section, as well as the proposals of Farber (1975) and Lewis and Feiring (1980), highlight several important dimensions that are relevant to a transactional perspective of family stress. First, it does appear likely that the relationships existing between subgroups within the family may contribute to the stress experienced, although more data are needed to determine the nature and the extent of this influence. Second, this relationship may be influenced by events external to the family as well as within it. The best example was provided by a study by Fraiberg (1974, 1975). Recall that when the mother was taught to look for alternative nonverbal signals, the mother's attachment to her infant was facilitated. Thus, appropriate support provided by individuals outside the family resulted in important changes in the ongoing relationship. Finally, an important temporal element is implied. As time elapses and roles are renegotiated and patterns of interaction change, the amount of stress is also likely to change.

Conclusion

Taken together, the findings presented thus far suggest that a range of "constitutional" characteristics may exert considerable influence on the amount and the nature of the stress experienced by families. Obviously, the characteristics presented thus far may represent only a small sample of the potential influences within the family. However, the relationship between these characteristics and stress may not necessarily be a direct one. As a result, it may be speculated that factors external to the family may influence both the amount and type of stress reported. Therefore, it is necessary to consider the environmental circumstances in which families function to determine their role in stress.

ENVIRONMENTAL FACTORS

The environment within which the family functions consists of a variety of elements, including socioeconomic circumstances and the amount of support available within the community. In this section, the relationship of each of these areas to the amount of stress reported by families is examined.

Socioeconomic Status

Perhaps the most widely recognized contributor to family stress involves the socioeconomic circumstances in which each family functions. Socioeconomic status is typically established by one or more indexes of occupation, income, or education, although a variety of potential stressors within the family's environment may be reflected as well.

One obvious source of stress is concern about the income that the family has available to meet the basic needs for food, shelter, medical care, and the like. Stress may also emanate from concerns about the neighborhood in which the family lives. Similarly, the sheer density of the population surrounding the family may, in and of itself, be a source of stress.

The results of studies in a number of different areas suggest that socioeconomic status may influence family functioning. For example, in recent years, socioeconomic status has been associated with various aspects of the home environment. Ramey, Mills, Campbell, and O'Brien (1975) compared the environments of infants who were considered "at risk" for sociocultural retardation with those of infants from the general population. Among other things, the sample from the general population scored more favorably on measures of maternal involvement, maternal warmth, organization of the environment, and absence of restriction and punishment. The authors suggested that one possible explanation for these findings was that, in disadvantaged environments, mothers may have to devote a substantial amount of time and energy simply to maintaining an intact household.

A more dramatic and clear-cut indicator of high levels of stress among families is the extent to which child abuse and neglect are found. Many reports have suggested a link between less advantageous socioeconomic conditions and family stress (Elmer, 1977; Garbarino, 1976; Garbarino & Crouter, 1978; Garbarino & Sherman, 1980; Light, 1973). These findings have been somewhat controversial because of the potential reporting biases inherent in the data. A summary of these concerns is discussed by Garbarino and Crouter (1978). However, in recent years, studies in which attempts have been made to control potential reporting bias have suggested that the relationship between abuse and socioeconomic status is not a spurious one. For example, in a recent study, Steinberg, Catalano, and Dooley (1981) observed monthly changes in the rate of abuse and neglect and in economic indicators such as the size of the work force and the rate of unemployment in two diverse metropolitan communities. Utilizing cross-correlational techniques over a 30-month period, these investigators found that increases in child abuse were preceded by periods of high job loss.

In addition, studies utilizing other indicators from which stress is often inferred also suggest that members of disadvantaged groups may be particularly vulnerable. For example, disproportionately high rates of medical and psychiatric disorders have long been found in deteriorating neighborhoods (Faris & Dunham, 1939).

The results of the studies cited thus far clearly provide support for the suggestion that families living in disadvantaged environments are often subject to considerable stress. Again, however, it is necessary to look beyond simple linear relationships between stress and socioeconomic conditions. Obviously, reports of child abuse and psychiatric disorder are not unique to families from disadvantaged families. Indeed, socioeconomic status appears to interact with a variety of other indicators to produce stress. The previously described study conducted by Bronfenbrenner and his colleagues (1977) illustrates this point. They found that families reporting the most stress were intact families at the lowest occupational levels who had more than one child. High levels of stress were also reported for single-parent families below the professional level, *and* for intact, upper-income families with only one child. Clearly, these findings suggest that although socioeconomic status does appear to influence stress, a true understanding of its impact must include a consideration of the particular characteristics of the family.

Support

One of the most consistent themes running throughout the literature concerned with family stress is focused on the role of social support (Cobb, 1976; McCubbin, 1979). In addition to the supports provided by ongoing relationships within the family, external support may be provided by friends, neighbors, extended family, and outside agencies. Studies with regard to the families of handicapped and nonhandicapped children alike suggest that the amount of external support a family receives may influence the amount of perceived stress. Some of the earliest findings in this regard were reported by Farber (1959) in his study of stress in families of severely handicapped children. He reported that frequent interactions with the wife's mother were associated with less stress, as were frequent contacts with the church. However, frequent interactions with the husband's mother were associated with more stress. In addition, Farber found that frequent associations with friends and neighbors were associated with more stress for the women whom Farber had labeled as poor marital risks. This finding is particularly relevant to the present discussion, as it suggests that the relationship between stress and support depends, at least in part, on specific characteristics of the women who were interviewed.

In recent years, evidence has begun to accumulate that further suggests that social support is particularly important *as it interacts* with other features of the family and the environment. For example, Crockenberg (1981) studied the influence of social support for the mother, infant irritability, and maternal responsiveness on the development of attachment in a sample of nonhandicapped infants and their mothers. She reported that the amount of social support reported by the mother when the infant was 3 months old was the best predictor of secure attachment at 1 year. Social support was *most* important for mothers of babies who were identified as irritable during the neonatal period.

Additional evidence of the importance of support is provided by the study by Bristol (1979) of factors relating to the stress reported by the mothers of autistic children. Bristol found that, in addition to specific characteristics of the child, the lack of informal supports was a major contributor to the amount of stress that the mothers reported.

In a similar vein, lack of social support appears to interact with other environmental factors to influence perceived stress. For example, in the pilot study by Bronfenbrenner and his colleagues, described earlier, lower levels of stress were found to be associated with the assistance that families received from friends, neighbors, and relatives, as well as the adequacy of the available child-care arrangements.

Further evidence of the importance of social support has been reported by Garbarino and Sherman (1980). These investigators selected a pair of low-income neighborhoods that were matched for socioeconomic level, but that were dramatically different in terms of the rate of child abuse and neglect. The neighborhood labeled as "high-risk" had a much higher rate of abuse than would have been expected based on demographic predictors. Conversely, the "low-risk" neighborhood had a much lower rate of abuse than would ordinarily have been predicted from demographic information. These authors interviewed a variety of individuals who were familiar with the neighborhood about various aspects of neighborhood life. These included neighborhood change, the availability of informal supports in the neighborhood, the quality of life, neighborhood involvement, child abuse, neighborhood appearance, the public image of the neighborhood, and the social characteristics of the neighborhood. The authors also interviewed a random sample of families from each neighborhood regarding stress and support. Generally, the responses suggested a more negative picture in all areas for the "high-risk" neighborhood. In particular, the parents in the "low-risk" neighborhood appeared to receive more of their support within the neighborhood. For example, they were more likely to exchange child supervision. In the "high-risk" neighborhood, there was less adequate child care, less reciprocal exchange, and less self-sufficiency.

In general, the findings presented with regard to social support are similar to the findings from studies investigating the relationship between other factors and family stress. The amount of support available to the family has been found to be associated with the amount of stress reported by its members. However, the relative influence of the available social supports may vary as a function of the other environmental circumstances and constitutional characteristics of the family.

Conclusion

The studies described in this section provide support for the argument that environmental factors are important contributors to the amount of stress experienced by families. However, just as it was difficult to consistently obtain a direct linear relationship between family stress and any single constitutional factor, it is also difficult to consistently obtain a direct linear relationship between family stress and any single environmental factor. It appears instead that the influence of any given factor depends on the relative influence that other environmental or constitutional factors are exerting at that time.

Although it seems intuitively obvious that the stress experienced by a family is influenced by factors both within and external to the family, more research is needed to establish the nature and the extent of this interaction. It may be more important, however, to note that, in the studies described thus far, stress is typically considered at one specified time rather than over time. Because many of the environmental and constitutional dimensions described earlier change on either an ongoing or an intermittent basis, it is important to consider studies in which the interaction of these factors is investigated longitudinally.

CHANGES IN STRESS OVER TIME

Perhaps the most distinguishing feature of the transactional model is that it emphasizes the interaction taking place between constitutional and environmental factors *over time*. This interplay is as important to the prediction of family stress as it is to the prediction of child development. However, relatively few studies have considered the amount of stress the family experiences from a longitudinal perspective. Studies of this nature are critical, both for purposes of prediction and in order to develop strategies that will effectively assist families as they cope with the birth of a handicapped child.

Despite the general paucity of research in which temporal factors have been considered, a few longitudinal studies have emerged in recent years. One particularly relevant study in this regard was conducted by Fotheringham, Skelton, and Hoddinott (1972). They examined the effect of mentally retarded children on their families by comparing the families of children who were institutionalized with the families of children who remained at home. The families were interviewed prior to institutionalization and one year later. Similarly, standardized tests of intelligence were administered prior to institutionalization and one year later. Significant declines in IQ were found for the children in both groups. The authors also found that the IQ changes in the institutional sample were not related to the families' overall level of functioning, nor to changes in the families' functioning over the year. However, for the families of children who remained at home, the child's decline in IQ was significantly related to a decline in family functioning over the year. These findings suggest that not only are the individual reactions of the parents to the child influenced by a decline in the child's rate of development, but the decline is also associated with an overall disruption in family functioning *over time*. These findings are particularly significant because they illustrate that the influence of a child's handicap changed *with time* as a result of environmental factors, although the causal nature of the change is impossible to determine.

Another study, conducted by Kogan, Tyler, and Turner (1974) illustrates these concurrent changes over time. These authors examined the longitudinal course of affective interactions between children with cerebral palsy and the adults working with them. Over a three-year period, a gradual decrease in the amount of warmth and acceptance that mothers displayed both during therapy and during play sessions was observed. The same trend occurred for the therapists. These authors also found a strong relationship between the mother's affect decrement and the child's deficit in gross motor development; again, however, the causal nature of the relationship cannot be determined. Similarly, there was a clear correlation between the size of the mother's affect decrement and whether or not the child was walking.

More recently, Burden (1980) explored the influence of a home-based intervention program on the stress reported by mothers of handicapped infants. Twenty mothers were interviewed at the beginning of home visits and two years later by means of the Malaise Inventory. Twelve mothers in another location, whose infants were not receiving home visits, were also interviewed at these times. The percentage of mothers who were identified as clinically depressed dropped over the two-year period for the mothers in the intervention group, whereas it increased for the mothers in the group who did not receive home visits. Although the nature of the data precludes overzealous interpretation, the findings support the importance of a transactional approach to family stress. Over time, the availability of support appeared to reduce at least some of the potentially adverse affects of a handicapping condition on the mother.

SUMMARY AND CONCLUSIONS

The studies presented in the preceding sections by no means represent an exhaustive review of the research concerning stress in the families of handicapped children. However, they can be used to demonstrate that these families are dealing with a complex array of influences, only one of which is the child's handicap, and that it is nearly impossible to consider the effect of one influence in the absence of the others. Moreover, these factors are likely to exert an ongoing influence on one another, so that the stress observed at one time may be very different from the stress observed at another time. Serious attempts to assist the family must take these ongoing processes into account. Doing so requires a research effort that goes beyond the measurement of these factors at a single point and that isolates a number of relevant dimensions. It is with this kind of information that we may begin to understand the impact of a child's handicap on the family system and to effectively assist families as they cope with stress.

REFERENCES

Beckman-Bell, P. *Characteristics of handicapped infants: A study of the relationship between child characteristics and stress as reported by mothers.* Unpublished doctoral dissertation, University of North Carolina, 1980.

Beckman-Bell, P. Child-related stress in families of handicapped children. *Topics in Early Childhood Special Education*, 1981, 1(3), 45–53.

Bell, R. Q. A reinterpretation of the direction of effects in studies of socialization. *Psychological Review*, 1968, 75, 81–95.

Bradshaw, J., & Lawton, D. Tracing the causes of stress in families with handicapped children. *British Journal of Social Work*, 1978, 8(2), 181–192.

Bristol, M. M. *Maternal coping with autistic children: Adequacy of interpersonal support and effect of child's characteristics.* Unpublished doctoral dissertation, University of North Carolina, 1979.

Bronfenbrenner, U. *Is early intervention effective?* (DHEW Publication No. OHD-74-25). Washington: D. C.: Department of Health, Education, and Welfare, 1974.

Bronfenbrenner, U., Avgar, A., & Henderson, C. *An analysis of family stresses and supports.* Unpublished manuscript, Ithaca, N.Y., 1977.

Burden, R. L. Measuring the effects of stress on the mothers of handicapped infants: Must depression always follow? *Child Health: Care and Development*, 1980, 6, 111–125.

Cobb, S. Social support as a moderator of life stress. *Psychosomatic Medicine*, 1976, 38(5), 300–314.

Crockenberg, S. B. Infant irritability, mother responsiveness, and social support influences on the security of infant–mother attachment. *Child Development*, 1981, 52(3), 857–865.

Cummings, S. T. The impact of the child's deficiency on the father: A study of fathers of mentally retarded and of chronically ill children. *American Journal of Orthopsychiatry*, 1976, 46, 246–255.

Cummings, S. T., Bayley, H., & Rie, H. Effects of the child's deficiency on the mother: A study of mother's of mentally retarded, chronically ill, and neurotic children. *American Journal of Orthopsychiatry*, 1966, 36, 595–608.

Elmer, E. A follow-up study of traumatized children. *Pediatrics*, 1977, 59, 273–279.

Elmer, E., & Gregg, C. D. Developmental characteristics of the abused child. *Pediatrics*, 1967, 40, 596–602.

Farber, B. Effects of a severely retarded child on family integration. *Monographs of the Society for Research in Child Development*, 1959, 24(2, Serial No. 71).

Farber, B. Family adaptations to severely mentally retarded children. In M. J. Begab (Ed.), *The mentally retarded and society.* Baltimore: University Park Press, 1975.

Farber, B., & Ryckman, D. B. Effects of severely mentally retarded children on family relationships. *Mental Retardation Abstracts*, 1965, 2, 1–17.

Faris, R., & Dunham, H. *Mental disorders in urban areas.* Chicago: University of Chicago Press, 1939.

Feiring, C., & Lewis, M. The child as a member of the family system. *Behavioral Science,* 1978, *23,* 225–233.

Fotheringham, J. B., Skelton, M., & Hoddinott, B. A. The effects on the family of the presence of a mentally retarded child. *Canadian Psychiatric Association Journal,* 1972, *17,* 283–290.

Fraiberg, S. Blind infants and their mothers: An examination of the sign system. In M. Lewis & L. A. Rosenblum (Eds.), *The effect of the infant on its caregiver.* New York: Wiley, 1974.

Fraiberg, S. Intervention in infancy: A program for blind infants. In B. Z. Friedlander, G. M. Serritt, & G. E. Kirk (Eds.), *Exceptional infants: Assessment and intervention* (Vol. 3). New York: Brunner/Mazel, 1975.

Freese, M. P., & Thoman, E. B. The assessment of maternal characteristics for the study of mother–infant interaction. *Infant Behavior and Development,* 1978, *1,* 95–105.

Gallagher, J. J., Beckman-Bell, P., & Cross, A. Families of handicapped children: Sources of stress and its amelioration. *Exceptional Children,* 1983, *50,* 10–19.

Garbarino, J. A preliminary study of some ecological correlates of child abuse: The impact of socioeconomic stress on mothers. *Child Development,* 1976, *47,* 178–185.

Garbarino, J., & Crouter, A. Defining the community context for parent–child relations: The correlates of child maltreatment. *Child Development,* 1978, *49,* 604–616.

Garbarino, J., & Sherman, D. High risk neighborhoods and high risk families: The human ecology of child maltreatment. *Child Development,* 1980, *51,* 188–198.

Holroyd, J. The Questionnaire on Resources and Stress: An instrument to measure family response to a handicapped member. *Journal of Community Psychology,* 1974, *2,* 92–94.

Holroyd, J., & McArthur, D. Mental retardation and stress on the parents: A contrast between Down's syndrome and childhood autism. *American Journal of Mental Deficiency,* 1976, *80,* 431–436.

Jeffcoate, J. A., Humphrey, M. E., & Lloyd, J. K. Disturbance in parent–child relationships following preterm delivery. *Developmental Medicine and Child Neurology,* 1979, *21,* 344–352.

Kogan, K. L., Tyler, N., & Turner, P. The process of interpersonal adaptation between mothers and their cerebral palsied children. *Developmental Medicine and Child Neurology,* 1974, *16,* 518–527.

Lewis, M., & Feiring, C. Direct and indirect interactions in social relationships. In L. Lipsitt (Ed), *Advances in Infancy Research* (Vol. 1). New York: Ablex, 1980.

Lewis, M., & Rosenblum, L. A. *The effect of the infant on its caregiver.* New York: Wiley, 1974.

Light, R. Abuse and neglected children in America: A study of alternative policies. *Harvard Educational Review,* 1973, *43,* 556–598.

McCubbin, H. I. Integrating coping behavior in family stress theory. *Journal of Marriage and the Family,* 1979, 237–244.

Osofsky, J. D., & Connors, K. Mother-infant interaction: An integrative view of a complex system. In J. D. Osofsky (Ed.), *Handbook of infant development.* New York: Wiley, 1979.

Parke, R. D. Perspectives on father-infant interaction. In J. D. Osofsky (Ed.), *Handbook of infant development.* New York: Wiley, 1979.

Parke, R. D., & Collmer, C. W. Child abuse: An interdisciplinary analysis. In E. M. Heatherington (Ed.), *Review of child development research* (Vol. 5). Chicago: University of Chicago Press, 1975.

Pearlin, L. I., & Schooler, C. The structure of coping. *Journal of Health and Social Behavior,* 1978, *19,* 2–21.

Rabkin, J. G., & Streuning, E. L. Life events, stress, and illness. *Science,* 1976, *194*(3), 1013–1020.

Ramey, C. T., Mills, P., Campbell, F. A., & O'Brien, C. Infant's home environments: A comparison of high-risk families and families from the general population. *American Journal of Mental Deficiency,* 1975, *80,* 40–42.

Richards, I. D., & McIntosh, H. T. Spina bifida survivors and their parents: A study of problems and services. *Developmental Medicine and Neurology* 1973, *15,* 292–304.

Robson, K. S., & Moss, H. A. Patterns and determinants of maternal attachment. *Journal of Pediatrics,* 1970, *77,* 976–985.

Rosenberg, S. A. *Family and parent variables affecting outcomes of a parent-mediated intervention.* Unpublished doctoral dissertation, George Peabody College for Teachers, 1977.

Rutter, M., Tizard, J., & Whitmore, K. *Education, health and behavior.* London: Longman, 1970.

Sameroff, A. J., & Chandler, M. J. Reproductive risk and the continuum of caretaking casuality. In F. D. Horowitz (Ed.), *Review of child development research* (Vol. 4). Chicago: University of Chicago Press, 1975.

Steinberg, L. D., Catalano, R., & Dooley, D. Economic antecedents of child abuse and neglect. *Child Development,* 1981, *52,* 975–985.

Tavormina, J. B., & Kralj, M. M. *Facilitating family dynamics: Family system issues in the overall management of the handicapped infant.* Paper presented at the meeting of the Exceptional Infant Symposium, Charlottesville, Va., December 1975.

Wise, S., & Grossman, F. K. Adolescent mothers and their infants: Psychological factors in early attachment and interaction. *American Journal of Orthopsychiatry,* 1980, *50,* 454–468.

13

Residential Environments and the Social Behavior of Handicapped Individuals

SHARON LANDESMAN-DWYER

INTRODUCTION

The theme of this book—social connections beyond the dyad—is one that is particularly relevant to the lives of handicapped individuals. From birth on, all children live in a world where their opportunities for social interaction and for learning depend on many others. For children with seriously handicapping conditions, this dependence on others constitutes one of the most salient features of their day-to-day lives. Compared with normal children, those who are handicapped are more likely to experience multiple disruptions in their primary dyadic relationships; yet theoretically, they are less able to adapt to the demands associated with environmental change.

Adaptation to environmental change requires perceptual awareness of the environment, cognitive processing of the relationships among elements in the environment (including oneself), and a willingness to explore alternative ways of responding. By definition, all organisms are adapted to their environments; the issue is how successfully they have

---

SHARON LANDESMAN-DWYER • Department of Psychiatry and Behavioral Sciences, University of Washington, Seattle, Washington 98195. This research was supported by the National Institute of Child Health and Human development (HD 11551 and HD 00346), the Department of Social and Health Services, State of Washington, and the Child Development and Mental Retardation Center (HD 02274) at the University of Washington.

adapted. For mentally retarded children, deficits in *adaptive behavior*, as well as poor performance on standardized measures of intelligence, are characteristic signs of their handicap (Grossman, 1977). The criteria for successful adaptation are interwoven with the norms of a given culture and include age-appropriate behavioral skills in home, school, and other social settings. Such criteria may be useful for clinical and administrative purposes, such as identifying children who need "special help" or support services, but such an *a priori* definition of successful adaptive behavior is limited.

The current definition of adaptive behavior (Grossman, 1977) implies that all children should naturally acquire the same set of behavioral skills in the process of adapting to their environments. This definition assumes that children will choose similar solutions to the demands and challenges they face, that is, that there is one best way to adapt. There is also the assumption that children experience similar worlds, living in culturally normative settings and attending to the same features of their environments.The implicit expectation is that standard environments will have uniform effects on development. In fact, the definition focuses only on specific outcomes and fails to consider the actual processes of adaptation within their natural contexts.

A social-ecological view reveals that handicapped children often live in environments that differ both objectively and subjectively from those of their nonhandicapped peers. Not only do the environments differ, but these children are extremely heterogeneous in terms of their sensorimotor abilities, cognitive processing strategies, social adeptness, activity levels, and personal preferences. The biases in the current definition of adaptive behavior thus preclude understanding some of the successful strategies that handicapped children may use to solve problems, particularly when the situations or the solutions are atypical or idiosyncratic.

Similar problems exist in most comparative studies of normal and handicapped children. Investigators dichotomize their subjects on the basis of culturally defined norms and then proceed to use measuring instruments designed to describe the development of children who experience "normal" environments. The theoretical significance of detecting differences or no differences is obscured by the untested assumptions inherent in the design and the methodology of such research. An alternative strategy is to expand the perspective on children's life situations. This expansion would necessitate constructing functional taxonomies of environments, so that they may be compared in their objective elements and in their subject-mediated or experiential features. If such taxonomies incorporated all individuals with whom a child interacts, then the potential would exist for detecting complex, nondyadic relationships and for analyzing special, atypical systems of interpersonal communication and adaptation.

Two fundamental axioms in social ecology are "that behavior cannot be studied apart from the environment in which it occurs" and "that substantial differences may occur in the behavior of the same individuals when they are in different milieus" (Moos, 1973). The basic interdependency of persons and environments appears to be an obstacle to separating out the critical influences on development. However, the opportunity to view an individual's total ecological niche and to study the ways in which he or she adjusts to different settings constitutes the "natural experiment" (Parke, 1979) that is life.

An important aspect of social ecology is its explicit value orientation and its concern with promoting maximally effective human functioning (Insel & Moos, 1974). In essence, social ecology is a young, idealistic subdiscipline in psychology that is seeking a merger of practical, philosophical, and theoretical perspectives; hoping to take everything important into account; and gleaning the best from the methodologies developed over the past century (cf. Stokols, 1981, 1982). As Moos (1973) proposed:

> Essentially, every institution in our society is attempting to set up conditions that it hopes will maximize certain types of behavior and/or vectors of development. . . . In this sense, it may be cogently argued that the most important task for the behavioral and social sciences should be the systematic description and classification of environments and their differential cost and benefits to adaptation. (p. 662)

A key question in mental retardation is "What residential environments foster optimal social and cognitive development?" (Landesman-Dwyer, 1981). Hundreds of thousands of retarded individuals live in group-care settings and need residential support services throughout their lives (Bruininks, Meyers, Sigford, & Lakin, 1981). Compared with the norms for natural families, most of these alternative living arrangements involve a greater dispersion of caretaking responsibilities among adults, more frequent changes in the group or family composition, and less opportunity for direct control by those who live and work there. In such settings, there are major difficulties in the formation and the maintenance of close dyadic relationships, including high staff turnover rates, frequent relocation of residents, intermittent and often irregular contact with parents or other relatives, and the past histories of both the residents and their careproviders. One-to-one interactions and reciprocated friendships typically occupy only a small proportion of time in these residents' lives (Landesman-Dwyer, Berkson, & Romer, 1979) and potentially exert less influence than do other variables in their environments. Appropriate study of the development of these individuals must advance beyond the dyad and inquire about the functional meaning of all social contacts.

The variables of potential significance for handicapped children include the physical and social resources in their immediate environments, which, in turn, are affected greatly by public policy decisions. Public policy controls residential programs via legislation, court decisions, licensing standards, funding allocations, and monitoring and evaluation criteria. The present Zeitgeist in mental retardation service-delivery has been shaped by three major factors: (1) public disclosure of past mistakes, notably the inhumane conditions prevalent in many traditional institutions (Center on Human Policy, 1979; Ferleger & Boyd, 1979); (2) encouraging results from model programs (e.g., Apolloni, Cappuccilli, & Cooke, 1980) and the application of behavioral research findings (Berkson & Landesman-Dwyer, 1977); and (3) the philosophy of normalization (Kugel & Shearer, 1976; Wolfensberger, Nirje, Olshansky, Perske, & Roos, 1972). Normalization ideology is built on the belief that all children, regardless of the nature or severity of their handicaps, will benefit from "conditions of everyday life which are as close as possible to the norms and patterns of mainstream society" (Nirje, 1976, p. 181).

Identifying the norms and patterns of mainstream society and then replicating these in the lives of handicapped individuals is not an easy task. The difficulty stems from at least three distinct sources. First, there is the genuine heterogeneity of lifestyles in American culture. The nuclear family, consisting of a father, a mother, and two or three children living in their own suburban home and remaining a stable group over the first 20 years of the children's lives, still captures the American ideal or prototype. In fact, the prevalence of this family unit has decreased steadily since World War II, and increasing numbers of children experience multiple shifts in where they live, in who their primary careproviders are, and in the composition of their sibling and peer groups. A variety of "new families" has emerged, including single-parent families, families in which the children are shared by their separated parents, combined households with two or more family groups, extended-family environments in which grandparents or other relatives assume caretaking responsibilities, and changeable or unpredictable family constellations where children live with different (nonrelated) adults who function in parental roles for varying lengths of time. To average all of these diverse family situations would yield a view of living environments that does not represent the actual experience of any members in our society. Alternatively, to select only the most frequently occurring or "typical" family unit—and to provide such a living arrangement for all handicapped children—would restrict the range of experiences and options unrealistically. Clearly, the concept of "normative" environments needs to be adjusted to geographic, ethnic, socioeconomic, and other factors relevant to the target group of individuals.

Second, the normative settings for nonhandicapped individuals are not static. Indeed, physical and social environments vary for normal individuals largely as a function of age. Age happens to be a convenient index of many important and correlated variables, including an individual's major developmental accomplishments, her or his degree of dependence on other family members, and society's expectations of the individual. For many handicapped individuals, especially those who are mentally retarded, age is a less powerful predictor of other life variables, notably mental age, social skills, and independence in daily living. This nonstatic nature of environments presents problems in designing a normative environment to meet the needs of a special group of children. Should the environment match the normative one for nonhandicapped children of similar biologic age, of similar mental age, or a combination? Moreover, when and how often should the specially designed environments be modified to replicate the naturally occurring changes in normal living environments?

Third, there has been a failure to distinguish between the topographical or structural features of environments and the behavioral or functional consequences of these features. The definition of normative conditions and life patterns is a hazy mixture of stimuli and responses. What is needed is a clear delineation of the objective components of children's living environments and the primary functional concomitants of these elements. If such an equation of living environments were formulated, then the effects of substituting or changing different elements could be tested systematically. Each environmental variable could be evaluated separately according to its structural and its functional features. Theoretically, this means that children could reside in settings that differ dramatically in the composition of the social group, but that appear to function similarly from the child's perspective. Alternatively, children in topographically identical family environments may have significantly different life experiences and developmental outcomes, depending on the child's interpretation of events and on the behavioral responses of other family members.

Despite the many unanswered questions, current social policy actively endorses providing smaller and more homelike living units for all handicapped persons. The research presented here was designed primarily to evaluate the impact of such settings on the daily behavior of retarded individuals. The data, however, provide an opportunity to explore social relationships beyond the dyad, because the social settings include large peer groups and multiple careproviders. In fact, if the behavior of individual subjects had been studied only in terms of specific dyadic relationships, many of the key changes in environment and in the individuals' behavioral repertoires would have been missed.

The Study of Group Living Contexts

By observing what happens in diverse living contexts, inferences may be made about the relative importance of social group variables and individual characteristics. Recent studies of retarded person in community group-care settings indicate that environmental variables predict an individual's current behavior better than do personal variables (see Landesman-Dwyer, 1981). These findings, however, are based on correlational studies and lack a longitudinal perspective on individuals who are transferred to new settings. The research presented here represents a prospective study of retarded subjects who were randomly assigned to alternative residential settings.

A traditional state institution was preparing for total physical renovation and for reorganization of its internal resources. The residents had been living in very large wards that were old, relatively barren, and understaffed according to federal standards. The settings were "institutional" in that they were regimented, lacked space for privacy, and showed little evidence that individual needs or preferences were recognized. The new duplexes to be built were designed to correct these environmental deficiencies and to improve the overall quality of life for the residents. Staff-to-resident ratios were to increase substantially, in part to encourage the formation of closer dyadic relationships.

The target population consisted of severely and profoundly retarded individuals who had no serious medical or behavioral problems.[1] All had some minimal self-help and communication skills, but more than half needed routine assistance in dressing or toilet use, and the majority were nonverbal or spoke only a few two- or three-word phrases. The subjects were selected after evaluating the entire population ($N = 606$) at the institution (Landesman-Dwyer & Schuckit, 1976; unpublished findings, 1978).

Initially, 180 subjects were matched in 60 trios, on the basis of sex,

---

[1] This target population represents the largest group of retarded individuals in need of lifelong residential support services (Landesman-Dwyer & Sulzbacher, 1981; Landesman-Dwyer, Schuckit, Keller, & Brown, 1977) and currently receiving services poorly matched to their needs (Eyman, 1976). The research presented here is part of an ongoing series of prospective, controlled studies of residential adaptation. The studies include observational assessment of these same subjects and matched controls in community-based settings, as well as a replication of this study in another traditional institution.

age, adaptive behavior or IQ level, and etiology. One subject from each trio was selected by the clinical staff to move to the small, homelike units. The remaining two were randomly assigned to old or new settings. This design was possible because the institutional renovation was to occur in several phases, and not all residents could be moved to the new duplexes at once. Because the randomly assigned and clinically selected subjects did not differ in their baseline demographic features and showed no significant differences in outcome measures, the findings presented here include all the subjects. (For full details regarding the study design and methods, see Landesman-Dwyer, in press.)

The final sample included 67 females and 80 males, for whom complete baseline and postmove data sets were obtained. Their mean age was 34.9 years ($sd = 13.2$ years) with prior length of institutionalization averaging 21.8 years ($sd = 11.6$ years). Their adaptive behavior levels ranged from those of normal 2.0- to 6.5-year-olds, with a mean developmental age of 3.4 years ($sd = 1.5$ years).

Monitoring Behavioral and Environmental Changes

The primary outcome measures for the residents were direct behavioral observations of day-to-day behavior, standardized ratings of adaptive behavior skills, health status indicators, and documented behavioral incidents. The baseline data were gathered over a one-year period prior to any resident relocation; the postmove data were collected for one year thereafter.

We gathered behavioral observations in three settings throughout the study: the living units, the dining halls, and the day training settings. These environments were selected to provide a range of resources and demands that characterized the places where the subjects spent the greatest amounts of time. Each subject was observed during a minimum of eight sessions every two weeks, a quantity that had been determined empirically as yielding a representative behavioral profile for this population (Landesman-Dwyer, in press). Each session consisted of six discrete observations taken at consecutive 30-second intervals. Each observational entry contained the following information about the subject's behavior: (1) the major ongoing activity (1 of 69 mutually exclusive and exhaustive codes); (2) vocalizations: (3) proximity to others; (4) expression of affect; (5) others with whom the subject was interacting; (6) the direction of the social interaction; and (7) the type of stereotypy, if any. In addition, each session was described in terms of the physical location, the time of day, the major ongoing activity, and the social resources available to the subject, that is, the number of others within sight of the subject. See Landesman-Dwyer and Watts (1979) for a detailed description of the observational procedures and definitions for each code.

Interrater agreement on the major codes exceeded 80% after two weeks of initial training. Every three months, interrater reliability was reassessed by having the six observers rotate in all possible pairs and simultaneously code the behavior of randomly selected subjects. To reduce the possibility that any systematic observer bias would influence the data, observers were assigned across subjects, times, days, and settings in a balanced manner.

The residential environments were characterized in terms of (1) physical space and resources; (2) staffing patterns, including the type of staff, their prior work and educational history, their job satisfaction, their attitudes toward mentally retarded individuals, their perception of their job responsibilities and autonomy, and their ideas for improving the institution; (3) the composition of the resident group, including heterogeneity of demographic and behavioral profiles and the number of changes in the group; (4) the degree to which the environments were institutional-oriented versus resident-oriented, according to the King, Raynes, and Tizard (1971) Revised Child Management Scale; (5) the amount of cognitive stimulation provided, as reflected on scores on the Caldwell Home Observation for Measurement of the Environment (HOME), using the 3- to 6-year-old form (comparable to the developmental level of the subjects); and (6) the availability of activities, based on documentation of all special events on and off the living units. Essentially, these measures focused on environmental resources and were collected several times during baseline and at repeated intervals after the subjects moved.

The findings from this study depict changes in group living contexts and their observed effects on the residents' behavior. Even during baseline, several basic principles about environment–behavior interactions appeared. More important, individual behavioral differences during baseline predicted alternative postmove adaptation strategies.

SETTING-SPECIFIC BEHAVIOR: EVIDENCE OF SENSITIVITY TO ENVIRONMENTAL RESOURCES AND DEMANDS

Initially, all the residents lived on large halls (40–60 residents) that had dormitory-style bedrooms, open bathing and toileting areas, large common living rooms, and clearly identified staff offices, coffee rooms, and storage areas. Although the old units appeared "institutional," they had an ambiance and an individuality, reflected in their decor and lighting, the availability of things to do, and the patterns of staff–resident interaction. Staff morale and behavior, as well as their attitudes toward mentally retarded people, varied significantly across the living units. Generally, the residents lived with others of similar adaptive behavior level, age, and temperament (e.g., aggressive versus not aggressive residents), although there were notable exceptions.

A key finding from the baseline observations is that severely and profoundly retarded residents did show setting-specific patterns of activity. That is, during mealtimes and training, the subjects displayed quite different behavioral profiles from those they showed in their living units. This finding indicates that despite severe retardation, these subjects were able to perceive and respond to the resources and the demands that distinguished these settings. The finding also provides evidence that the behavioral code was capturing some of the appropriate dimensions of adaptation, an important methodological issue for interpreting the results of any longitudinal and ecological study.

Figure 1 highlights the behavioral profiles of the residents across the settings,[2] plus their social choice patterns. In the home (ward) environments, the residents spent nearly 33% of the observed periods "doing nothing," that is, not observably engaged in any activity and not attending to their immediate environments. Another 22% of the sampled time involved simply looking at someone or something or locomoting, which was combined into a summary category of exploration. All object-related activity comprised 12% while general social interaction occupied only 6% of their observed behavior. Less than 2% of the observations fell into any of the special affiliative codes (e.g., sharing resources, assisting others, showing affection, or praising others). In the home setting, the residents indicated a slight preference for peers in social choice situations, rather than for staff members.

In the training setting, the subjects showed two large differences in their behavior. They had much higher rates of object-related activity—more than double that on the wards—and they interacted far more with staff members (more than 20% of the periods) than with peers (about 6%). These findings match the demands of the training setting. What is surprising, however, is that signs of special affiliative behavior increased by a factor of 2, indicating that the quality of the social interactions was altered when the residents were expected to be more active and more task-oriented.

The types of staff-mediated demands differed markedly across the

---

[2] For presentation in this chapter, the original behavioral codes have been combined into a few broad categories. Multivariate data analyses of changes over time and across settings were conducted on the original codes.

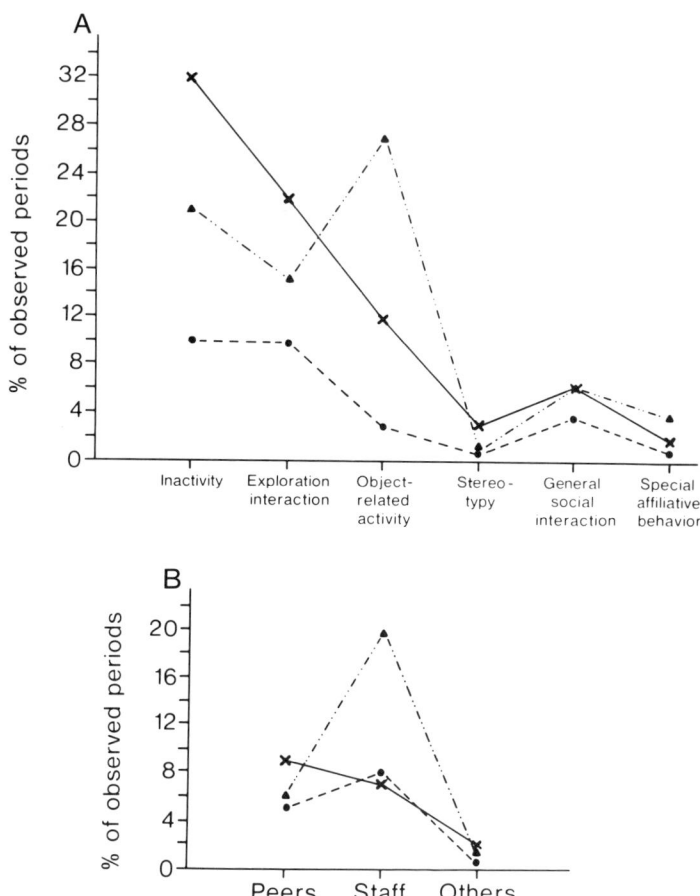

FIG. 1. Baseline behavioral profiles as a function of settings. (A) major activity; (B) social choice; (×- - -×) home (ward); (▲—··—▲) training setting; (●- - -●) dining hall.

three baseline settings. In their training programs, the residents were encouraged by the staff to be actively involved with either objects or other persons for an average of 11% of the observed periods. In contrast, they received such demands from the staff only 1% of the time when on the wards and 2% during mealtimes. The training environments further provided a large number and variety of physical resources appropriate for teaching specific skills and for being manipulated independently by the residents. Another setting difference was the staff's educational level, which was significantly higher for the training staff than for the direct-care staff on the living units. In addition, the training staff had been in their positions for a much shorter time, received higher wages, and had a narrower range of specific job responsibilities than did the direct-care attendants.

Traditional Wards versus New Living Units

The new living units were 14-bed duplexes (six to eight residents per side). The residents had single or double bedrooms, places for their own clothes and personal possessions, and private bathing and toileting areas. Each side had its own kitchen (although the meals were prepared in a centralized kitchen), dining area, and small living room. The furniture was more homelike and colorful. The duplexes were attractive single-story brick homes, surrounded by sidewalks, streets, and homelike yards.

The management practices on the new duplexes were significantly less institutional, according to scores on the revised King, Raynes, and Tizard Resident Management Scale. The maximum score possible was 30, reflecting the most institutional type of regime. None of the new duplexes had scores higher than 14, whereas all but 2 of the 15 old halls had scores above this level. The old units had a mean score of 18.2 ($sd = 7.1$) compared to 7.6 ($sd = 2.0$) for the new units, a highly significant difference ($F = 43.8$, $df = 1/34$, $p < .001$).

Although the old and new units also differed on the Caldwell HOME Scale ($F = 22.9$, $df = 1/34$, $p < .001$), their mean scores were closer together and at the lower (i.e., less stimulating) end of the scale: the new duplexes averaged 21.6 ($sd = 2.7$), the old halls 16.3 ($sd = 4.0$). In comparison, homes for normal 4-year-olds in this area had a mean score of 68.4 ($sd = 5.7$) (Landesman-Dwyer, Ragozin, & Little, 1981). Interestingly, the two highest HOME scores were in old halls, one of which was fairly institutional in its management practices, whereas the other was not.

In addition to differences in the physical resources and management practices, the old and new halls contrasted dramatically in their social contexts. Besides the obvious drop in the average number of residents living on each unit, there were significant changes in the average size of the resident group observed and in the number of staff present and visible. During the postmove period, the average size of the resident group was 10.8 ($sd = 5.0$) on the modified old wards, and 3.65 ($sd = 1.0$) on the new duplexes ($F = 42.6$, $df = 1/34$, $p < .001$). Figure 2 depicts the social resources observed on both new and old units. On the old halls, the residents were alone (group size of 1) about 5% of the observed periods and spent 47% of their time in groups of 10 or more residents.

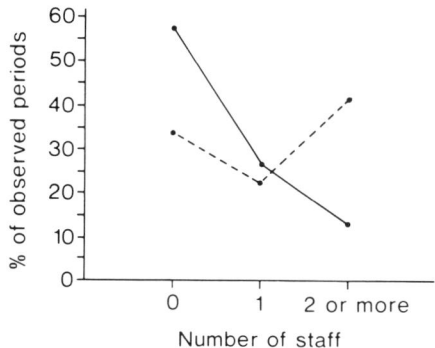
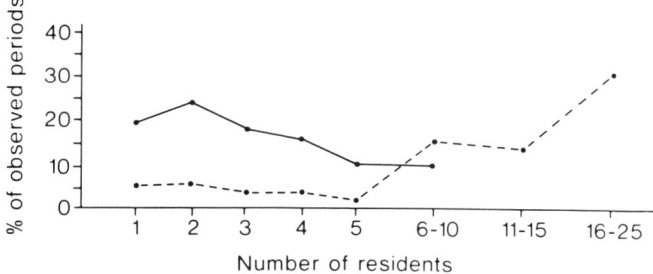

FIG. 2. Social resources in old wards versus new duplexes; (●———●) new duplexes; (●– – – ●) old wards.

A very different picture emerged in the new settings, where the residents spent 19% of their time alone, 23% with only one other person, and only 22% with five or more individuals. In essence, the residents in the new duplexes experienced two events that seldom occurred during their years of living on large wards: being totally alone and being with only one person at a time.

Although the formal assignment of the staff to the new duplexes resulted in a more than fourfold increase in the number of staff per resident, this change did not lead to an increase in the *functional* presence of staff members near the residents. As Figure 2 shows, the staff members were *not* present during 58% of the coded periods on the new duplexes, a rate much higher than that of 35% on the old units. Moreover, when a staff member was observed on the new units, he or she was much more likely to be alone, whereas staff members usually were in pairs or trios on the traditional halls.

To what extent did these changes in physical and social resources affect the behavioral profiles of residents over time? To answer this question, a variety of statistical and descriptive analyses may be applied. A multivariate analysis of covariance (Huck & McLean, 1975) was used[3] to minimize capitalizing on chance and to control for baseline differences in subjects. The dependent measures were the 10 major behavior categories that accounted for 80%–85% of the residents' observed activities. The independent measures were Group (control or old ward versus experimental or new duplexes) and Time (in four three-month segments during the postmove year).

The overall significant main effects were Group ($F = 4.31$, $df = 1/145$, $p < .001$) and Time ($F = 1.79$, $df = 3/435$, $p < .01$). The results from subsequent univariate analyses of variance, however, indicated that the subjects in the old and new units differed significantly only in their television watching ($F = 22.14$, $df = 1/145$, $p < .001$) and stereotyped/idiosyncratic behavior ($F = 12.27$, $df = 1/145$, $p < .001$).

Although the controls spent an average of 33% of their time inactive, compared with 25% for the experimentals, this difference was not greater than that observed during the baseline period, when all the subjects lived on the old wards. For all subjects, there was a small decrease (2%–3%) in awake-inactivity after major changes at the institution. Despite the fact that the experimental subjects were in novel environments, their exploratory and looking behavior did not increase significantly either above their own baseline rate or in comparison to that of the controls. Similarly, although the opportunities to engage in activities with objects were somewhat greater on the new units, as reflected in Caldwell HOME scores, these were not sufficient to induce major behavioral changes.

As Figure 3 shows, the largest group difference was the increased amount of television watching by the experimental subjects—averaging 10% of their time or twice the amount for the controls and a 3.5% increase over their own premove rate. Concerning stereotyped and idiosyncratic behavior, Figure 3 indicates that the residents in both settings showed an increase over baseline. The control subjects, however, had significantly more variable patterns and increased even more than did the experimentals—averaging 8.6% during the postmove year compared with 7.6% for the controls.

---

[3] Even for the carefully matched subjects who were randomly assigned, there were significant group differences on 4 of the 20 major behavior codes that comprised more than 85% of their behavior on the living unit. This underscores the need to conduct prospective studies and to consider individual change patterns, as conventional subject matching strategies do not necessarily ensure that the initial *behavior* of groups is comparable.

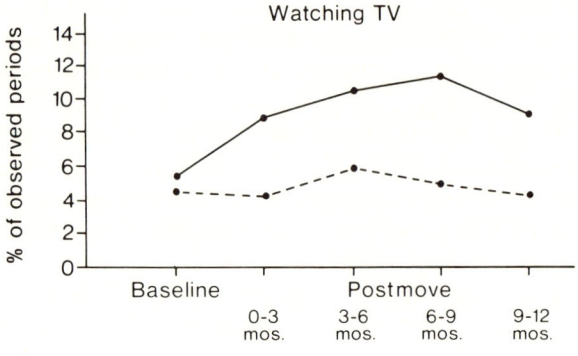
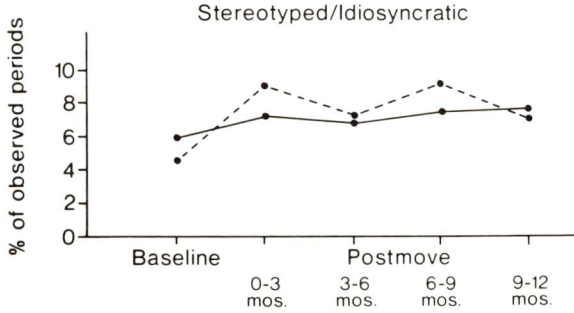

FIG. 3. Significant behavioral changes for experimental and control subjects; (●———●) experimental subjects; (●– – –●) control subjects.

Is There Evidence of Person–Environment Interaction?

An important issue is to what degree the individual characteristics of the subjects influenced their behavioral patterns. Correlations were computed for the major behavior codes and the subject variables of adaptive behavior, age, length of institutionalization, and sex. Adaptive behavior level correlated significantly with only five types of behavior and showed a similar pattern during both the baseline and the postmove periods. Adaptive behavior scores correlated positively with social interaction (.42), object-related activity (.24), and watching television (.21), and negatively with inactivity (−.35) and stereotyped or idiosyncratic behavior (−.35). None of these correlations, however, accounted for more than 20% of the variance. (When entered as a covariate in the multivariate analysis of observed behavior, adaptive behavior level did not alter the findings reported above concerning the effects of moving to the new duplexes.) Age correlations indicated that the older residents were more inactive (.28) whereas younger ones demonstrated more stereotyped and idiosyncratic behavior (−.24). Length of institutionalization, even though correlated significantly with age (.68), showed minimal association with observed behavior, correlating positively only with inactivity (.20).

Because this analysis did not reveal any significant interactions between individual characteristics and environmental responses, an alternative way of classifying individuals, described below, was tried.

Clustering Subjects on the Basis of Observed Behavioral Patterns

Although these retarded adult subjects were relatively homogeneous in their demographic characteristics and IQ range, they revealed great diversity in their daily patterns of observable behavior—even within the baseline institutional condition. Initially, 16 behavioral measures were selected. All were judged to be important for social behavior or adaptive functioning. When they were factor-analyzed, as shown in Table 1, six primary factors emerged, accounting for 95% of the variance. Each subject was assigned factor scores. These were used to perform a cluster analysis by Ward's method (1963) with relocation (Everitt, 1980). Five distinct behavioral clusters of subjects emerged: Table 2 presents their mean factor scores.

There is no best method for deciding how many factors to use, how to compute factor scores, or what is the optimal number of clusters. The program for clustering generates from 2 to 10 different clusters. By studying the computer printouts and by reviewing which subjects were sorted into the clusters, I concluded that five clusters effectively separated the subjects whom I knew to be "different," whereas four or fewer clusters

TABLE 1. Varimax Rotated Factor Matrix (After Rotation with Kaiser Normalization) for Major Behaviors, Baseline Period

Factor #	Eigenvalue	% of Variance	Highest loading variables
1	2.94	33.3	Verbalization (.94), general social (.91)
2	1.53	17.4	Idiosyncratic (.95), stereotypy (.89)
3	1.33	15.1	Physical contact (.95), special affiliative (.48), social caretaking (.29), being supervised (.28)
4	1.20	13.6	Inactive (−.86), TV/radio/stereo (.45), focused activity (.42), unfocused activity (.31)
5	.78	8.8	Locomote (.86)
6	.62	7.0	Look (.78)

TABLE 2. Mean Factor Scores for Five Behavioral Clusters[a]

	Cluster 1 $\bar{X}$ (sd)	Cluster 2 $\bar{X}$ (sd)	Cluster 3 $\bar{X}$ (sd)	Cluster 4 $\bar{X}$ (sd)	Cluster 5 $\bar{X}$ (sd)
Factor 1					
Social and verbal	−.26 (.44)	−.27 (.49)	2.43[b] (1.11)	−.30 (.40)	−.23 (.19)
Factor 2					
Stereotyped/ idiosyncratic	−.17 (.43)	−.29 (.36)	−.20 (.14)	−.40 (.50)	2.57[b] (.92)
Factor 3					
Receives supervision/ contact	1.49[b] (.83)	−.50 (.52)	−.03 (.62)	−.41 (.44)	−.13 (.65)
Factor 4					
Object-related activity	.18 (.70)	.76 (.75)	.03 (.70)	−1.18[b] (.58)	−.30 (.77)
Factor 5					
Locomotion	−.16 (.42)	.35 (1.15)	−.08 (.60)	−.34 (.59)	−.50[b] (.69)
Factor 6					
Looking	.33[b] (.94)	−.08 (.80)	−.03 (.59)	−.03 (.85)	−.24 (.37)
Number of subjects per cluster	26	47	13	31	13

[a] Based on Ward's method with relocated clustering (Everitt, 1980). Note: The same clusters resulted when the analysis started with 10 randomized groups.
[b] Highest mean value for each factor across clusters.

combined too many people into a very large and somewhat nondescript group, and more than five clusters added only a few extremely small groups with unusual profiles. I also applied a different statistical program that created subject clusters from a randomized start. Because the two methods converged almost exactly when five clusters were chosen (Everitt, 1980) (i.e., more than 95% of the subjects clustered in the same group), but not at other levels, this similar sorting provided further justification for the final choice of five clusters. Above all, the characteristic profiles of the subjects in these five clusters seemed to make sense behaviorally and clinically.

Cluster 1 consisted of 26 residents who received lots of supervision and physical contact from the staff and who mostly looked around and had only minimal interactions with objects of other people in their environments. Cluster 2 was the largest group, with 47 residents who were not extremely high or low on any behavioral factor. These subjects did show more object-oriented activity than did other clusters, although the rate was still low. In contrast, Cluster 3 contained the 13 most social and verbal individuals, who showed few maladaptive behaviors and not much interest in object-related activities. The 31 subjects in Cluster 4 represented the least active group; they had high negative loadings on the factor of object-related behavior. Finally, Cluster 5 contained the 13 subjects whose behavioral profiles consisted mostly of stereotyped and idiosyncratic behavior and who showed little evidence that other people or objects were part of their functional environments.

These subject clusters also differed significantly from one another in the vast majority of the remaining behavior codes (i.e., those not entered into the factor analysis). Some of the social characteristics of these clusters are highlighted in Tables 3 and 4.

Besides their initial baseline differences in average rates of social interaction, the subjects in these five clusters continued to differ significantly during each of the four postmove periods. The differences occurred in the proportion of time they interacted with peers and with staff and in their rates of initiating and mutually engaging in social contact with others. Prior to the move, the Cluster 1 subjects, who received much direct supervison from staff, had the highest rates of receiving contact from others. After the move, however, this distinction disappeared. In fact, the subjects in Cluster 1 experienced proportionately more of a decrease in attention from others than did those in any other cluster. For most subjects in the other clusters, the rates of receiving attention from others increased after their living environments were changed. Perhaps the Cluster 1 residents were vulnerable to a loss of attention because of a disruption in particular friendships or dependency relationships. Certainly, some regression to the mean may be expected, although such an effect tends to be rather weak in the entire data set.

The cluster analysis was performed primarily to see if the residents differed in their response to the living environments, that is, whether the principle of person × environment interaction applied to these sub-

TABLE 3. Cluster Differences in Peer and Staff Interaction during Pre- and Postmove Periods

Subject cluster	Mean percentage of periods interacting with			
	Peers		Staff	
	Pre	Post	Pre	Post
1	12.2	10.2	12.3	9.9
2	6.5	7.2	5.3	7.8
3	21.9	18.0	12.0	15.8
4	6.2	5.8	5.6	6.8
5	3.0	2.1	3.6	4.5

TABLE 4. Cluster Differences in Initiation of Social Contact during Pre- and Postmove Periods

	Mean percentage of periods in which subject					
	Initiated		Received		Engaged mutually	
Subject cluster	Pre	Post	Pre	Post	Pre	Post
1	5.4	5.3	7.6	5.2	3.1	2.8
2	3.1	4.5	2.6	3.9	2.1	2.3
3	15.5	16.1	5.2	5.7	8.9	7.7
4	1.4	2.0	2.7	3.8	1.1	1.1
5	1.2	1.6	2.5	3.0	.3	.2

jects and their living situations. Based on repeated measures analyses of variance, performed separately for each cluster, for the 10 major behavior codes, the answer is yes: the clusters did differ in their patterns of behavioral change. Two of the clusters showed significant main effects of treatment (old wards versus new duplexes) on several behaviors; two other clusters changed significantly over time, regardless of their living environments, and one cluster showed no significant change over time or across settings.

Figure 4 summarizes the cluster × environment × time interactions, displayed in terms of proportionate changes in behavior from the baseline (premove) period to the end of the postmove period (average of observations from 9–12 months). Significant three-way interactions were

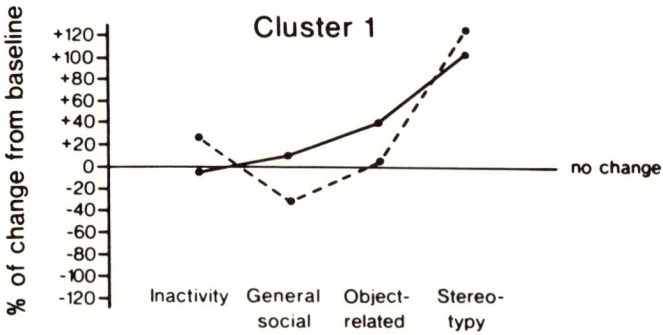

FIG. 4. Changes in behavior categories as a function of subject clusters and environments. (*) Comparisons were not computed when baseline rates were less than .10% for a given cluster of subjects; (●———●) experimental subjects; (●– – –●) control subjects.

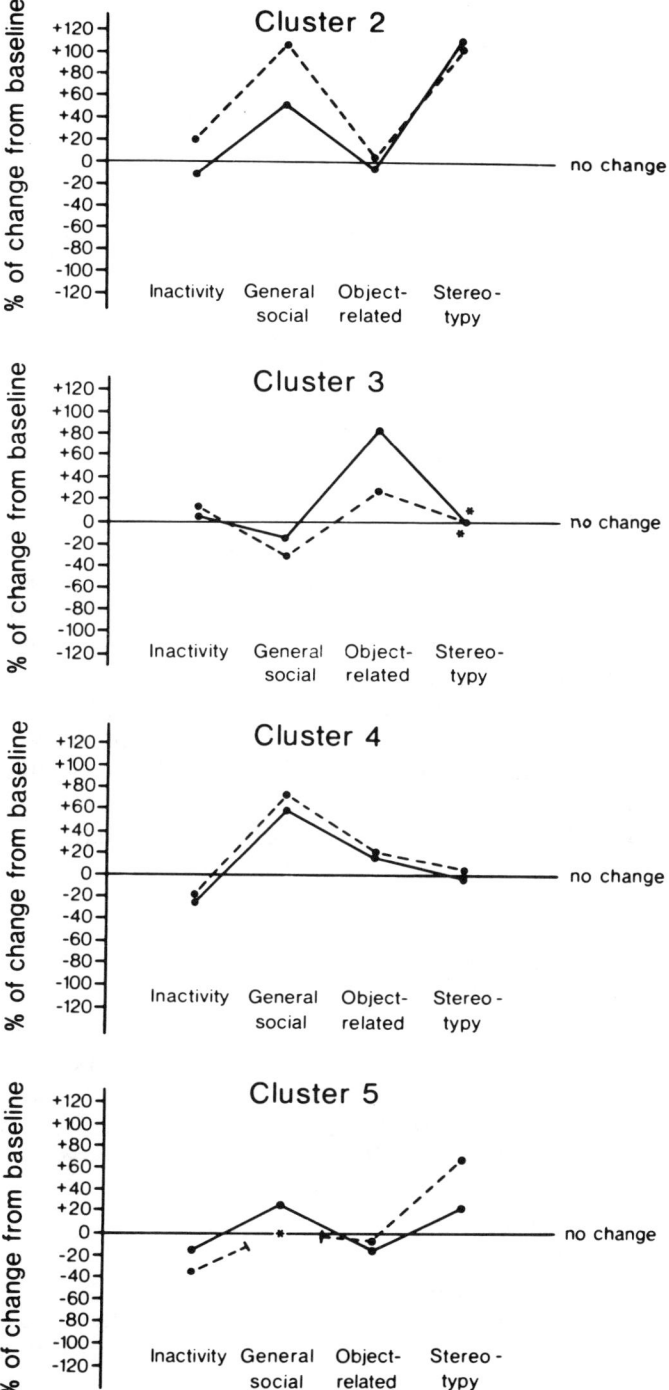

detected in a repeated-measures multivariate analysis of variance for general social interaction, household activities (a subcategory of object-related behavior), and stereotyped behavior. The experimental subjects in Clusters 1 and 2 showed significantly more positive increases in pro-social behavior than did the controls from these same clusters. In contrast, the Cluster 4 subjects in both old and new living units showed significant increases in their general social behavior.

Object-related activities increased the most for the experimental subjects in Clusters 1 and 3. Stereotyped behavior, however, showed a very different person × environment interaction effect. Only for the experimental subjects in Cluster 5, the cluster already distinguished by high baseline rates of stereotypy and idiosyncratic behavior, did a positive effect of moving to the new duplexes appear. Specifically, Cluster 5 residents in the new units did not have as great an increase in stereotypy as the controls in the old wards did. For other clusters, the pre- and postmove levels of stereotypy were similar regardless of where they lived.

These results indicate that topographically similar environments may have markedly different effects on individuals, depending on their initial behavioral profiles. Subjects' baseline behavior is hypothesized to be a valid index of their perceptual awareness of and dependence on resources in the immediate environment. Accordingly, changes in environmental resources would be expected to influence behavior only to the extent that they are perceived by the subjects. The exact nature and amount of behavioral change are, in turn, mediated by the subjects' availability of appropriate responses, that is, the diversity of the baseline repertoire and its suitability to the particular changes that have occurred.

Understanding Social Interactions "Beyond the Dyad"

The study reported here demonstrates what happens when individuals living in large group settings move to smaller settings or continue to live in groups that are modified socially. The subject population consisted of adults whose functional and cognitive abilities were comparable to preschool children's levels, and who had spent the majority of their lives in situations that were nondyadic and that were highly variable in terms of peer and staff group members.

In the new duplexes, where the assigned staff-to-resident ratios had been enriched considerably, there was no evidence that this change in social resources led to increased interactions among staff members and residents. In fact, the experimental residents actually spent significantly more time totally alone or without any staff person present than they had in the old halls. This finding is consistent with other observations that the exact staffing patterns do not determine interaction frequency or style (e.g., Grant & Moores, 1977; Harris, Veit, Allen, & Chinsky, 1974), and with one prospective study in an institution, in which Knight, Weitzer, and Zimring (1978) reported that in smaller settings staff often isolate themselves from the residents. In this natural experiment, the staff did not receive new job demands related to social interaction with the residents. When taught new patterns of behavior however, staff can change—and this change, in turn, alters residents' behavior (e.g., Schinke & Landesman-Dwyer, 1981).

The opportunities for dyadic interactions among peers did increase considerably in the new duplexes, yet the behavior of most residents was not significantly altered. As expected, the one group of individuals who showed selectivity in their social interactions with staff, and who appeared to be quite dependent on staff attention, was influenced by moving to the new duplexes. These individuals had a decrease in interactions with staff, but they substituted positive interactions with objects in their environments and increased their interactions with peers. The subjects who were "highly social" initially continued to be so, even though many of their social partners had been changed. Apparently, these subjects were able to use prior skills of social initiation and responsiveness to adapt to changes in their social group. This finding does not imply that no special friendships were disrupted, or that these subjects are incapable of developing dyadic relationships that may matter to them. Rather, the finding suggests that, for individuals with a long history of disruptions in group living arrangements, learning to adapt includes developing the ability to interact with newcomers as well as with familiar individuals.

Interestingly, for those subjects who appeared to be the most inactive and the least interested in their environments, the changes to somewhat smaller groups and the presence of more staff did contribute to increases in positive social behavior—in both the old and the new settings. The dyadic situations were clearly not the key mediating variable for this positive change because they differed markedly in the two types of residential environments. Moreover, the social interactions were more likely to occur when these particular residents were in group settings of three or more, rather than when they were in one-to-one situations.

At this stage, empirical data are not available to help make precise predictions about how to "match" different types of individuals and different environments. The paradigm of person × environment interaction is not a new one, but translating it into a meaningful theory about behavioral adaptation has been slow (Bem & Allen, 1974; Hunt, 1975;

Pervin & Lewis, 1978). For any effective analysis of environments, the function of all of the key elements must be considered, as well as the differences among individuals in what they perceive to be important, interesting, challenging, or stimulating features of their living environments. There is no evidence that dyadic situations are different, in and of themselves, in ways that have pronounced effects on the behavior of most handicapped persons.

ACKNOWLEDGMENTS

I extend special thanks to those who helped collect and analyze the observational data: Judy Bly, Al Cory, Christine Curtis, Karla Fredericksen, Pamela Garnett, Kathy Kipp Hauck, Kevin Isherwood, Jean Jameson, Susanne Keller, Margaret Knowles, Charles Lund, Katherine Mabbatt, Victor Morin, Darcy Polanecsky, Judith Schuckit, Sorena Southerland, and Ann VanderStoep.

REFERENCES

Apolloni, T., Cappuccilli, J., & Cooke, T. P. (Eds.) *Achievements in residential services for persons with disabilities: Toward excellence.* Baltimore: University Park Press, 1980.

Bem, D. J., & Allen, A. On predicting some of the people some of the time: The search for cross-situational consistencies in behavior. *Psychological Review*, 1974, 81, 506–520.

Berkson, G. B., & Landesman-Dwyer, S. Behavioral research in severe and profound mental retardation (1955–1974). *American Journal of Mental Deficiency*, 1977, 81, 428–454.

Bruininks, R., Meyers, C. E., Sigford, B. B., & Lakin, K. C. (Eds.) *Deinstitutionalization and community adjustment of mentally retarded people.* Washington, D.C.: American Association on Mental Deficiency, Monograph No. 4, 1981.

Caldwell, B. *Home observation for measurement of the environment, preschool version: Instruction manual.* Fayetteville: University of Arkansas, n.d.

Center on Human Policy. *The community imperative: A refutation of all arguments in support of institutionalizing anybody because of mental retardation.* Syracuse, N.Y.: Syracuse University, 1979.

Everitt, B. *Cluster analysis* (2nd ed.). New York: Wiley, 1980.

Eyman, R. K. Trends in the development of the profoundly mentally retarded. In C. C. Cleland, J. D. Swartz, & L. W. Talkington (Eds.), *The profoundly mentally retarded*, proceedings of the Second Annual Conference on The Profoundly Mentally Retarded, Austin, Texas: The Hogg Foundation, 1976.

Ferleger, D., & Boyd, P. A. Anti-institutionalization: The promise of the Pennhurst case. *Stanford Law Review*, 1979, 31, 101–135.

Grant, G. W. B., & Moores, B. Resident characteristics and staff behavior in two hospitals for mentally retarded adults. *American Journal of Mental Deficiency*, 1977, 82, 259–265.

Grossman, H. J. (Ed.). *Manual on terminology and classification in mental retardation* (1977 revision). Washington, D.C.: American Association on Mental Deficiency, 1977.

Harris, J. M., Veit, S. W., Allen, G. J., & Chinsky, J. M. Aide-resident ratio and ward population density as mediators of social interaction. *American Journal of Mental Deficiency*, 1974, 79, 320–326.

Huck, S. W., & McLean, R. A. Using a repeated measures ANOVA to analyze the data from a pretest-posttest design: A potentially confusing task. *Psychological Bulletin*, 1975, *82*, 511–518.

Hunt, D. E. Person–environment interaction: A challenge found wanting before it was tried. *Review of Educational Research*, 1975, *45*, 209–230.

Insel, P. M., & Moos, R. H. Psychological environments: Expanding the scope of human ecology. *American Psychologist*, 1974, *29*, 179–188.

King, R. D., Raynes, R. V., & Tizard, J. *Patterns of residential care: Sociological studies in institutions for handicapped children*. London: Routledge and Kegan Paul, 1971.

Knight, R. C., Weitzer, W. H., & Zimring, C. M. (Eds.). *Opportunity for control and the built environment: The ELEMR project*. Amherst: University of Massachusetts, 1978.

Kugel, R. B., & Shearer, A. (Eds.). *Changing patterns in residential services for the mentally retarded*. President's Committee on Mental Retardation. Washington, D.C.: U.S. Government Printing Office, 1976.

Landesman-Dwyer, S. Living in the community. *American Journal of Mental Deficiency*, 1981, *86*, 223–234.

Landesman-Dwyer, S. The changing structure and function of institutions. A search for optimal group care environments. In S. Landesman-Dwyer & P. Vietze (Eds.), *Living with retarded people*. Baltimore: University Park Press, in press.

Landesman-Dwyer, S., & Schuckit, J. J. *Preliminary findings of the survey of state institutions for the mentally retarded*. Olympia, Washington: Department of Social and Health Services, 1976, Library of Congress No. 76-620025.

Landesman-Dwyer, S., & Sulzbacher, F. Residential placement and adaptation of severely and profoundly retarded individuals. In R. Bruininks, C. E. Meyers, B. B. Sigford, & K. C. Lakin (Eds.), *Deinstitutionalization and community adjustment of mentally retarded people*. Washington, D.C.: American Association on Mental Deficiency, Monograph No. 4, 1981.

Landesman-Dwyer, S., & Watts, G. *Manual for behavioral observations of severely and profoundly retarded individuals*. Olympia, Washington: Department of Social and Health Services, 1979.

Landesman-Dwyer, S., Schuckit, J. J., Keller, L. S., & Brown, T. W. A prospective study of client needs relative to community placement. In P. Mittler (Ed.), *Research to practice in mental retardation, Vol 1: Care and intervention*. Baltimore: University Park Press, 1977.

Landesman-Dwyer, S., Berkson, G. B., & Romer, D. Affiliation and friendship of mentally retarded residents in group homes. *American Journal of Mental Deficiency*, 1979, *83*, 571–580.

Landesman-Dwyer, S., Ragozin, A. S., & Little, R. E. Behavioral correlates of prenatal alcohol exposure: A four-year follow-up study. *Neurobehavioral Toxicology and Teratology*, 1981, *3*, 187–193.

Moos, R. H. Conceptualizations of human environments. *American Psychologist*, 1973, *28*, 652–665.

Nirje, B. The normalization principle. In R. B. Kugel & A. Shearer (Eds.), *Changing patterns in residential services for the mentally retarded*. President's Committee on Mental Retardation, 1976, Library of Congress No. 76-20170.

Parke, R. D. Interactional designs. In R. B. Cairns (Ed.), *The analysis of social interactions: Methods, issues, and illustrations*. New York: Wiley, 1979.

Pervin, L. A., & Lewis, M. (Eds.). *Perspectives in interactional psychology*. New York: Plenum Press, 1978.

Schinke, S. P., & Landesman-Dwyer, S. Training staff in group homes serving mentally retarded persons. In P. Mittler (Ed.), *Frontiers of knowledge in mental retardation, Vol. 1: Social, educational, and behavioral aspects*. Baltimore: University Park Press, 1981.

Stokols, D. Group × place transactions: Some neglected issues in psychological research on settings. In D. Magnusson (Ed.), *Toward a psychology of situations*. Hillsdale, N.J.: Erlbaum, 1981.

Stokols, D. Environmental psychology: A coming of age. In A. Kraut (Ed.), *G. Stanley Hall Lecture Series* (Vol. 2). Washington, D.C.: American Psychological Association, 1982.

Ward, J. H. Hierarchical grouping to optimize an objective function. *Journal of the American Statistical Association*, 1963, *58*, 236–244.

Wolfensberger, W., Nirje, B., Olshansky, S., Perske, R., & Roos, P. (Eds.). *The principal of normalization in human services*. Toronto: National Institute on Mental Retardation, 1972.

Author Index

Italic numbers indicate pages where complete reference citations are given.

Abbott, S., 48, *57*
Abernathy, V., 175, *188*, 266, *274*
Abrahams, B., 165, 166, *188*
Abramovitch, R., 24, 25, *34*
Adams, B. N., 148, *156*, 163, 178, 180, *188*
Ainsworth, M. D. S., 17, *34*, 62, 71, 82, *84*, 195, 196, 197, 199, 200, 202, *218*, *219*, *220*, 253, *274*, *278*
Alan Guttmacher Institute, 177, *189*
Albrecht, R., 170, 175, *189*
Aldous, J., 23, *34*, *191*
Allen, A., 319, 320, *320*
Allen G. J., 319, *321*
Als, H., 138, *157*
Anderson, B. J., 13, 39, 63, *88*, 136, *159*
Anderson, C. W., 24, 28, *35*
Alexander, R. D., 92, 104, *106*
Altman, S. A., 96, *106*
Altmann, D., 95, *106*
Antonucci, T. C., 181, *191*
Apolloni, T., 302, *320*
Aquilino, W., 60, *88*
Arbel, T., 19, *36*

Arend, R., 132, 133, *156*, *159*, 183, *192*, 200, 219, 220, 253, *277*
Arezzo, D., 242, *250*
Aries, P., 34, *35*
Arling, G. L., 110, *126*, 134, *159*
Arnold, F., 66, *84*
Arnold, P., 105, *106*
Asch, S. E., 28, *35*
Asher, S. R., 165, *189*
Atchley, R. C., 164, *189*, *194*
Attneave, C. L., 141, *156*
Aug, R. G., 80, *84*
Avgar, A., 286, *296*

Badger, E., 176, *189*
Baker, C. T., 4, *12*
Bales, R. F., 16, *35*, 137, *159*, 241, *250*
Balijian, H., 132, *157*
Ballou, J., 104, *107*
Baltes, P., 256, *274*
Ban, P., 54, *57*, 61, *84*, 137, *158*
Bane, M. J., 77, 81, *84*
Barclay, A. G., 80, *84*
Barnes, J. A., 140, *156*
Barnett, C. R., 17, *40*
Barret, R., 71, *84*
Barry, H., 44, *56*
Barry, W. A., 138, *156*
Barton, K., 265, *276*

Bauman, K., 262, *278*
Bausano, M., *57*
Bayley, H., 282, *296*
Becker, J. M. T., 25, *35*
Beckman-Bell, P., 282, 284, 286, 287, *296*, *297*
Beckwith, L., 215, *219*
Bee, H. L., 2, *11*
Bell, N. W., 17, *40*
Bell, R. Q., 56, *56*, 178, *189*, 256, 275, 288, *296*
Bell, S. M., 199, 200, *218*, *219*, 253, *274*
Belmont, L., 67, *84*
Belsky, J., 13, 24, *35*, 62, 74, 75, 76, *84*, 162, 167, 174, 176, 182, *189*, 253, 256, 260, 263, 265, 267, 273, 275, 277, *278*
Bem, D. J., 320, *320*
Bengston, V. L., 162, 163, 167, 178, 180, *189*, *194*
Berardo, F. M., 148, *156*, 178
Berg, I., 69, *84*
Berg, M., 225, 238, *249*
Berger, M., 169, 185, *189*
Berghorn, F. J., 169, *189*
Berger, J., 25, *35*
Bergman, A., 215, *220*

323

Berkson, G. B., 301, 302, 320, 321
Bernard, J., 257, 275
Biben, M., 100, 106
Biller, H. B., 80, 84
Birigen, Z. C., 165, 190
Birney, E. C., 99,. 108
Bittman, S., 176, 189
Black, D., 164, 190
Blanchard, R. W., 80, 84
Blau, P., 19, 35
Blehar, M. C., 17, 24, 34, 35, 62, 74, 84, 196, 199, 200, 213, 219
Block, Jack, 134, 156
Block, Jeanne, 134, 156
Blomfield, J. M., 68, 85
Blood, R. O., Jr., 16, 35
Blumberg, L., 178, 189
Boehm, C., 48, 57
Boll, E. S., 14, 35, 61, 84
Booth, A., 243, 249
Borgatta, E. F., 16, 35
Boriskin, J., 263, 276
Borod, J., 69, 84
Bossard, J. H., 61, 84
Bott, E., 140, 141, 144, 146, 155, 156
Bowlby, J., 82, 84, 132, 157, 195, 219
Boyd, E., 256, 263, 276
Boyd, P. A., 302, 321
Bradley, R. H., 182, 183, 189, 190
Bradley-Johnson, S., 24, 40
Bradshaw, J., 285, 287, 289, 296
Brady, C. A., 97, 99, 106
Bradwein, R. A., 81, 84
Brassard, J. A., 140, 157, 162, 163, 167, 182, 189, 267, 275
Brazelton, T. B., 138, 157
Bright, T., 80, 84
Brim, O. G., 257, 275
Bringle, R., 67, 89
Bristol, M. M., 282, 284, 285, 292, 296
Broderick, C. B., 17, 35
Bronfenbrenner, U., 15, 26, 35, 43, 57, 138,

Bronfenbrenner, U. (Cont.) 157, 162, 163, 189, 257, 258, 275, 282, 286, 291, 296
Brooks, J., 9, 11, 20, 22, 25, 35, 39, 41, 75, 87, 132, 159, 196, 220, 224, 249
Brossard, J. H. S., 14, 35
Brown, B., 80, 85
Brown, C. A., 81, 84
Brown, J. S., 178, 189
Brown, K., 98, 106
Brown, T. W., 304, 321
Brownlee, E., 74, 86
Brugger, K., 104, 107
Bruinicks, R., 301, 320
Bulatao, R. A., 66, 84
Burden, R. L., 295, 296
Burke, P. J., 16, 35
Burns, D., 189
Busch-Rossnagel, N., 256, 277
Buss, D. M., 263, 275

Cain, R. L., Jr., 13, 39, 63, 88, 136, 159
Cairns, R. B., 5, 11, 176, 192
Caldwell, B. M., 74, 85, 182, 183, 189, 190, 253, 275, 306, 320
Campbell, A., 257, 275
Campbell, F., 74, 88, 291, 298
Campos, J. J., 17, 18, 30, 35, 36, 38
Cancian, F., 48, 57
Caplan, G., 266, 267, 275
Caplow, T., 16, 35
Cappuccilli, J., 302, 320
Carew, J. V., 253, 275
Carey, W. B., 32, 33, 35
Carlson, R., 191
Carroll, J., 261, 275
Carter, C. S., 105, 106
Cassell, J., 266, 267, 275
Catalano, R., 291, 298
Cattell, R., 265, 276
Center on Human Policy, 302, 320
Centers, R., 16, 35

Chandler, M. J., 271, 278, 282, 283, 298
Charlesworth, W. R., 30, 35
Chase-Lansdale, L., 71, 86, 209, 219, 220
Chavez, A., 263, 275
Cherlin, A. J., 176, 185, 189
Cherry, L., 256, 263, 275
Chibnick, M., 48, 57
Chilman, C. S., 14, 35
Chinsky, J. M., 319, 321
Chivers, D. J., 98, 99, 106
Christen A., 98, 106
Cicirelli, V. G., 136, 157
Claby, D. A., 242, 249
Clarke-Stewart, K. A., 18, 30, 36, 61, 63, 68, 73, 85, 136, 138, 157, 172, 189, 215, 219, 253, 275
Clinger, J. B., 24, 37
Coates, D. L., 61, 62, 85, 86
Cobb, S., 180, 190, 266, 267, 275, 292, 296
Cochran, M. E., 74, 85
Cochran, M. M., 140, 157, 162, 163, 167, 182, 189, 267, 275
Cohen, L. J., 18, 36
Cohen, S. E., 73, 85
Cohen, W., 235, 250
Cohler, B. J., 180, 181, 189
Cole, C., 278
Coleman, J. S., 21, 26, 36
Colletta, N. D., 80, 85, 266, 276
Collmer, C. W., 22, 39, 126, 127, 174, 182, 192, 208, 220, 263, 273, 278, 285, 297
Connell, D. B., 201, 219
Connolly, K., 225, 233, 234, 250
Connors, K., 288, 297
Constantine, J. M., 23, 36
Constantine, L. L., 23, 36
Cook, K. S., 19, 36
Cooke, T. P., 302, 320

Author Index

Cookson, D., 69, 85
Coombs, L. L., 242, 249
Coopersmith, S., 254, 276
Corsini, D., 74, 87
Corter, C., 24, 34
Costanzo, P. R., 19, 38
Cowen, E., 132, 157
Cox, M., 81, 86, 176, 191, 244, 249, 265, 276
Cox, R., 81, 86, 176, 191, 244, 249, 265, 276
Cramer, J. C., 24, 36
Crockenberg, S. B., 79, 85, 183, 190, 272, 296
Cromwell, R. E., 16, 36
Cromwell, V. L., 16, 36
Croog, S. H., 169, 190
Crook, C. K., 20, 40
Cross, A., 282, 297
Crouter, A., 291, 297
Cummings, E. M., 24, 36
Cummings, S. T., 282, 284, 296
Cusumano, D., 80, 84
Cutler, N. E., 167, 189
Cutrona, C., 165, 190

D'Andrade, R. G., 48, 57
Daniels, P., 170, 190
Darley, J. M., 29, 36, 38
Davenport, R. K., 110, 113, 118, 119, 120, 121, 122, 126, 127
Dawe, H. C., 16, 36
Deutch, M., 80, 85
Dewsbury, D. A., 96, 99, 105, 106
Dickstein, S., 200, 219
Dielman, T., 265, 276
Dodsworth, R. O., 110, 126
Dooley, D., 291, 298
Douglas, J. W. B., 68, 85
Droegemueller, W., 126, 127
Dudley, D., 96, 98, 99, 106
Dunham, H., 291, 297
Dunn, J., 18, 24, 36, 38, 83, 85, 136, 157, 165, 190

Earp, J. A., 182, 193
Easterbrooks, M. A., 28, 36, 62, 85, 200, 211, 213, 215, 219, 254, 263, 268, 277
Eckerman, C. O., 30, 36, 39
Edwards, C. P., 27, 36, 45, 58, 179, 190
Egeland, B., 200, 221
Eisenberg, J. F., 92, 105, 106, 107
Eisenman, R., 69, 84
Elardo, R., 182, 190
Elder, G. H., 163, 185, 186, 190, 256, 276
Elmer, E., 126, 126, 285, 291, 296
Emde, R. N., 17, 30, 35, 38
Emerson, P. E., 61, 88
Emerson, R. M., 19, 27, 36
Ensminger, M. E., 182, 191
Epple, G., 92, 95, 98, 106
Erikson, E., 195, 219, 257, 276
Erikson, K. T., 21, 36
Erlanger, H., 261, 276
Espenshade, T. J., 66, 85
Estes, D., 200, 201, 219, 221
Everitt, B., 313, 320
Evers, C., 235, 250
Eyman, R. K., 304, 320
Eysenck, H. J., 69, 85

Falbo, T., 16, 36
Farber, B., 282, 284, 288, 289, 292, 296
Faris, R., 291, 297
Farran, D. C., 26, 36, 74, 85
Fearnly, W., 69, 84
Fein, G., 73, 85, 266, 277
Feinman, S., 17, 20, 28, 30, 31, 36, 37, 138, 139, 157
Feiring, C., 2, 4, 7, 11, 16, 18, 19, 27, 30, 31,

Feiring, C. (*Cont.*)
37, 38, 59, 62, 63, 64, 67, 68, 76, 79, 82, 85, 86, 87, 135, 136, 137, 138, 139, 140, 142, 144, 152, 153, 155, 157, 158, 159, 167, 173, 192, 266, 276, 288, 289, 297
Feldman, H., 276
Feldman, S. S., 61, 85, 165, 166, 187, 188, 190
Ferleger, D., 302, 321
Festinger, L., 17, 21, 28, 37
Field, T. M., 18, 19, 23, 26, 37, 176, 190
Fisek, K., 25, 35
Fjellman, S., 48, 57
Foote, N., 191
Fossey, D., 123, 126
Fotheringham, J. B., 294, 297
Fowlkes, M. A., 169, 185, 189
Fox, E. M., 81, 84
Fraiberg, S., 288, 289, 297
Frame, G. W., 92, 106
Frame, L. H., 92, 94, 99, 103, 106
Frankel, D. G., 19, 37
Freedle, R., 4, 11
Freeman, H., 74, 86
Freese, M. P., 286, 297
French, J. R. P., Jr., 180, 181, 190
Freud, S., 195, 219
Friederich, W., 263, 276
Frommer, E., 261, 276
Furman, W., 133, 157
Furstenberg, F. F., 176, 186, 190

Gaensbauer, T., 17, 35
Galdikas, B. M. F., 122, 126
Gallagher, J. J., 282, 297
Gallas, H. B., 170, 190
Galligan, R., 136, 159
Gallimore, T., 46, 48, 57
Galper, A., 173, 190

Gamble, W., *278*
Gamson, W. A., 16, *37*
Gans, H. J., 14, 21, *37*
Garbarino, J., 22, *37*, 182, 190, 208, *219*, 273, 291, 293, 297
Gavish, L., 105, *106*
Gecas, V., 256, *276*
Getz, L. L., 105, *106*
Gewirtz, J., 256, 263, *276*
Gibson, F., 176, *190*
Gibson, H., 265, *276*
Gilford, R., 164, *190*
Gilliam, G., 182, *190*
Glick, C. P., 163, 164, *190*
Glueck, E., 69, *86*
Glueck, S., 69, *86*
Goetsch, G. G., 16, *37*
Goldberg, S., 61, *87*, 155, *157*, 253, 254, 263, 273, *276*, 277
Goldberg, W. A., 174, *190*
Golden, M., 74, *86*
Goldman, J. A., 19, *37*
Goodall, J., 123, *126*
Goode, M., 253, *275*
Gordon, E. W., 177, *194*
Gordon, M., 178, *190*
Gottman, J. M., 7, *11*, 138, *159*, 162, 165, 189, *193*
Gove, F. L., 132, *156*, 200, *219*
Goy, R. W., 124, *128*
Granovetter, M. S., 16, 22, 23, 24, *37*
Grant, G. W. B., 319, *321*
Greenbaum, C. W., 138, *157*
Greenwood, A., 30, *40*
Greenwood, P. J., 95, *106*
Gregg, C. D., 285, *296*
Grossi, M., 74, *86*
Grossman, F. K., 285, *298*
Grossman, H. J., 300, *321*
Grossman, J. C., 69, *84*
Grunebaum, H. V., 180, 181, *189*
Gunnar, M. R., 30, *37*
Guttman, D., 257, *276*

Hagestad, G. O., 163, 178, *191*
Hall, F., 261, 262, *276*
Hamilton, W. D., 103, *106*
Hansen, D. A., 79, *86*
Harcourt, A. H., 123, 124, *126*
Hardy, J. B., 176, *191*
Hare, A. P., 16, *37*
Hare-Mustin, R. T., 163, *191*
Hareven, T. K., 164, 185, 186, *191*
Harkness, S., 43, 45, *57*
Harlow, H. F., 9, *11*, 27, 40, 60, 75, *86*, *88*, 110, *126*, 132, 133, 134, *158*, *159*, 160
Harlow, M. K., 60, *86*, 110, *126*, 132, *158*
Harmon, R. J., 215, *219*
Harper, L., 56, *56*
Harris, G., 144, *159*
Harris, J. M., 319, *321*
Hartung, T. G., 96, 99, 105, *106*
Hartup, W. W., 9, *11*, 133, *157*, 224, *249*, 256, *276*
Havighurst, R. J., 180, *191*
Haviland, J., 30, *37*
Hawkins, C., 25, *40*
Hay, D. F., 18, *37*
Haythorn, W. W., 20, *40*
Hearn, J. P., 95, *106*
Heath, D. H., 138, *158*
Heider, F., 17, *37*
Heinicke, C., 27, *37*
Henderson, C., 17, 35, 286, *296*
Hentig, H. V., 170, *191*
Hess, B. B., 169, 181, *191*
Hess, J. P., 110, *127*
Hess, R. D., 2, *11*, 255, *276*
Hetherington, E. M., 80, 81, *86*, 176, 187, *191*, 244, *249*, 265, 266, *276*

Hickey, L. A., 179, *191*
Hickey, T. H., 179, 187, *191*
Hill, R., 79, *86*, 180, *191*
Hoage, R. J., 97, 98, 99, 103, *107*
Hock, E., 24, *37*, 71, 73, 74, *86*
Hoddinott, B. A., 294, *297*
Hofferth, S. L., 23, *39*, 72, *86*
Hoffman, L. W., 66, 72, 75, *86*, 254, 267, *276*
Hogan, R., 253, *278*
Holland, P. W., 141, *158*
Hollingshead, A. B., 202, *219*
Holroyd, J., 282, 284, 286, *297*
Honig, A. S., 74, *85*
Horenstein, D., 126, *127*
Horr, D. A., 122, *127*
Hosking, J., 262, *278*
Huck, S. W., 311, *321*
Hultsch, D. K., 257, *277*
Humphrey, M. E., 287, *297*
Hunt, D. E., 320, *321*

Ignatoff, E., 176, *190*
Indik, B. P., 16, *37*
Ingram, J. C., 98, 103, *107*
Insel, P. M., 301, *321*
Isso, L. D., 132, *157*

Jacklin, C. N., 25, *40*, 150, *158*, 159
Jacobs, B. S., 17, *37*
Jacobson, S. W., 30, *38*
James, R. M., 25, *40*
Jantz, R. K., 173, *190*
Jaskir, J., 152, *158*
Jeffcoate, J. A., 287, 289, *297*
Jeffries, V., 182, *193*
Johnson, S., 265, *277*
Jordan, V. B., 180, 187, *191*

Kagan, J., 14, 26, *40*, 74, *86*, 257, *275*

Author Index

Kahana, B., 164, 178, 180, 187, *191*
Kahana, E., 164, 178, 180, 187, *191*
Kahn, R. L., 181, *191*
Kahn, S., 176, *194*
Kalish, R. A., 179, *191*
Katz, E., 21, *36*
Katz, M., 46, *57*
Kaufman, I. C., 131, *159*
Kaye, K., 30, *38*
Klein, R. P., 215, *219*
Keim, A. M., 16, *40*
Kellam, S. G., 182, *191*
Keller, L. S., 304, *321*
Kelman, H., 28, *38*
Kempe, C. H., 126, *127*
Kemper, T., 265, 267, *277*
Kendrick, C., 18, 24, *36*, *38*, 83, *85*, 136, *157*, 165, *190*
Kennell, J. H., 134, *158*, 185, *191*
Kessen, W., 266, *277*
Killen, M., 30, *38*
Kimmel, D., 265, *277*
King, R. A., 242, *250*
King, R. D., 306, *321*
King, T. M., 176, *191*
Kirchshofer, R., 109, 110, 113, 116, 118, *127*
Klaus, M. H., 134, *158*, 185, *191*
Kleiman, D. G., 91, 92, 93, 94, 95, 96, 97, 103, 104, 105, *107*
Klinnert, M., 30, 31, *38*
Knight, R. C., 319, *321*
Knoll, C. E., 178, *190*
Kogan, K. L., 295, *297*
Kohn, M. L., 25, *38*, 267, *277*
Komarita, S. S., 16, *38*
Komarovsky, M., 181, *191*
Konner, M., 46, *57*, 144, *158*
Kornfein, M., *57*
Kornhaber, A., 164, 165, 169, 172, 173, 174, 177, 179, *191*
Kortlandt, A., 123, *127*
Koslowski, B., 138, *157*

Kotelchuck, M., *86*
Krajl, M. M., 282, *298*
Kreitzberg, V. S., 17, 18, 24, *39*, 152, 155, *158*
Kreutzer, M. A., 30, *35*
Krolick, G., 75, *88*
Kugel, R. B., 302, *321*

Lakin, K. C., 301, *320*
Lakin, M., 19, *38*
Lakin, M. G., 19, *38*
Lamb, M. E., 5, 6, *11*, 18, 24, 26, 27, 28, 30, *36*, *38*, *57*, 62, 63, 68, 71, 83, *85*, *86*, 136, 138, *158*, 162, 165, 170, 172, *191*, 196, 197, 200, 201, 209, 211, 212, 213, 215, 218, *219*, 220, 221, 253, 254, 263, 268, *277*, *278*
Lambert, W. W., 46, *57*, 175, *192*
Landau, R., 138, *157*
Landesman-Dwyer, S., 301, 302, 304, 305, 306, 309, 319, *320*, *321*, *322*
Landis, P. H., 69, *86*
Lando, B., 24, *34*
Lang, L., 23, *39*
Laslett, P., 163, *191*
Latané, B., 29, *38*
Lawton, D., 286, 287, 289, *296*
Leavitt, L. A., 215, 218, *221*
Leavitt, M. J., 20, *38*
Leavitt, S., 242, *250*
Leckie, M. S., 2, *11*
Lee, G. R., 141, 146, 150, *158*, 163, 169, 177, 178, 180, 181, *191*, *192*
Leibowitz, A., 70, 72, *86*
Leiderman, G. F., 51, *57*
Leiderman, P. H., 17, 40, 51, *57*
Leinhardt, S., 141, *158*
Lenington, S., 96, *106*
Lenski, G., 34, *38*

Lenski, J., 34, *38*
Lerner, R. M., 60, *88*, 256, 258, 264, 267, *277*
Leslie, G. R., 148, *158*
Levin, H., 67, 195, *220*
Levine, R., 45, *57*
Levine, S., 169, *190*
Levinger, R. L., 16, *39*
Lewis, L. D., 29, *36*
Lewis, M., 2, 3, 4, 5, 7, 9, *11*, 13, 15, 16, 17, 18, 19, 20, 22, 24, 25, 26, 27, 30, 31, *35*, *36*, 37, *38*, *39*, 41, 54, 57, 59, 61, 62, 63, 64, 67, 68, 69, 75, 76, 78, 79, 82, 83, *84*, *85*, *86*, 87, 132, 133, 134, 135, 136, 137, 138, 139, 140, 142, 144, 148, 152, 153, 155, *157*, *158*, 159, 162, 163, 167, 173, 175, 179, *190*, *192*, 196, *220*, 224, 249, 253, 254, 255, 256, 263, *275*, *277*, 288, 289, *297*, *320*, *322*
Lewis, N. G., 162, *192*
Lieberman, A. F., 183, *192*, 199, 200, *219*, *220*, 225, 244, 245, 249
Liebow, E., 26, *39*
Light, R., 291, *297*
Lipsitt, L., 256, *274*
Lipson, A., 169, *190*
Little, R. E., 309, *321*
Litwak, E., 164, 178, *192*
Lloyd, J. K., 287, *297*
Lobitz, G., 265, *277*
Loda, F. A., 182, *193*
Longfellow, C., 266, *279*
Lunn, S. F., 95, *107*
Lynn, D. B., 78, 79, *87*
Lytton, H., 18, 20, *39*

Maccoby, E. E., 25, *40*, 67, 150, *158*, 159, 195, *220*
MacDonald, R., *191*

Mack, D. S., 98, *106*
MacKinnon, J. R., 122, *127*
Magalam, J. J., 178, *189*
Mahler, M. A., 215, *220*
Main, M., 138, *156*, 183, *192*, 200, 201, 202, 211, *220*, 253, *277*
Malcolm, J. R., 92, 95, 96, 97, 98, 99, 100, 103, 105, *106*, *107*
Marjoribanks, K., 67, *87*
Markus, G. B., 67, *89*, 152, *160*
Markus, H., 67, *89*, 152, *160*
Marolla, F. A., 67, *84*
Martin, J. A., 20, *39*
Martinez, C., 263, *275*
Maruyama, G., 247, *249*
Masters, J. C., 195, 200, *220*
Matas, L., 132, 133, *159*, *193*, 200, 202, 210, *220*, 253, *277*
Mattson, M. L., 16, *41*
Mayhew, B., 16, *39*
McArthur, D., 282, 284, *297*
McCall, R., 254, *277*
McCord, J., 80, *87*
McCord, W., 80, *87*
McCubbin, H. I., 292, *297*
McDermott, J. L., 105, *106*
McDevitt, S. C., 32, 33, *35*
McFarland, D. D., 16, *37*
McGhee, L. J., 30, *36*
McGrew, W. C., 19, *39*
McIntosh, H. T., 282, *298*
McIntosh, S., 66, *88*
McLean, R. A., 311, *321*
Medrich, E. A., 225, 238, *249*
Menzel, H., 21, *36*
Merton, R. K., 27, 28, *39*
Meyer, G. A., 166, *193*
Meyers, C. E., 301, *320*
Michalson, L., 9, *11*, 25, *39*, 75, *87*, 132, *159*, 196, *220*, 224, *249*

Miller, C. E., 16, *39*
Miller, L. C., 122, *127*
Miller, N., 247, *249*
Miller, S. J., 164, *194*
Mills, L. A., 176, *192*
Mills, P., 291, *298*
Mills, T. M., 16, *35*
Minturn, L. A., 46, *57*, 175, *192*
Minuchin, S., 17, *39*
Mitchell, R., 266, 267, *277*
Moehlman, P. D., 100, 102, *107*
Mogey, J., 169, *192*
Moltz, H., 1, *12*
Monane, J. H., 62, *87*, 135, *159*
Mondell, S., 262, *277*
Moore, D., 16, *38*
Moore, K. A., 23, *39*, 72, *86*
Moore, J. W., 163, *192*
Moore, T., 74, 75, *87*
Moores, B., 319, *321*
Moos, R. H., 301, *321*
Moran, D. L., 181, *189*
Morgan, G. A., 17, *39*, 215, *219*, *220*
Moskowitz, D., 74, *87*
Moss, H. A., 17, *37*, 256, 263, *277*, 288, *298*
Most, R., 253, *275*
Mueller, C. W., 79, *88*
Mueller, E. C., 75, 83, *87*, 196, *220*
Munroe, L., 45, 46, 48, *57*
Munroe, R., 45, *57*
Murchinson, N., 81, *87*

Nadler, R. D., 110, 111, 112, 114, 115, 116, 122, 126, *127*
Nash, A., 18, *37*
Nash, S. C., 165, 166, 187, *188*, *190*
Neal, A., 26, *39*
Nelson, J. I., 16, *39*
Nelson, U. L., *12*
Neugarten, B. L., 162, 163, 170, 172, 178, 188, *192*, 257, *277*
Nimkoff, M. F., 164, *192*

Nirje, B., 302, *321*, 322
Norman, R. Z., 25, *35*
Norton, A. J., 77, *86*
Nuttal, J., 24, *40*
Nye, F. I., 75, *86*, 266, *277*
Nyman, B. A., 2, *11*

O'Brien, C., 291, *298*
O'Donnell, L., 225, 236, 238, *249*, *250*
Olds, D. L., 184, 185, *192*
O'Leary, S., 162, *193*
Olshansky, S., 302, *322*
Olson, D. H., 16, *36*
Orlansky, H., 253, *277*
O'Shea, G., 261, *276*
Osofsky, J. D., 256, 263, *278*, 288, *297*
Owen, M. T., 71, *86*, 209, *219*, *220*
Owens, D. M., 261, *278*

Palerson, M., 69, *84*
Parke, R. D., 7, *12*, 15, 22, 27, 30, *39*, 126, *127*, 136, 138, *159*, 162, 165, *167*, 172, 174, 175, 182, 187, *192*, *193*, 208, *220*, 256, 263, 273, *278*, 285, 288, *297*, 301, *322*
Parry, M. H., 30, *40*
Parsons, T., 137, *159*, 163, *193*, 241, *250*
Pascoe, J. M., 182, *193*
Pastor, D. L., 18, *39*, 200, 211, *220*, 245, *250*, 253, *278*
Pattison, M., 141, *159*
Pawlby, S., 26, *37*, 261, *276*
Pearlin, L. I., 169, *193*, 285, *297*
Pedersen, F. A., 6, *12*, 13, 18, *39*, 61, 63, 64, 68, 78, 80, *87*, *88*, 136, *157*, *159*, 162, 174, *193*, 196, *220*, 253, 265, *278*, *279*

Pedersen, J., 18, *37*, 215, *219*
Pederson, A., 132, *157*
Peplau, L. A., 16, *36*
Perske, R., 302, *322*
Pervin, L. A., 134, *159*, 320, *322*
Piaget, J., 30, *39*
Pine, F., 215, *220*
Pleck, J., 23, *39*
Plemons, J. K., 257, *277*
Plimpton, E. H., 225, *250*
Policare, H., 74, *86*
Pollitt, E., 263, *278*
Pollock, G., 69, *84*
Pope, H., 79, *88*
Porton, I., 98, *99*
Powell, D. R., 184, *193*, 266, 267, *278*
Power, T. G., 7, *12*, 138, *159*, 162, 172, *193*
Pulliam-Krager, H., 17, *35*
Putney, E., 20, *40*

Rabkin, J. G., 282, 285, *298*
Radke-Yarrow, M., 242, *250*
Ragozin, A. S., 309, *321*
Rake, D. F., 133, *157*
Ralls, K., 104, *107*
Ramey, C. T., 26, *36*, 74, 85, *88*, 263, *279*, 291, *298*
Rasa, O. A. E., 98, *107*
Rathbun, G., 92, 94, 97, *107*
Raven, B. H., 16, *35*
Raynes, R. V., 306, *321*
Reed, R. H., 66, *88*
Reese, H., 256, *274*
Regan, R. A., 69, *89*
Rehberg, R. Q., 68, *88*
Reichler, M., 265, 267, *277*
Reiss, D., 163, *193*
Reiss, P. J., 144, *159*
Reynolds, F., 123, *127*
Reynolds, V., 123, *126*
Rheingold, H. L., 30, *39*
Ricciuti, H. N., 17, 24, *39*, 74, *88*, 215, *220*

Richards, I. D., 282, *298*
Richardson, A. H., 148, *158*
Rie, H., 282, *296*
Riegel, K., 166, *193*, 258, *278*
Riegel, R. M., 166, *193*
Rigler, D., 22, *40*
Rijksen, H. D., 122, *127*
Robertson, J. F., 177, *193*
Robins, E., *278*
Robinson, J. P., 71, 72, 88, 187, *193*
Robson, K. S., 61, *87*, 162, *193*, 288, *298*
Rockwell, R., 185, 186, *190*
Rodgers, W. L., 178, *190*
Rodin, J., 29, *38*
Rodman, P. S., 122, *127*
Rodrigues, A., 16, *35*
Rogers, C. M., 110, 113, 118, 119, 120, 121, *126*, *127*
Rohner, E., 261, *278*
Rohner, R., 175, 181, *193*, 261, *278*
Rollins, B. C., 136, *159*
Romer, D., 301, *321*
Rood, J. P., 92, 95, 99, 101, 102, *107*
Roos, P., 302, *322*
Roseborough, M., 16, *35*
Rosen, B. C., 68, *88*
Rosenberg, M., 26, 27, *39*
Rosenberg, S. A., 285, *298*
Rosenblatt, P. C., 136, *159*
Rosenblum, L. A., 1, 2, 3, 5, *11*, *12*, 13, *39*, 60, 75, 82, *87*, 131, 136, *158*, *159*, 167, *192*, 196, *220*, 225, *250*, 255, 256, *277*, 288, *297*
Rosenbluth, L., 74, *86*
Rosow, I., 169, *193*
Rosser, C., 144, *159*
Rossetti-Ferreira, M., 263, *278*
Rossi, A. S., 28, *39*

Rubenstein, J., 253, *279*
Rubin, Z., 223, 226, *250*
Rueveni, V., *193*
Rupenthal, G. C., 134, *159*
Rutter, M., 265, *278*, 286, *298*
Rykman, D. B., 284, *296*

Sackett, G. P., 134, *159*
Salamon, S., 16, *40*
Saltz, R., 184, *193*
Sameroff, A. J., 263, 271, *278*, 282, 283, *298*
Samuels, H. R., 18, *40*
Sander, L., 60, *88*
Sanders, M., 262, *278*
Sanger, W., 14, *35*
Sangree, W. H., 48, *57*
Savin-Williams, R. C., 16, *40*
Sawin, D., 256, *278*
Scanzoni, J., 150, *159*
Schachter, F., 73, 74, 75, *88*
Schaefer, E. S., 79, *88*, 262, *278*
Schaeffer, S., 62, *87*, 133, 134, *159*
Schafer, D. E., 169, *189*
Schaffer, H. R., 20, 30, *40*, 61, *88*
Schaller, G. B., 123, *127*
Schiavo, R. S., 235, 239, *250*
Schinke, S. P., 319, *322*
Schlegel, A., 44, *56*
Schooler, C., 169, *193*, 267, *277*, 285, *297*
Schubert, J. B., 24, *40*
Schuckit, J. J., 304, *321*
Schwartz, T., 48, *57*
Schwarz, J., 74, 75, *87*, *88*
Schwarzweller, H. K., 178, *189*
Sears, R., 67, 69, *88*, 195, *221*
Seefeldt, C., 173, *190*
Seeman, M., 26, *39*
Serock, K., 173, *190*
Shanas, E., 22, *40*, 164, 167, 177, *193*

Shearer, A., 302, *321*
Sherif, M., 28, *40*
Sherman, D., 22, *37*, 208, 219, 291, 293, *297*
Shipman, V. C., 2, *11*, 255, *276*
Shipp, D. A., 176, *191*
Shweder, R. A., 48, *57*
Siegel, E., 262, *278*
Sigford, B. B., 301, *320*
Silver, H. K., 126, *127*
Silverman, F. N., 126, *127*
Simmel, G., 15, *16*
Simpson, H. R., 68, *85*
Skelton, M., 294, *297*
Sloman, J., 225, *250*
Smith, B., 74, *88*
Smith, E. W., 176, 181, *193*
Smith, J. W., 181, *194*
Smith, P. K., 19, *40*, 225, 233, 234, *250*
Smith, S., 20, *40*
Smythe, N., 92, *108*
Snow, M. E., 25, *40*, 215, *219*
Sojit, C. M., 17, *40*
Solomon, S. K., 235, 239, *250*
Sontag, L. W., 4, *12*
Sorce, J., 30, *38*
Spanier, G. B., 60, *88*, 267, *277*, *278*
Speicher, J. L., 178, *191*
Spinetta, J. J., 22, *40*
Sroufe, L. A., 62, *88*, 132, 133, 134, *156*, *159*, *160*, 183, *192*, *194*, 200, 210, 215, *219*, *220*, 221, 244, *250*, 253, 261, 262, *277*, *278*, *279*
Stack, C. B., 14, 21, 22, *40*
Staines, G. L., 23, *39*
Staples, R., 181, *194*
Starr, M., 139, *157*
Stayton, D. J., 199, *219*, *220*, 253, 274, *278*
Steele, B. F., 126, *127*
Steeve, G. H., 169, *189*

Steinberg, L. D., 24, *35*, 74, 75, 76, *84*, 291, *298*
Stephenson, G. R., 5, *11*
Stern, D. N., 138, *160*
Stevenson, M. B., 212, 215, 218, *221*
Stewart, K. J., 123, 124, *126*, *127*
Stickland, R. G., 75, *88*
Stocking, S. H., 242, *250*
Stokols, D., 301, *322*
Straus, M. A., 261, *278*
Streib, G. F., 167, 170, *194*
Streissguth, A. P., 2, *11*
Stringer, S., 176, *190*
Stodbeck, F. L., 16, 25, 35, *40*
Struening, E. L., 282, 285, *298*
Stueve, C. A., 225, 238, *250*
Sugiyama, Y., 123, *127*
Sulzbacher, F. MacL., 304, *321*
Suomi, S. J., 5, *11*, 27, *40*, 75, *88*, 126, *127*, 134, *159*, *160*
Super, C., 45, *57*
Sussman, M. B., 163, *194*
Svanum, S., 67, *89*
Svejda, M. J., 30, *38*
Svendsen, G. E., 99, *108*
Szeleniji, I., 164, *192*

Tannenbaum, J., 74, *85*
Tavormina, J. B., 282, *298*
Taylor J., 62, *89*, 138, *157*, 266, *276*
Teger, A. I., 29, *36*
Terhane, K. W., 67, 68, 69, *89*, 152, *160*
Tharp, R., 48, *57*
Thoman, E. B., 17, *40*, 286, *297*
Thomas, J. A., 99, *108*
Thompson, R. A., 200, 201, 213, 215, 218, 219, 221, 253, *278*
Thurber, E., 80, *87*

Tilson, R. L. 91, *108*
Tinsley, B. R., 172, 175, *193*, *194*
Tizard, J., 286, *298*, 306, *321*
Tolan, W. J., 176, *189*, 200, *220*, 256, *275*
Tomasini, L., 200, *220*
Tomilin, M. I., 113, 118, *128*
Toms-Olson, J., 266, *278*
Townsend, P., 178, *194*
Trickett, E., 266, 267, *277*
Trivers, R. L., 19, *40*
Troll, L. E., 162, 163, 164, 167, 177, 178, 180, 186, 187, *194*
Tronick, E., 138, *157*
Trost, M. A., 133, *157*
Tuckman, J., 69, *89*
Tulkin, S. R., 14, 26, *40*, *278*
Turnbull, C., 22, *40*
Turner, B., 180, *194*
Turner, J., 182, *191*
Turner, P., 295, *297*
Turner, R. H., 68, *89*
Tyler, F., 262, *277*
Tyler, N., 295, *297*

Uhlenberg, P., 164, *194*
Umeh, B. J., 215, *219*
Unger, D., 182, *194*, 266, 267, *278*
Uzgiris, I. C., 30, *38*
Uzoka, A. F., 163, *194*

Vallance, R., 69, *84*
Vandell, D. L., 25, *40*, 75, 83, *87*, 196, *220*
Van den Berghe, P. L., 16, *40*
Vander Veen, F., 265, *277*
Van Egeren, L. F., 2, *11*
van Lawick, H., 92, 95, 99, *106*, *108*
van Lawick-Goodall, J., 123, *127*
Vaughn, B., 200, 201, 203, 204, 209, *221*

Author Index

Veit, S. W., 319, *321*
Vietze, P., *189*
Vogel, E. F., 17, *40*
Von Bertalanffy, L., 62, *89*

Wagner, G., 48, *57*
Wagner, S. S., 96, *106*
Wahler, R. G., 169, *194*
Wainwright, W., *279*
Walberg, H. J., 67, *87*
Wall, S., 17, *34*, 62, *84*, 197, *219*
Wandersman, A., 176, *194*
Wandersman, L. P., 176, *194*
Ward, J. H., 313, *322*
Ward, M., 261, 262, *278*
Waring, J. M., 169, 181, *191*
Waters, E., 17, *34*, 62, *84*, *88*, *133*, *160*, 183, *194*, 196, 200, 201, 202, 210, 211, *219*, *220*, *221*, *244*, *250*, *253*, *279*
Watts, G., 306, *321*
Webster, M. A., 16, 27, *40*
Weingarten, K., 170, *190*
Weinraub, M., 7, *11*, 13, 15, 20, 22, 27, 39, *40*, *41*, 72, 83, 86, *87*, *89*, 136, 138, *159*, 162, 167, 174, *192*
Weinstein, K. K., 162, 170, 172, 178, 188, *192*
Weisbard, C., 124, *128*

Weisner, T. S., 46, 48, 53, *57*
Weiss, J. C., 181, *189*
Weitzer, W. H., 319, *321*
Welcher, D. W., 176, *191*
Wellman, H. W., 200, *220*
Werner, E. E., 181, *194*
Westby, D. L., 68, *88*
Westheimer, I., 27, *37*
Weston, C. R., 183, *192*, 201, *220*
Whatley, J. L., 30, *36*
Whiting, B. B., 45, 46, 47, *57*, *58*, 163, *194*
Whiting, J. W. M., 46, *58*, 163, *194*
Whitmore, K., 286, *298*
Whyte, W. F., 26, *41*
Widmayer, S. M., 176, *190*
Williams, R. M., 16, *41*
Wilmott, P., 14, 22, *41*, 144, *160*
Wilson, C. D., 26, *38*, 255, *277*
Wilson, E. O., 92, *108*, 142, *160*
Wilson, K. S., 25, *40*
Wilson, S. C., 99, 100, *108*
Wilsson, L., 99, *108*
Winch, R. F., 178, *194*
Wippman, A., 133, *160*
Wippman, J., 183, *194*, 200, *221*, *244*, *250*, *253*, *279*
Wise, S., 285, *298*
Wiseman, R. F., 169, *189*

Wittenberger, J. F., 91, *108*
Wolfe, D. M., 16, *35*
Wolfensberger, W., 302, *322*
Wolff, K. L., 15, *41*
Wolkind, S., 261, *276*
Wolters, H. J., 101, *108*
Woodward, K. L., 164, 165, 169, 172, 173, 174, 177, 179, *191*
Wrangham, R. W., 123, *128*
Wright, C. M., 74, *85*

Yarrow, L. J., 63, *88*, *159*, *253*, *279*
Yaschine, T., 263, *275*
Yerkes, R. M., 113, 118, *128*
Yogman, M. W., 138, *157*, 172, *194*
Yorburg, B., 163, *194*
Young, G., 9, *11*, 25, *39*, 75, *87*, 132, *159*, 196, *220*, 224, *249*
Young, L., 126, *128*
Young, M., 14, 22, *41*, 144, *160*

Zahn-Waxler, C., 242, *250*
Zalk, S. R., 176, *189*
Zajonc, R. B., 67, 73, *89*, 152, *160*
Zelditch, M., Jr., 25, *35*
Zeskind, P., 263, *279*
Zigler, E., 177, *194*
Zimring, C. M., 319, *321*
Zur-Szpiro, S., 266, *279*

Subject Index

Adaptive behavior, 299–300
Attachment
  and day care, 74, 76
  and family circumstances, 202–210
  measurement of, 197–200
  and peer relations, 211, 244–245
  as primary relationship, 132–133, 134, 195–197, 211
  security of, 196–197, 210, 217–218
  and stranger sociability, 210–217
  temporal stability of, 200–203
  See also Mother–child dyad
Attachment theory, 196–197
  See also Epigenetic model

Biological model, 1–2
Birth order, 17–18
Birth rate, 65–67

Child abuse, 22
  and family stress, 291
  in gorillas, 110, 113–118, 126
  and peer behavior, 134–135
Cognitive development. *See* Intelligence
Critical period, 134, 210
Cultural effects, 52–56
  See also Kisa; Nairobi
Cultural niche, 43–47
  effects on families, 47–56
  measures of, 46

Day care, 23, 24, 73–75, 76
  selection of, 233–234
Direct effects, 138–139
  of grandparents, 167–169, 170–173, 181–185
  See also Indirect effects

Divorce
  effect on child, 79, 81
  effect on family, 61
  See also Single-parent family
Dyad
  effect of social connectedness, 20–23
  See also Attachment; Father–child dyad; Mother–child dyad; Teacher–pupil dyad
Dyadic interaction
  in infant research, 13–14

Emotional development, 62
  See also Attachment
Environment, normative, 302–303
Environmental effects
  on handicapped children, 306–318
Epigenetic model, 131–135

Family
  effects on attachment, 203, 210
  nuclear versus extended, 101–104
  as a social system, 16–17, 59–60, 62–65
  See also Single-parent family
Family interaction, 3–4, 136–137
  birth order effects on, 17–18
  cultural niche effects on, 47–56
  divorce effects on, 61. *See also* Single-parent family
  fertility rate effects on, 61, 65–69
  maternal employment effects on, 61, 64–65, 70–77
  quantity versus quality of, 61–62
  as small group process, 16–17
  See also Indirect effects
Family size
  and child's personality, 69
  and intelligence, 67–69

Family system, *See* Family; Family interaction
Father
  absence of, 78–80. *See also* Single-parent family
  effects of, 6
    on animal development, 95–100. *See also* Indirect effects
Father–child dyad
  divorce effects on, 80
  maternal employment effects on, 71, 72, 76
Fertility, 61, 65–69
Friendships
  effects on parents, 236–237
  parental effects on, 223–249
  parental values about, 226–230, 245–248
  See also Peer relationships

Gorillas
  mother–child interaction in, 109–118
Grandparents, 163–166
  children's perceptions of, 178–180, 187
  direct effects of, 167–169, 170–173
  effect on development, 181–185
  indirect effects of, 167–169
  mediating factors, 177–181
  measurement of influence, 166–167, 185–188
Group processes, 6–7, 15–20
  See also Family interaction

Handicapped children
  environmental effects on, 299–320
  and family stress, 281–296
Helpers, 100–101, 102, 103, 104
  See also Substitute caregivers
HOME scale, 182, 306
Human Relations Area Files, 44

Incest taboos, 104–105
Indirect effects, 4, 13, 30, 63–64, 138–139
  of animal fathers, 95–96, 97–98, 114, 116–117
  in family interactions, 18
  of grandparents, 167–169, 173–176, 181–185
  of human fathers, 78–80, 82, 136
Individual differences. *See* Birth order; Cultural effects; Sex differences; Social class; Temperament

Intelligence
  and family size, 67–69
  and father absence, 80
  and maternal employment, 73
Intervention, 294–295
  See also Day care; Residential environments

Kisa, 48–51
  See also Cultural effects

Language development, 3–4, 73
Learning model, 2–4

Maternal employment, 61, 70
  effects on attachment, 74, 209
  effects on development, 23–24, 70, 73–74
  effects on social network, 152–153
Measurement
  of attachment, 197–200
  of cultural niche, 46
  of dyadic interactions, 4–6
  of grandparent effects, 185–188
  of group interaction, 6–7, 18–20
  of residential environmental effects, 300–312
  of social network, 140–142
Modeling
  and friendships, 242–244
  See also Indirect effects; Social referencing
Models of development, 1–4, 131–139, 282–283
Monogamy
  and sociality, 92–95
  evolution of, 91–93
Mother–child dyad, 1–6, 195–197
  in chimpanzees, 118–121, 122–123
  in epigenetic model, 131–132
  in gorillas, 109–118, 121, 123–126
  in orangutans, 122–123
  in social network systems model, 134
  See also Attachment; Maternal employment; Parents
Multiple roles
  and relationships, 23–28

Nairobi, 51–52
  See also Cultural effects
Normative environment, 302–303

Subject Index

Parent–child dyad, 62–63
   See also Father–child dyad;
      Mother–child dyad
Parents
   as coaches, 224, 225, 239–242
   as decision-makers, 224, 225, 230–237
   effects on peer relations, 223–249
   as models, 224, 225, 242–244
   as providers, 224, 225, 244–245
   relationships between. See Indirect
      effects
   as social organizers, 224, 225, 237–238
Parent values
   about children's friendships, 226–230,
      245–248
Paternal care. See Father; Father–child
      dyad; Indirect effects
Pattison Psychosocial Network
      Inventory, 141
Peer relationships, 25, 113
   and attachment, 211, 244–245
   divorce effects on, 81
   maternal employment effects on, 73,
      75, 76
   parental effects on, 223–249
   See also Friendships
Personality
   and family size, 69
   and father absence, 80
   and maternal employment, 74
Prenatal/Early Infancy Project, 184

Relationships. See Father–child dyad;
      Friendships; Grandparents;
      Mother–child dyad; Parent–child
      dyad; Peer relationships; Siblings
Residential environments
   effects on handicapped, 299–320
Revised Child Management Scale, 306

Sex differences, 81, 148–150, 241
Second-order effects. See Indirect effects
Siblings, 73
Single-parent families, 61, 77–82
Social competence, 134
Social ecology. See Cultural niche
Social influences
   types of, 15
Social network, 13–34, 83, 142–146,
      154–156

Social network (Cont.)
   age changes in, 153–154
   birth order differences in, 150–152
   effects on mother–child dyad, 82. See
      also Indirect effects
   measurement of, 140–142
   of parents, 235–237
   SES differences in, 146–148
   sex differences in, 148–150
Social network systems model, 135–137
Social referencing, 28–33
Social class
   effects on attachment, 203–204,
      207–210
   effects on relationships, 26
   and stress, 290–292
Status
   effects on relationships, 25–26
Stranger–infant interaction, 17
   and attachment, 210–217
   temperament effects on, 31–33, 218
   See also Indirect effects
Strange situation procedures, 196,
      197–200, 212–213
Stress, 282
   changes over time, 294–295
   effects on attachment, 207
   in families of handicapped children,
      281–296
   and intervention, 294–295
   social class differences, 290–292
Substitute caregivers, 27–28, 73, 76
   See also Day care; Helpers; Residential
      environments
Support system
   and stress, 292–293

Teacher–pupil dyad, 2–4
Temperament
   and stranger interactions, 31–33, 218
Trait theory, 133–135
Transactional model, 282–283

Urban migration, 48

Values
   and childrearing, 24–25

Working mothers. See Maternal
      employment